STUDY GUIDE

Earth: Portrait of a Planet

STUDY GUIDE

Earth: Portrait of a Planet

Stephen Marshak

Rita Leafgren
UNIVERSITY OF NORTHERN COLORADO

 W·W· Norton & Company · NEW YORK · LONDON

Copyright © 2001 by W. W. Norton & Company, Inc.

All rights reserved

ISBN 0-393-97660-2 (pbk.)

W. W. Norton & Company, Inc., 500 Fifth Avenue, New York, NY 10110
http://www.wwnorton.com

W. W. Norton & Company, Ltd., Castle House, 75/76 Wells Street, London W1T 3QT

1 2 3 4 5 6 7 8 9 0

CONTENTS

Preface: How to Get the Most from This Study Guide	vii
Chapter 1 | Cosmology and the Birth of Earth	1
Chapter 2 | Journey to the Center of the Earth	11
Chapter 3 | Drifting Continents and Spreading Seas	20
Chapter 4 | The Way the Earth Works: Plate Tectonics	31
Chapter 5 | Patterns in Nature: Minerals	45
Chapter 6 | Up from the Inferno: Magma and Igneous Rocks	55
Chapter 7 | A Surface Veneer: Sediments and Sedimentary Rocks	67
Chapter 8 | Change in the Solid State: Metamorphic Rocks	82
Chapter 9 | The Wrath of Vulcan: Volcanic Eruptions	93
Chapter 10 | A Violent Pulse: Earthquakes	103
Chapter 11 | Crags, Cracks, and Crumples: Crustal Deformation and Mountain Building	121
Chapter 12 | Deep Time: How Old Is Old?	133
Chapter 13 | A Biography of Earth	153
Chapter 14 | Squeezing Power from a Stone: Energy Resources	166
Chapter 15 | Riches in Rock: Mineral Resources	181
Chapter 16 | Unsafe Ground: Landslides and Other Mass Movements	190
Chapter 17 | Streams and Floods: The Geology of Running Water	201
Chapter 18 | Restless Realm: Oceans and Coasts	218
Chapter 19 | A Hidden Reserve: Groundwater	235
Chapter 20 | An Envelope of Gas: Earth's Atmosphere and Climate	249
Chapter 21 | Dry Regions: The Geology of Deserts	264
Chapter 22 | Amazing Ice: Glaciers and Ice Ages	275
Chapter 23 | Global Change in the Earth System	290

PREFACE: HOW TO GET THE MOST FROM THIS STUDY GUIDE

The basic purpose of the guide is to help you learn more geology and do better in your class. Each chapter has three main parts to it:

1. **Guide to Reading** sections present an overview of the major ideas introduced in each text chapter. If you start out with "the big picture," it's less likely you'll get lost in the details and miss the point of it all. Do read it before you read the chapter, and it's not a bad idea to reread it after reading the chapter as a review of those important concepts you should have grasped.

2. **Learning Activities** ask you to apply what you've learned. Because people learn in different ways and different materials lend themselves to different approaches, I've provided a variety of activities. Sometimes you'll be asked to do very mundane things: fill in the blank, match descriptions with names, label diagrams, or work problems. Sometimes you'll be required to apply your knowledge more creatively, by writing short essays or creating your own summaries of topics. There will always be a key terms/vocabulary exercise in this section because you must understand what individual words mean before you can hope to understand what ideas they are trying to convey when they get used with each other.

3. **Practice Tests** challenge you to put it all together and provide you with the opportunity to test yourself before your teacher does. Check your answers against the annotated feedback to ensure that you not only know which answer is correct but *why*.

General words of advice:

- The more *time* you spend with the subject matter, the better you'll know it. Even if you don't do all of the guide activities, any time spent will increase your knowledge.
- Your *active involvement* in doing exercises will help you tremendously. In learning, as in many areas of life, being a passive spectator just doesn't get you as far as does active participation.
- *Do read your text*—an obvious suggestion, but often overlooked. Read the entire chapter. Continuity counts. Piecemeal looking for answers to these exercises just doesn't equal reading the chapter.
- *Go to class.*
- *Start early and study often.* Great athletes work out every day. Concert pianists practice daily. And A students exercise their minds daily.

Whether you're an avid science student looking forward to the course, science shy, or somewhere in between, do remember that geology literally means "study of the earth." This course covers a lot of material. Doing a good job will take a lot of your time, but in return you'll have a better understanding of your world and a deeper appreciation for how awesome a place Earth really is.

CHAPTER 1 | Cosmology and the Birth of Earth

GUIDE TO READING

If you've always thought that the study of Earth is limited to identifying rocks and minerals, you're in for some surprises, starting with this chapter. Before you begin reading it, look back at Table 1 in the Prelude of the text, which lists the many subdivisions of geology. Notice how geology claims as subject matter just about anything Earth related. It's a very eclectic subject (composed of a broad selection of topics from many different sources). With this is mind, begin reading Chapter 1, and note how it very logically starts at the beginning of everything. Everything is literally the Universe.

For thousands of years humans have sought to understand the Universe they're part of—its beginnings, structure, functioning, and future. Today we term these studies *cosmology*. Many of the earliest ideas were fanciful, revolving around gods and goddesses, though there were also investigations of a more scientific nature. People asked the right questions, made careful observations of natural phenomena, used common sense and ingenuity, and came up with quite accurate answers to some basic yet profound questions. Human history and this chapter are both filled with many of these milestones of understanding. Included in this chapter are discussions of the following:

- the discovery that planets are wanderers, different from stars
- Toscanelli's influence on Columbus, which convinced him the world was round
- Kepler's work that showed planetary orbits are elliptical
- Galileo's use of the telescope to study Earth's distant neighbors
- Newton's calculations about gravity and the laws of motion
- Eratosthenes' calculations that gave Earth's correct circumference
- the use of parallax to figure distances to stars
- Ptolemy's belief in an Earth-centered Universe
- Copernicus's belief in a Sun-centered Universe
- the discovery of galaxies, including our own Milky Way
- Doppler's explanation of wavelengths and frequencies altered by moving sources
- the correlation between the Doppler effect and the red shift of the expanding Universe
- the big bang theory of the beginning of the Universe
- the realization that stars have beginnings, lifetimes, and deaths
- the process of element formation in stars
- the development of our round Earth and our planetary system
- the origin of our Moon

The vastness of the Universe defies comprehension so, as your author explains, scientists created special units like the light year to make the mathematics about it more manageable and frequently use analogies to make its large-scale processes more accessible.

By chapter's end you may feel you've had a crash course in ancient astronomy. In a way you have, but you also have established a firm foundation for your study of geology.

LEARNING ACTIVITIES

Completion

Test your recall of new vocabulary terms.

 1. the study of the Universe—its beginning, development, structure, functioning and future

_____ 2. the systematic analysis of natural phenomena

_____ 3. all space and matter

_____ 4. the idea that Earth is the center of the Universe and everything revolves around it

_____ 5. the idea that the Sun is the center of the Universe and everything revolves around it

_____ 6. the apparent motion of a distant object against an even more distant background when viewed from a new position

_____ 7. the distance light travels in 1 year (roughly 6 trillion mi)

_____ 8. the change in the perceived frequency of sound waves (pitch of the sound) or the perceived wavelength of light waves (color) when the source moves toward or away from the observer

_____ 9. the note in the musical scale that a sound corresponds to

_____ 10. the number of wave crests (high points) or wave troughs (low points) that pass a fixed reference point in a given time

_____ 11. the distance between successive crests (or successive troughs)

_____ 12. the shift of the perceived color of light waves toward the red end of the visible spectrum (lower energy, longer wave length) as the source moves away

_____ 13. the shift of the perceived color of light waves toward the blue end of the visible spectrum (higher energy, shorter wave length) as the source moves closer

_____ 14. the theory that every galaxy is moving away from every other galaxy at great velocity

_____ 15. the amount of mass (matter) in a given volume (i.e., weight per volume)

_____ 16. groups of many billions of stars

_____ 17. our galaxy, a flattened spiral with curving arms, all in motion

_____ 18. a theory of how the Universe began

_____ 19. a cloud in space composed of gas and dust

_____ 20. the apparent outward force experienced by a body moving in a circle

_____ 21. the process in which intense heat causes small atoms to stick together and become larger atoms, and in so doing, to liberate energy

_____ 22. the central core of a spinning gas ball which has collapsed, been flattened by centrifugal force, and continues to collapse and get extremely dense and hot and which is the precursor of a star

_____ 23. a substance that can't be broken down into a simpler substance by any ordinary physical or chemical procedure

_____ 24. the cataclysmic explosion of a dying star

_____ 25. the residual ring of gas and dust surrounding a protostar turned star

_____ 26. tiny pieces of rock and metal formed from the dust specks of a planetary nebula

_____ 27. planet-sized bodies that develop when enough planetesimals accrete (collide and stick together)

_____ 28. solid chunks that are "left over" planetesimals, that were not incorporated into the original protoplanets but fell onto the protoplanets (and today fall onto the planets) when their orbits bring them close enough

_____ 29. a stream of protons and electrons flowing outward from the Sun

_____ 30. the inner, Earth-like planets (Mercury, Venus, Earth, and Mars)

_____ 31. the outer, Jupiter-like gas-giant planets (Jupiter, Saturn, Uranus, and Neptune)

_____ 32. the imaginary plane in space, traced out by the orbits of the planets, in which all of the planets (except Pluto) are found

Matching Questions

BELIEFS OR ACCOMPLISHMENTS

Match the belief or accomplishment with the person or group responsible for it.

a. astronomers in 1838
b. Nicolaus Copernicus (1473–1543)
c. C. J. Doppler (1803–1853)
d. Eratosthenes (276–194 B.C.)
e. Jean Foucault (1819–1868)
f. Galileo (1564–1642)
g. Greek astronomers, 3,000 years ago
h. Edwin Hubble

i. Edwin Hubble and Milton Humason (1929)
j. Isaac Newton (1642–1727)
k. Ptolemy (100–170 C.E.)
l. Paulo Toscanelli (1430)

_____ 1. Earth is a flat disk in the center of the celestial sphere, in which stars hang above Earth and there is a scary world below. Numerous gods and goddesses are the causes of astronomical occurrences.

_____ 2. Earth is a sphere that sits motionless at the center of the heavens. Everything else moves in circular orbits around it. Calculations supporting this idea predict planet wanderings.

_____ 3. Influenced Columbus; convinced him the Earth was spherical and he could sail around it and come back to where he started.

_____ 4. Wrote *De revolutionibus* and released the book just prior to his death. It supported the heliocentric concept.

_____ 5. Used a telescope to see the phases of Venus and the moons of Jupiter. His observations showed the geocentric theory could not be correct, and this got him into big trouble.

_____ 6. Derived the law of gravity and the laws of motion for objects in space.

_____ 7. Applied the principles of pendulum motion to prove that Earth rotates on its axis.

_____ 8. Calculated Earth's circumference, on the basis of the angles of incidence of the Sun's rays at different locations.

_____ 9. These were the first persons to apply the parallax method to determine the distance to a nearby star.

_____ 10. Discovered the following: "If the source of sound waves is moving away, the frequency of the waves goes down and therefore the pitch of the sound goes down. If the source is moving closer, the perceived frequency and therefore the perceived pitch go up."

_____ 11. Applied the Doppler effect to light waves from a distant galaxy and deduced the galaxy was moving away.

_____ 12. Developed the expanding Universe theory.

SIGNIFICANT NUMBERS

Match the number with its description.

a. 237,100 mi
b. 100 billion
c. 300 billion
d. 15 billion light years
e. 93,000,000 mi
f. 4.6 billion years
g. 10–20 billion years
h. 100, 000 light years
i. 24,865 mi
j. 1.3 light s
k. 8.3 light min
l. 6 trillion mi
m. 186,000 mi/s

_____ 1. number of miles in 1 light year
_____ 2. distance from Earth to the Sun (in miles)
_____ 3. distance from Earth to the Sun (in respect to speed of light)
_____ 4. age of Earth
_____ 5. age of the Universe
_____ 6. distance from Earth to the Moon (in miles)
_____ 7. distance from Earth to the Moon (in respect to speed of light)
_____ 8. distance to the edge of the visible Universe
_____ 9. circumference of Earth
_____ 10. speed of light
_____ 11. diameter of the Milky Way
_____ 12. number of stars in our galaxy
_____ 13. number of galaxies in the visible Universe

Short-Answer Questions

Answer the following with a few words, numbers, or phrases:

1. What very *un*scientific reasoning caused ancient astronomers to believe all astronomical bodies traveled in circular orbits? _____

2. What did Kepler (1571–1630) have to say about orbital paths? _____

3. Fill appropriate numbers in the following equation to show what a light year involves and how its value is calculated:

   186,000 mi/s × _____ s/min × _____ min/h × _____ h/day × _____ day/year = _____ mi/year

   (Handle the units of measurement properly as you go along to show the answer really does come out as miles per year.)

   To the closest trillion, how many miles are there in 1 light year? _____

4. When you gaze at the star Alpha Centauri (which, by the way, you'd have to do from the Southern Hemi-

sphere), you're looking at the closest star to Earth besides the Sun. It's 4.3 light years away. You see it now, but you're looking at the star as it was how many years ago? _____

5. How distant a star can a person see without the aid of a telescope? _____
_____
(In other words, how many years back in time can you look with the naked eye?)

6. Telescopes let you see much farther. The best modern ones allow you to look how far away? _____ light years. How far back in time is this? _____

7. What is the name of our galaxy? _____

8. How many stars are in it? _____

9. Name the galaxy that is our closest neighbor. _____

10. How far away is it? _____

11. Compare your answers for questions 5, 9, and 10. Do you need a telescope to see our nearest neighbor galaxy? _____

12. How can you tell the difference between stars and planets just by observing the night sky often and thoughtfully? _____

13. What force causes stars to pull together into systems called galaxies? _____

14. The size of a star and the length of its lifetime are related. For about how long will a star the size of our Sun burn? _____

15. You're sitting on some railroad tracks and hear a train whistle in the distance. The train is out of sight behind a bend. The pitch of the whistle is going down. Are you in danger of being hit by that particular train? _____ What science effect are you staking your life on? _____

Supply your own short answers or select from the choices offered in parentheses.

BIG BANG

1. Where was everything that was to become the Universe at the instant of beginning? _____

2. Why was there no matter in existence at that instant? _____

3. At age 1 s, temperature 5 billion°, what formed? _____

4. What did the Universe look like at age 1 million years? _____

5. As the gas distribution of the Universe started to vary, what happened to the original nebula? _____

6. What force caused revolving nebulae to collapse and spin faster? _____

7. What force caused these rapidly revolving nebulae to flatten into disks with bulbous central cores? _____

8. What is the name for these bulbous central cores? _____

9. Protostars eventually became dense enough and hot enough for colliding hydrogen atoms to release energy and stick together. What is the name of this process? _____

UNEXPECTED WAVELENGTHS AND FREQUENCIES

When the light from stars is observed through prisms that separate it into its component colors, the wavelengths are found to be shifted _____ (toward the red, toward the blue) end of the spectrum, and the frequencies are _____ (higher than, lower than) normal. This is known as the _____ (red shift, blue shift) and is an indication the Universe is _____ (expanding, collapsing). When this same thing happens to sound waves, the phenomenon is known as the _____ effect.

ELEMENT FACTORIES

1. What were the only two elements in existence in the early Universe? _____

2. After all the hydrogen of a star has been used up, what happens to the helium atoms? _____

3. Generally speaking, small atoms fuse to form larger atoms. For example, _____ (magnesium atoms, carbon atoms) might fuse to form _____ (carbon atoms, magnesium atoms).

4. Compared to our Sun, how large are the stars that are capable of forming the really large atoms? _____

SOLAR SYSTEM FORMATION

1. Rings of gas and dust surrounding protostars are called what? _____

2. These rings eventually develop into small particles of rock and metal called what? _____

3. The small rock and metal particles clump together to become what? _____

4. What blew most of the light, gaseous elements away from the inner planets? _____

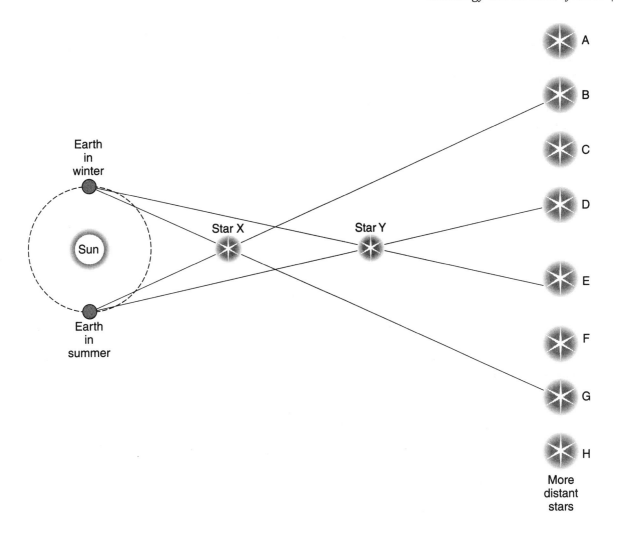

Making Use of Diagrams

View star X from Earth in summer. What star is in the background, along your line of sight? _____

View star X from Earth in winter. Now what star is in the background? _____

Star X appears to have moved from star _____ to star _____ when viewed from these different positions.

View star Y from Earth in summer. What star is in the background? _____

View star Y from Earth in winter. Now what star is in the background? _____

Star Y appears to have moved from star _____ to star _____ when viewed from these different positions.

Which star apparently moved farther, X or Y? _____

It's obvious from the diagram that star _____ is farther from Earth than star _____ is and the difference in their apparent motions is in agreement with what the _____ method would predict.

Short-Essay Questions

1. How did Ptolemy, who lived approximately 2,000 years ago, think the Universe was organized?

2. What more-enlightened viewpoint did Copernicus present about 500 years ago?

3. Where is Earth located in the Milky Way? In what direction is it moving? How fast is it going?

4. Why is there variety in the ages of existing stars?

5. Can our Sun ever become a supernova? Explain your answer.

6. Why do large stars have shorter life times than small stars?

7. How did the Moon form?

8. Planetesimals are jagged and irregular in shape; planets are nearly spherical. Explain why.

9. Our Sun isn't big enough to have formed the heavier elements, yet Earth has all 92 naturally occurring elements. Where did they come from?

10. You've read how the Doppler effect reveals when sound or light sources are moving toward or away from an observer. It does the same when applied to a radar image. Ordinary radar can recognize low-pressure air masses that signify a storm; Doppler radar can examine both edges of the storm mass and thereby tell a trained observer whether the storm is a tornado. What must Doppler radar be "seeing" that allows it to decide this?

Practice Test

1. The heavy elements on Earth
   a. were created by fusion in our Sun and blown here by the solar wind.
   b. were formed in a massive star and disbursed by its supernova event.
   c. were formed by fusion in first-generation stars.
   d. were in existence at the moment of the big bang.
   e. All of the above are true statements.

2. Imagine yourself in a universe that is exhibiting a blue shift. This must mean
   a. it's expanding much faster than our Universe.
   b. it's expanding but in a jerky fashion.
   c. it's collapsing.
   d. it's in a steady state, neither expanding nor collapsing.
   e. There's no such thing as a blue shift of light waves.

3. Parallax
   a. is the apparent motion of an object against more distant objects when viewed from different positions.
   b. is a method used in special radar instruments to observe tornadoes.
   c. was important evidence for the expanding Universe theory.
   d. shows that the farther away an object is, the more it seems to move against a distant background.
   e. All of the above are true statements.

4. Select the FALSE statement: The Doppler effect
   a. says the frequency of sound waves received is greater if the wave source is moving toward you than if it's moving away.
   b. says the pitch of a sound will get higher if the source is receding.
   c. is relevant to sound but can also be applied to the behavior of light waves.
   d. is the basic explanation for the red shift of our Universe.
   e. is the basic reason for the belief in an expanding Universe.

5. Identify the FALSE statement: The big bang
   a. occurred between 10 and 20 billion years ago.
   b. cataclysmically exploded matter outward.
   c. began with all matter and energy concentrated in a single point.
   d. theory states that at the instant of explosion, everything had so much thermal energy even the smallest atomic pieces couldn't stick together.
   e. is the explanation for how our solar system developed.

6. Earth is round because
   a. erosion over time has worn down the jagged edges of its planetesimals.
   b. its interior rock is warm enough to flow slowly in response to gravity.
   c. solar wind has shaped it.
   d. the gravitational tug of the Moon has worn down the jagged edges.
   e. the gravitational tug of the Sun has worn down the jagged edges.

7. Our Moon
   a. was formed by a collision of Earth with a Mars-sized protoplanet.
   b. was a separate small planet before it was captured by Earth's gravity.
   c. is 93 million mi away from Earth.
   d. is 8.3 light years away from Earth.
   e. has never been visited by humans.

8. Identify the FALSE statement: Stars
   a. create elements by fusing the nuclei of small atoms into the nuclei of larger atoms.
   b. have limited amounts of fuel and therefore don't exist forever.
   c. all explode cataclysmically when they die and contribute their matter to future star generations.
   d. the size of our Sun burn for about 10 billion years.
   e. begin as protostars, which fire up when they collapse and get denser and hotter.

9. Our solar system
   a. is centrally positioned in the Milky Way galaxy.
   b. formed 10 billion years ago.
   c. has a massive, first-generation star at its center.
   d. completes one rotation around the center of the Milky Way every 250 million years.
   e. All of the above are true statements.

10. If the star Alpha Centauri, 4.3 light years distant from the Earth, were to explode this instant, we would not be aware of this event until 4.3 years from now.
    a. True
    b. False

11. Mars, Jupiter, Saturn, and Venus are the Jovian planets.
    a. True
    b. False

12. Fusion reactions happen at a slower rate in bigger stars than in smaller ones.
    a. True
    b. False

13. The element carbon, so important to life, was formed in first-generation stars
    a. True
    b. False

14. Planets can be distinguished from stars because their light doesn't twinkle as light from stars does.
    a. True
    b. False

15. Johannes Kepler showed the planets followed elliptical, not circular orbits.
    a. True
    b. False

16. Galileo's discoveries of the phases of Venus and the moons of Jupiter supported the geocentric Universe concept.
    a. True
    b. False

17. Because of Earth's rotation, a swinging pendulum rotates about a vertical axis once every 24 h.
    a. True
    b. False

18. Galaxies are evenly distributed throughout the Universe.
    a. True
    b. False

19. In the early Universe the only two elements in existence were carbon and hydrogen.
    a. True
    b. False

20. The Sun is roughly 6 trillion mi (1 light year) away from Earth.
    a. True
    b. False

21. Modern telescopes have allowed us to see roughly 15 billion years back into the past.
    a. True
    b. False

22. Our Sun is a first-generation star.
    a. True
    b. False

23. The solar wind is a stream of protons and electrons streaming outward from the Sun.
    a. True
    b. False

24. The solar wind blew light, gaseous elements out of the inner solar system.
    a. True
    b. False

25. All the planets except Jupiter travel their orbits within the ecliptic.
    a. True
    b. False

ANSWERS

Completion

1. cosmology
2. science
3. Universe
4. geocentric Universe concept
5. heliocentric Universe concept
6. parallax
7. light year
8. Doppler effect
9. pitch
10. frequency
11. wavelength
12. red shift
13. blue shift
14. expanding Universe theory
15. density
16. galaxies
17. Milky Way
18. big bang theory
19. nebula
20. centrifugal force
21. nuclear fusion
22. protostar
23. element
24. supernova
25. planetary nebula
26. planetesimals
27. protoplanets
28. meteors
29. solar wind
30. terrestrial planets
31. Jovian planets
32. ecliptic

Matching Questions

BELIEFS OR ACCOMPLISHMENTS

1. g
2. k
3. l
4. b
5. f
6. j
7. e
8. d
9. a
10. c
11. i
12. h

SIGNIFICANT NUMBERS

1. l
2. e
3. k
4. f
5. g
6. a
7. j
8. d
9. i
10. m
11. h
12. c
13. b

8 | Chapter 1

Short-Answer Questions

1. Circles were considered the most perfect of geometric forms and the Universe was a perfect creation.
2. Planets have elliptical, not circular, orbits.
3. 60; 60; 24; 365$\frac{1}{4}$; 5,869,713,600,000; 6 trillion mi
4. 4.3 years
5. about 2.2 million light years distant
6. 15 billion; 15 billion years
7. the Milky Way
8. over 300 billion
9. Andromeda
10. over 2.2 million light years distant
11. no
12. Planets move relative to the stars and to each other; stars remain fixed relative to each other.
13. gravity
14. 10 billion years
15. no; Doppler effect

BIG BANG

1. Packed into one point, so small it occupied no volume at all.
2. Too hot; thermal energy kept the atomic pieces from sticking together.
3. protons and neutrons, the basic building blocks of atoms
4. a hot cloud of gas (98% hydrogen and 2% helium)
5. Clumped into nebulae that separated from each other, with emptier space between them.
6. gravity
7. centrifugal force
8. protostars
9. nuclear fusion

UNEXPECTED WAVELENGTHS AND FREQUENCIES

toward the red; lower than; red shift; expanding; Doppler

ELEMENT FACTORIES

1. hydrogen and helium
2. Helium atoms fuse to form larger atoms.
3. carbon atoms; magnesium atoms
4. 10 to 30 times the mass of the Sun

SOLAR SYSTEM FORMATION

1. planetary nebulae
2. planetesimals
3. protoplanets
4. solar wind

Making Use of Diagrams

star B; star G; B; G
star D; star E; D; E
star X
Y; X; parallax

Short-Essay Questions

1. Ptolemy believed in the geocentric Universe concept, in which Earth sat motionless at the center of the Universe and all other heavenly bodies orbited in perfect circles around it. He supported his belief with mathematical calculations that seemed to predict planet wanderings. The church favored Ptolemy's idea, and it became the only safe viewpoint to express in the Western world for more than a thousand years.
2. Copernicus reintroduced the heliocentric (Sun-centered) concept of the motion of heavenly bodies in a book, *De revolutionibus,* published just days before he died.
3. The Milky Way galaxy is a flattened spiral, 100,000 light years across, with giant curving arms rotating about its center. Earth sits near the outer edge on one of these arms and rotates around the center about once every 250 million years. Relative to the galactic center, it's traveling about 200 km/s.
4. Stars are of different sizes, burn (nuclear fusion) at different rates, and consequently have different life spans. Therefore at any given moment in time there are many generations of stars in existence, from the new, young ones to the old, dying stars.
5. Our Sun can never become a supernova because it's not big enough. Only very massive stars are dense enough and therefore hot enough to create the cataclysmic explosion known as a supernova.
6. Large stars are denser and therefore hotter than small stars. More heat produces faster fusion, so large stars burn faster, use up their fuel faster, and die sooner.

7. On the basis of rock samples brought back by mission Apollo astronauts, current thought is that the Moon formed when a Mars-sized protoplanet hit Earth tens of millions of years ago and blasted away much of Earth's rocky mantle. This debris orbited Earth and coalesced to become the Moon.

8. Planetesimals keep jagged, irregular shapes as long as they remain small, cool, and rigid. As they coalesce, they become part of a larger body. When this body reaches a critical size, a little smaller than our Moon, its insides get warm enough to allow rock to soften and slowly flow under the influence of gravity. The new shape is one that allows the force of gravity to be the same at all points on its surface, in other words, a sphere.

9. All earth elements came from some star, including but not limited to our Sun. Our solar system consists of material scattered into space by a supernova event of a previous-generation massive star. We are all literally made of stardust.

10. The Doppler effect tells whether the wave source is moving toward or away from the observer and how fast it's moving. If one side of the air mass is moving away and the other side is moving toward, the air mass is rotating. If you know the speed of rotation, you can tell whether the air mass is a tornado.

Practice Test

1. b The formation of elements larger than iron can only take place under conditions achieved during supernova explosions.

2. c When a light source moves toward you, it becomes blue as it shifts to a higher frequency; receding light shifts to red, a lower frequency.

3. a Astronomers observe the position of a nearby star relative to a backdrop of very distant stars at times that are 6 months apart, measure the angle, and use trigonometry to complete their calculation of the distance.

4. b The Doppler effect states that the pitch will get lower as the source recedes.

5. e The big bang theory is the explanation for how our Universe began.

6. b A sphere is a shape that permits the force of gravity to be the same at all points on its surface.

7. a Samples from the Apollo mission confirmed that the Moon was formed from a piece of Earth's rocky mantle.

8. c Only massive stars become supernovas.

9. d Our solar system lies near the outer edge of the Milky Way, it formed less than 5 billion years ago, and our Sun is a third- or fourth-generation star.

10. True By the definition of light year, the light we see from Alpha Centauri left the star 4.3 years ago, so we are looking at the past; to see what is happening on Alpha Centauri now, we must wait 4.3 years.

11. False Jupiter, Saturn, Uranus, and Neptune are the Jovian planets; Mercury, Venus, Earth, and Mars are the terrestrial planets.

12. False Large stars, because of their greater mass, become denser and therefore hotter than small stars; fusion reactions are accelerated; and the large stars burn out faster.

13. False Hydrogen, the lightest element in the Universe, was formed by first-generation stars.

14. False Planets move in relation to each other and to stars; stars remain fixed relative to one another.

15. True By showing that the planets follow elliptical orbits, Kepler showed that Ptolemy's calculations, the bulwark of the geocentric hypothesis, were wrong.

16. False Galileo's discoveries demonstrated that not all heavenly bodies orbit Earth.

17. True Foucault's demonstration of this in the middle of the nineteenth century finally proved Earth rotated on its axis.

18. False There are some regions of the Universe where galaxies cluster together, and others that have no galaxies.

19. False Hydrogen and helium were the only two elements that existed in the early Universe.

20. False The Sun is 93,000,000 mi (about 8.3 light min) from Earth.

21. True The edge of the visible Universe is about 15 billion light years distant; this means that light traveling to Earth from the edge of the Universe began its journey over 10 billion years before Earth even existed.

22. False Our Sun is less than 5 billion years old and is likely a third- or fourth-generation star.

23. **True** When the Sun began to produce fusion energy, it also produced a solar wind, a stream of particles that had enough energy to escape the Sun's gravity.

24. **True** The solar wind left the inner planets (Mercury, Venus, Earth, and Mars) relatively free of light, gaseous elements such as hydrogen and helium.

25. **False** Pluto is the exception, not Jupiter.

CHAPTER 2 | Journey to the Center of the Earth

GUIDE TO READING

Chapter 1 dealt with the long ago and far way, presenting theories about the origin and development of the Universe. Chapter 2 brings you into the present and much closer to home with an overview—and an inner view—of planet Earth. It starts miles high above Earth in the vacuum of interplanetary space, then zooms you down through Earth's magnetic field, magnetosphere, and Van Allen radiation belts, pausing in the troposphere to comment on the obvious topography of Earth and the great amount of hydrosphere covering Earth's surface. The journey continues, diving down below the ocean surface and progressing through Earth's crust, mantle, and core.

Most of the chapter deals with the ocean bottom and inner earth, as preparation for plate tectonics theory, presented in Chapters 3 and 4. The author discusses Earth's composition (organic chemicals, minerals, glasses, rocks, metals, melts, and volatiles) and layers (oceanic and continental crust, the Moho, oceanic and continental lithosphere, asthenosphere, upper mantle, transition zone, lower mantle, outer core, and inner core.)

Few humans have visited the ocean bottom, and no human has physically been more than 2 mi below Earth's land surface. How do we know what it's like inside the Earth? Your author describes scientists' efforts to reach deep into Earth and discusses the various approaches and scientific reasoning that have provided answers to questions about composition, structure, and conditions within Earth. These include the use of clues obtained from measuring Earth's density and shape and the study of earthquake waves, seismic velocities, and meteorites.

There are lots of terms and many numbers involved in this survey of Earth. Try not to get mired down in lists of rock types or thickness of layers. Instead concentrate on the thought processes necessary to analyze something you can neither see nor touch and on the truly amazing world that exists under your feet. John Milton's underworld of the 1600s and Jules Verne's fanciful journey of the 1800s were tame compared to the real thing. They imagined exotic versions of environments on Earth's surface, all places within the realm of human experience. The real interior of Earth is beyond any human's experience. It's a place of awesome pressures and temperatures, much closer than the stars but just as unreachable. Truth can certainly be stranger than fiction.

LEARNING ACTIVITIES

Completion

Test your recall of new vocabulary terms.

_____ 1. an area that contains very little matter per given volume

_____ 2. icy remnants of the material that formed our solar system, some of which orbit the sun in very elliptical paths

_____ 3. any region that is affected by the force of a magnet

_____ 4. anything that has a magnetic field and possesses both a north and a south pole

_____ 5. imaginary lines along which magnetic particles in the field would align

_____ 6. the region between Earth's surface and its magnetic field lines

_____ 7. areas about 10,500 and 3,000 km out from Earth that are composed of solar wind particles and cosmic rays (largely hydrogen nuclei)

_____ 8. the envelope of gas encompassing Earth

_____ 9. force per unit area; in respect to air, the "push" that squeezes molecules of air closer together

_____ 10. the lowest layer of Earth's atmosphere, just above its surface, in which convection occurs and weather exists

_____ 11. the circulation of a fluid, like air, in which warm less dense fluid rises and cold more dense fluid sinks

_____ 12. variations in elevations of Earth's surface; the "ups and downs" of the land

_____ 13. all of Earth's surface that's covered by water: oceans, lakes, and streams

_____ 14. a graph that plots the percentage of all of Earth's solid surface, submarine as well as subaerial, against its surface elevation

_____ 15. vibrations of Earth caused by the rapid release of energy as rocks break within Earth

_____ 16. simply means "earthquake"

_____ 17. a fracture in the Earth along which sliding occurs

_____ 18. the speed at which earthquake waves travel

_____ 19. the boundary at which an abrupt change in seismic velocity occurs

_____ 20. the rate of change in temperature as you go downward into Earth

_____ 21. a meteoroid that was large enough to survive its trip through the atmosphere and land on Earth's surface

_____ 22. the water that fills microscopic pores (open spaces) that exist in the top several kilometers of both oceanic and continental crust

_____ 23. the resistance of a material to flow (A high degree means it flows with difficulty, like cold molasses; a low degree means it flows freely and smoothly, like water.)

Matching Questions

CATEGORIES OF EARTH MATERIALS

a. alloy
b. crystal
c. glass
d. grain
e. igneous rocks
f. lava
g. magma
h. melts
i. metals
j. metamorphic rocks
k. mineral
l. organic chemicals
m. precipitate
n. rocks
o. sedimentary rocks
p. volatiles

Choose from the list of terms above.

_____ 1. carbon-containing compounds found in living organisms

_____ 2. a naturally occurring inorganic solid with an orderly internal structure (crystalline)

_____ 3. the solid formed when atoms that have been in solution come together and don't stay dissolved

_____ 4. a single bit of mineral that exhibits a very orderly structure, with planar surfaces and specific angles between them

_____ 5. a fragment of a once larger crystal or crystal group

_____ 6. a form of solid that froze from its liquid state so quickly it didn't have time to develop an orderly internal arrangement

_____ 7. aggregates (mixtures) of mineral crystals or grains, and masses of natural glass

_____ 8. rocks formed when hot molten rock cools and hardens

_____ 9. rocks formed by the natural cementing together of grains of previously existing rock (sediments) or by the precipitation of mineral material out of water solution

_____ 10. changed rocks, formed from previously existing rocks under the influence of heat and pressure

_____ 11. solids composed of a category of atoms known as metals (copper, iron, etc.)

_____ 12. a mixture formed by the melting together of two or more metals (example: bronze)

_____ 13. solid materials that are hot enough to assume their liquid state

_____ 14. molten rock beneath Earth's surface

_____ 15. molten rock that has exited Earth and flowed on its surface

_____ 16. materials that change to their gaseous state with just a slight rise in temperature

ROCKS AND MINERALS

Choose from the list of terms below.

a. basalt
b. gabbro
c. granite
d. intermediate
e. mafic
f. peridotite
g. silica
h. silicic
i. silicate minerals
j. silicate rocks
k. ultramafic

_____ 1. a silicic igneous rock with large crystals

_____ 2. a mafic igneous rock with small crystals

_____ 3. a mafic igneous rock with large crystals

_____ 4. an ultramafic igneous rock with large crystals, that makes up the mantle

_____ 5. a mineral or rock composition between silicic and mafic

_____ 6. a composition high in iron and magnesium, low in silica

_____ 7. a composition extremely rich in iron and magnesium, extremely low in silica

_____ 8. the chemical compound silicon dioxide (SiO_2)

_____ 9. rich in silica

_____ 10. minerals composed of silica and some other elements

_____ 11. rocks composed of silicate minerals

LAYERS OF THE EARTH

Choose from the list of terms below.

a. asthenosphere
b. continental crust
c. inner core
d. lithosphere
e. lower mantle
f. mantle
g. oceanic crust
h. outer core
i. transition zone
j. upper mantle

_____ 1. the portion of the outermost layer of Earth that underlies the continents

_____ 2. the portion of the outermost layer of Earth that underlies the oceans

_____ 3. the rigid, nonflowable Earth layer that consists of crust and the uppermost part of the mantle and starts at the surface of Earth and goes down to a depth of 100–150 km

_____ 4. the softer, flowable layer below the lithosphere, which starts at 100–150 km

_____ 5. the entire layer under the crust and above the core, whose upper boundary is called the Moho

_____ 6. the upper portion of the mantle, which extends from the Moho down to a depth of 440 km

_____ 7. the middle portion of the mantle, from 440 km down to 670 km

_____ 8. the lower portion of the mantle, from 670 km down to the core-mantle boundary

_____ 9. the outer part of Earth's inner layer, from 3,900 km down to 5,150 km

_____ 10. the innermost part of Earth's inner layer, from 5,150 km down to Earth's center at 6,370 km

Short-Answer Questions

Answer the following with a few words, numbers, or phrases:

1. What are the three sources of the few atoms that do exist in interplanetary space? _____

2. For how long have scientists been aware that Earth acts like a giant magnet? _____

3. What causes a comet's tail to form as it approaches the sun? _____

4. What deflects solar wind particles and keeps most of them from hitting Earth? _____

5. How do the Van Allen radiation belts benefit life on Earth? _____

6. a. The atmosphere has two major components and several minor ones. Which gas makes up roughly 78% of the atmosphere? _____
   b. Which gas makes up about 21%? _____
   c. List several gases that make up the remaining 1%. _____

7. The higher you get above Earth's surface, the _____ (greater, less) the weight of the overlying column of air and therefore the _____ (higher, lower) the air pressure.

8. Why is rocket propulsion a necessity for aircraft that fly higher than 25 km? _____

9. Most land surface lies no higher than _____ km above sea level. Most sea floor lies between _____ km and _____ km deep.

10. List the four elements that make up 90% of all Earth matter, and the percentage of each, in descending order. _____

11. Are topographic differences limited to the dry land surface of Earth? Briefly support your answer. _____

12. Are all carbon-containing compounds organic? Explain. _____

13. What are the three main rock groups? _____

14. The deepest humans have set foot in Earth is in a gold mine in Africa. How deep is this? _____

15. Which rocks are denser, silicic rocks or mafic rocks? _____

16. a. What is the numerical value of the geothermal gradient in the upper part of the crust? _____
    b. As you go deeper into Earth, what happens to the gradient? _____

17. Calculations (not observations) show temperatures at Earth's center to be just slightly less than the temperature of the Sun's surface. What is this temperature? _____

18. The crust is not just the cooled surface of the mantle. What is its origin? _____

19. a. If (a big if) you could drive, at highway speed of 100 km/h (60 mph), from Earth's surface all the way through the oceanic crust to the crust-mantle boundary, how long would it take? _____
    b. If you could continue going on down through the mantle and core all the way to Earth's center, how much longer would this take? _____

20. How do scientists recognize where the crust ends and the mantle begins? _____

21. Why is the crust-mantle boundary called the Moho? _____

22. List three subdivisions of oceanic crust and give the composition of each. _____

23. Approximately how much thicker is continental crust than ocean crust? _____

24. What fills the tiny pore spaces of upper crustal rock, both continental and oceanic varieties? _____

25. Mantle rock is solid, not liquid, but some of it does flow. How fast does it flow? _____

26. Geologists have collected data that show a blotchy pattern of warmer and cooler areas of the mantle. Analysis of the pattern leads them to conclude what? _____

27. Geologists believe Earth's core is composed of an iron alloy. What elements compose this alloy? _____

28. Give two lines of investigation geologists used to decide Earth's core is divided into two regions, an outer liquid iron alloy and an inner solid iron alloy. _____

29. How can Earth's inner core be solid when it is so extremely hot? _____

30. The term "lithosphere" refers to the outer rigid layer of Earth. In terms of crust/mantle/core classification, what is the lithosphere composed of? _____

31. Is the bottom of the lithosphere above or below the Moho? _____

32. a. What is the maximum thickness of the ocean lithosphere? _____
    b. What is the maximum thickness of the continental lithosphere? _____

33. Name the layer of rock, capable of flowing, that is immediately below the lithosphere. _____

34. a. In terms of Earth layers, what is significant about the temperature 1,250°C? _____
    b. Can rock melt at this temperature? _____

Making Use of Diagrams

Use the diagram provided to answer the following questions pertaining to the layers of Earth:

1. On the diagram, fill in the names of the layers marked a, b, and c.

2. Identify the following layers of Earth:
   a. The _____ is the boundary layer at 70 km.
   b. The region from 70 to 440 km defines the _____.

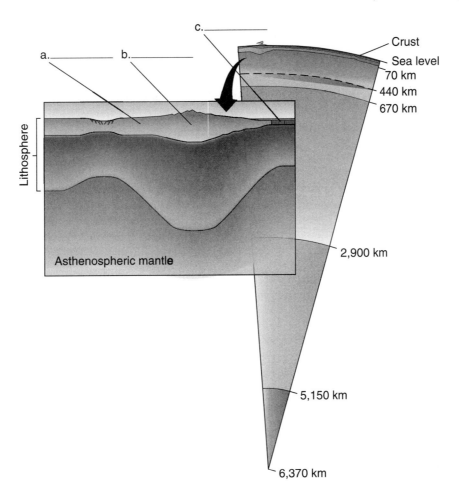

c. The region from 440 to 670 km defines the
   _____.

d. The region from 670 to 2,900 km defines the
   _____.

e. The region from 2,900 to 5,150 km defines the
   _____.

f. The region from 5,150 to 6,370 km defines the
   _____.

Short-Essay Questions

1. *Into Thin Air* is a recent popular nonfiction book about climbing Mt. Everest. Offer some facts to show this is an appropriate title.

2. Humans have barely penetrated Earth's surface, yet scientists believe they know much about Earth's interior. Briefly discuss how they've used three lines of investigation (Earth's density, Earth's shape, and earthquakes) to study Earth's interior.

3. We don't have samples, but scientists believe they know the composition of Earth's interior. List four lines of evidence they've used to determine this.

4. What is drilling mud and what is its purpose?

5. Discuss Project Mohole, the Deep Sea Drilling Project, and the Ocean Drilling Project.

6. Discuss deep drilling on land for the purpose of scientific research. Your answer should include where they're drilling, how far they've gotten, and why they haven't gone deeper.

Practice Test

1. The two most common gases of Earth's atmosphere and their percentages are
   a. carbon dioxide (CO_2) 21%, and oxygen (O_2), 78%.
   b. water (H_2O), 21%, and oxygen (O_2), 78%.
   c. oxygen (O_2), 21%, and carbon dioxide (CO_2), 78%.
   d. nitrogen (N_2), 21%, and oxygen (O_2), 78%.
   e. nitrogen (N_2), 78%, and oxygen (O_2), 21%.

2. Which of the following would you consult to predict the effect of sea-level changes on shapes of land areas?
   a. seismic velocity graphs
   b. hypsometric curve
   c. ship's log of the *Glomar Challenger*
   d. readings of a magnetosphere
   e. All would be relevant.

3. Choose the list that shows rock composition decreasing in silica and increasing in iron and magnesium.
   a. granite, basalt, peridotite
   b. peridotite, basalt, granite
   c. basalt, peridotite, granite
   d. granite, peridotite, basalt
   e. The question is invalid; they're all equally rich in iron and magnesium.

4. Boundaries between Earth layers
   a. are defined by abrupt changes in earthquake wave velocities.
   b. are defined by seismic velocity discontinuities.
   c. are places where there's abrupt change in the density of the rocks.
   d. have been discovered over the last 200 years.
   e. All of the above.

5. Which of the following statements is FALSE? Peridotite
   a. is rare at Earth's surface.
   b. is the most abundant rock of Earth.
   c. is ultramafic.
   d. is very rich in iron and magnesium, very poor in silica.
   e. None of the above.

6. Which of the following statements is FALSE?
   a. The Moho marks the boundary between the lithosphere and the asthenosphere.
   b. Continental crust can be four times as thick as ocean crust.
   c. Continental crust is less mafic than ocean crust is.
   d. Ocean crust is denser than continental crust is.
   e. Ocean crust consists of basalt and gabbro; continental crust consists of a great variety of rock types.

7. Which of the following statements concerning density is TRUE?
   a. Earth's density is a little over 5 g/cm³.
   b. The density of crustal rock is between 2.2 and 2.5 g/cm³.
   c. Mafic rock is more dense than silicic rock.
   d. Changes in density can be detected by changes in seismic velocity.
   e. All of the above are true statements.

8. Which of the following is FALSE? Convection currents
   a. cause hotter fluid to rise and cooler fluid to sink.
   b. occur in a pan of boiling water.
   c. occur in the lithosphere, bringing new rock up to replace eroded rock.
   d. occur in the mantle, from the asthenosphere down.
   e. occur in the atmosphere and are an important part of weather phenomena.

9. Which of the following statements about the crust is FALSE?
   a. The crust is the top of the mantle that has cooled and hardened.
   b. The upper portion of the crust contains microscopic pores that contain groundwater.
   c. The crust plus the upper part of the mantle make up the lithosphere.
   d. There are two different types of crust: oceanic and continental.
   e. The crust has never been penetrated down to the crust-mantle boundary.

10. Roughly 50% of Earth's surface is covered by hydrosphere.
    a. True
    b. False

11. The Van Allen radiation belts serve as a trap for cosmic rays and thus shield life on Earth from excessive radiation.
    a. True
    b. False

12. The troposphere is the highest layer of atmosphere, beyond which is the vacuum of space.
    a. True
    b. False

13. Iron, oxygen, silicon, and potassium make up most of Earth's matter.
    a. True
    b. False

14. A rock is a solid organic compound in which atoms are arranged in an orderly pattern (crystalline); a mineral is an aggregate (mixture) of rocks.
    a. True
    b. False

15. Basalt, gabbro, and peridotite are all mafic, but basalt has small crystals and the others have large crystals.
    a. True
    b. False

16. The deepest well ever drilled penetrates the upper half of the continental crust.
    a. True
    b. False

17. Earthquakes are generated when rock moves along faults.
    a. True
    b. False

18. The core accounts for most of the volume of Earth.
    a. True
    b. False

19. The geothermal gradient increases as you go deeper into the Earth.
    a. True
    b. False

20. Earth's core is composed of iron alloy and gold.
    a. True
    b. False

21. The inner core is so hot it's liquid, and therefore some seismic waves can't travel through it.
    a. True
    b. False

22. The *Glomar Challenger* and *Joides Resolution* are both ships that have been used in ocean floor drilling projects.
    a. True
    b. False

23. All mantle rock eventually becomes hot enough to melt, then flow at the rate of several kilometers per year.
    a. True
    b. False

ANSWERS

Completion

1. vacuum
2. comets
3. magnetic field
4. dipole
5. magnetic field lines
6. magnetosphere
7. Van Allen radiation belts
8. atmosphere
9. pressure
10. troposphere
11. convection
12. topography
13. hydrosphere
14. hypsometric curve
15. earthquakes
16. seismic
17. fault
18. seismic velocity
19. seismic velocity discontinuity
20. geothermal gradient
21. meteorite
22. groundwater
23. viscosity

Matching Questions

CATEGORIES OF EARTH MATERIALS

1. l
2. k
3. m
4. b
5. d
6. c
7. n
8. e
9. o
10. j
11. i
12. a
13. h
14. g
15. f
16. p

ROCKS AND MINERALS

1. c
2. a
3. b
4. f
5. d
6. e
7. k
8. g
9. h
10. i
11. j

LAYERS OF THE EARTH

1. b
2. g
3. d
4. a
5. f
6. j
7. i
8. e
9. h
10. c

Short-Answer Questions

1. the nebulae that formed our solar system; atmospheres of other planets; the Sun by the solar wind

2. William Gilbert, Queen Elizabeth's physician, proposed the idea around 1600.

3. The solar wind blows gas out of the icy (but warming and evaporating) comet body.

4. Earth's magnetic field

5. They trap cosmic rays and thus lower the amount of radiation reaching Earth; this is good for life here.

6. a. nitrogen   b. oxygen   c. carbon dioxide, neon, methane, ozone, carbon monoxide, sulfur dioxide

7. less; lower

8. There's not enough oxygen above this level to allow a jet engine to function.

9. 1; 2.5; 4.5

10. iron, 35%; oxygen, 30%; silicon, 15%; magnesium, 10%

11. No, there are submarine plains, ridges, mountains, and troughs.

12. No, only those that occur in living organisms are considered organic; simple ones like pure carbon (C), carbon dioxide (CO_2), carbon monoxide (CO), lime (CaO), and calcium carbonate ($CaCO_3$) aren't organic.

13. igneous, sedimentary, and metamorphic

14. 3.5 km, or about 2 mi

15. mafic rocks

16. a. 20–30°C/km   b. Decreases to 10° or less.

17. 4,300°C

18. elements brought to Earth's surface as magma

19. a. 5 min   b. 63 h

20. There's a seismic velocity discontinuity there.

21. in honor of Andrija Mohorovicic, who discovered the seismic velocity discontinuity there in 1909

22. on top: a sediment layer composed of clay grains and plankton; middle layer: basalt; bottom layer: gabbro

23. Continental crust is about 4 times as thick as ocean crust, but there is some variation.

24. groundwater

25. very slowly; just a few cm per year, which is comparable to the growth of fingernails

26. The mantle is convecting, like simmering water; hot rock slowly rises, cooler rock slowly sinks.

27. iron, oxygen, nickel, silicon, and sulfur

28. how seismic waves refract as they pass through Earth and the fact that a certain type of seismic wave can't pass through the outer core

29. Incredible confining pressure on the atoms, simply because they exist amid all the Earth material, doesn't allow them to move around (be liquid) even with all the thermal energy they possess.

30. the crust and the uppermost rigid part of the mantle

31. below

32. a. up to about 100 km (60 mi)   b. up to about 150 km (80 mi)

33. the asthenosphere

34. a. It's the temperature at the boundary between lithosphere and asthenosphere; rock hotter than this is soft enough to flow.
    b. No, it must be much hotter to melt.

Making Use of Diagrams

1. a. continental crust, normal   b. continental crust, thickened   c. oceanic crust

2. a. Moho   b. upper mantle   c. transition zone
   d. lower mantle   e. outer core   f. inner core

Short-Essay Questions

1. At sea level air pressure averages 14.7 lb/in^2, also known simply as 1 atmosphere. Atop Mt. Everest the air pressure is 0.3 atmospheres, which is too little air for humans to survive on for more than a few hours.

2. From Earth's density, creative analysis of a displaced plumb bob in a surveying expedition in the Andes in the late 1700s correctly gave the average density of Earth as greater than 5 g/cm^3. Crustal rocks average between 2.2 and 2.5 g/cm^3. Therefore Earth's interior rocks must be significantly denser than crustal rocks, and Earth's interior cannot be filled with a lot of open space. From Earth's shape, if Earth's density were concentrated toward its surface, centrifugal force would cause Earth to be a flattened disk. It isn't, so much of Earth's mass must lie close to its core. Earth's land surface does not respond to the Moon's gravitational pull and rise and fall significantly, as does the ocean surface (tides), so Earth is not a molten ball enveloped by a thin, outer solid crust. From earthquakes, studies of seismic waves show abrupt changes in velocities as they reach different depths. This indicates the rock type changes significantly, so these depths are considered layer boundaries.

3. a. lab experiments to simulate lava and thus theorize about magma within Earth
   b. studies of unusual rock chunks that have been carried up from depth in magma
   c. studies of seismic wave velocities, both under lab conditions and within the Earth
   d. studies of meteorites to theorize what planetesimals of early Earth were like

4. Drilling mud is a mixture of water and clay that's piped down into a drill hole, leaves the drill stem through holes in the drill bit, and works its way back up to the surface between drill stem and drill hole sides. It cools the drill stem, which is hot due to friction, and flushes out pulverized rock.

5. Project Mohole was an attempt in the early 1960s to drill all the way through the ocean crust and into the Moho. It never made it. The Deep Sea Drilling Project followed in the late 1960s. The drilling ship *Glomar Challenger* crisscrossed the sea drilling hundreds of holes, from 300 m to 2 km deep, in water depths up to 5 km. The cores obtained provided much valuable scientific information. The Ocean Drilling Project was the successor to the Deep Sea Drilling Project. It, too, did research drilling from the deck of a larger ship, the *Joides Resolution*. The Moho has never been reached.

6. Scientists are drilling on land on the Kola Peninsula, northern Russia. They've reached a depth of 12 km, which is about 7 mi. With today's technology it's impossible to drill deeper, because the drill bits melt and cuttings can't be flushed out. Even if technology is developed to do the job it might not be done because it's extremely expensive.

Practice Test

1. e Earth's atmosphere consists of 78% nitrogen, 21% oxygen, with minor amounts of argon, carbon dioxide, neon, methane, ozone, carbon monoxide, and sulfur dioxide.

2. b The hyposometric curve shows at a glance what percentage of land lies at what elevation.

3. a Granite is a silicic rock; basalt is a mafic rock; and peridotite is an ultramafic rock.

4. e Statements a to d are all true.

5. e Statements a to d are all true.

6. a The Moho marks the boundary between the crust and the mantle.

7. e Statements a to d are all true.

8. c Lithosphere is rigid and doesn't flow; convection happens below the lithosphere.

9. a The crust is not simply cooled mantle; rather, it consists of rocks created from elements brought to the surface in magma.

10. False Roughly *70%* of the Earth's surface is covered by hydrosphere.

11. True Van Allen radiation belts do trap cosmic rays and shield Earth from excessive radiation.

12. False The troposphere is the closest layer in Earth's atmosphere, where atmospheric convection and weather occur.

13. False Potassium does not belong in the list; magnesium does.

14. False Rock is an aggregate of minerals; a mineral is a solid inorganic compound.

15. True Basalt is a fine-grained mafic rock; gabbro and peridotite are coarse-grained mafic rock.

16. False Continental crust is at least 35 km thick; the deepest well, in northern Russia, is only 12 km deep.

17. True Movement along faults is the most common cause of major earthquakes.

18. False The mantle accounts for most of the volume of Earth.

19. False Although the temperature increases with depth, the geothermal gradient decreases from about 20–30°C/km near the surface to 10°C/km or less in the core.

20. False Earth's core is partly composed of iron alloy but not gold.

21. False The outer core is liquid and impedes travel of some seismic waves.

22. True The *Glomar Challenger* was used in the Deep Sea Drilling Project of the 1960s. The *Joides Resolution* was used in the more recent Ocean Drilling Project.

23. False Mantle rock does flow, but it remains solid and flows very slowly (about as fast as your fingernails grow, which is a very few centimeters per year).

CHAPTER 3 | Drifting Continents and Spreading Seas

GUIDE TO READING

This chapter sets the stage for an explanation of *plate tectonics*, which is explained in detail in Chapter 4. In a sense, this chapter is also setting the stage for the rest of the text, because plate tectonics theory, a relative newcomer to geologic thought, supplies the fundamental explanation for so many geologic processes that it has become a unifying principle in the modern study of geology.

The author begins by drawing a picture of Alfred Wegener who, in the early 1900s, proposed the idea of *continental drift*, the idea that the continents have moved around in relation to one another. His arguments in support of continental drift included:

- the fit of the continents
- paleoclimatic studies that showed evidence of past glaciation; coal deposits, ancient reef deposits, ancient sand dunes, and salt beds that make no sense in today's world climate belts
- the occurrence of the same fossils on lands now separated by oceans
- the matching of geologic units (distinct assemblages of rocks) on lands now separated by oceans

For several reasons, Wegener's ideas were not accepted for decades. First, although he was a scientist, he was not a geologist. Second, accepting his ideas would have meant huge changes in geologic thought. And third, he couldn't supply an explanation of how and why continents moved. New discoveries after his death eventually proved that continents do move. The new areas of study involved:

- changes in the earth's magnetic field over time (paleomagnetism)
- changes in the sea floor—the shape of its surface, types and ages of its rock, heat flow within it, and sea-floor earthquakes—all of which support the idea of sea-floor spreading

The author spends considerable time developing a historical context for the theory of plate tectonics. Why? Because:

- working through the reasons for accepting new theories, such as the meaning and significance of paleomagnetism, provides practice in good scientific thinking
- the gradual acceptance of plate tectonics theory provides an excellent example of the process by which scientific knowledge advances as new evidence and better instruments and techniques are introduced
- plate tectonics was a revolutionary idea that caused profound changes in the study of geology and thus merits your thoughtful study and understanding

LEARNING ACTIVITIES

Completion

Test your recall of new vocabulary terms.

_____ 1. the name Wegener used for his supercontinent (It's Greek for "all land.")

_____ 2. Wegener's name for his theory of shifting continents

_____ 3. the geologic interval that began 245 million years ago, at the end of the Paleozoic

_____ 4. the geologic interval that preceded the Mesozoic, between 540 and 245 million years ago

_____ 5. the geologic interval between Earth's formation 4.6 billion years ago and the beginning of the Paleozoic 545 million years ago

_____ 6. rivers, or sheets, of ice that move slowly across the land

_____ 7. unconsolidated surface material that may vary in size, from sand, silt, and clay to pebbles and boulders

_____ 8. the glacial sediment carried on glacial ice

_____ 9. evidence of life in the geologic past (preserved relics like shells and bones)

_____ 10. times in Earth's history when large areas of land were covered by glaciers

_____ 11. the theory that an electric current generates magnetism and that the movement of an electrical conductor in a magnetic field produces electricity

_____ 12. patterns of lines that represent an area influenced by magnetic forces

_____ 13. the places on Earth where the spin axis intersects Earth's surface

_____ 14. the places on Earth near but not identical to the geographic poles, where Earth's dipole intersects the surface

_____ 15. the record, preserved in rock, of Earth's past magnetic fields

_____ 16. the imaginary line that connects one magnetic pole to another magnetic pole, which can be extremely small (the size of a mineral grain or even an atom) or can refer to the entire Earth

_____ 17. the angle difference between the direction a compass needle points (magnetic north) and the direction of geographic north

_____ 18. the angle of tilt of a compass needle mounted to allow vertical as well as horizontal motion

_____ 19. an iron-rich magnetic mineral

_____ 20. an instrument that measures the strength of Earth's magnetic field

_____ 21. the positions of Earth's magnetic poles at different times in the geologic past

_____ 22. a curving line that shows the supposed progressive change in the position of Earth's magnetic pole (The polar-wandering concept was proven false, so this term is meaningless.)

_____ 23. what the polar-wander path was renamed after scientists learned poles don't wander

_____ 24. the shape of the sea-floor's surface

_____ 25. the procedure that bounces sound waves off a distant surface, then uses the length of time it takes for the sound to return to calculate distance (echo sounding)

_____ 26. the igneous rock formed when iron-rich lava cools and hardens

_____ 27. a cross section of the sea-floor bottom produced by continuous sonar scans across an area

_____ 28. broad, flat, deep regions of the ocean floor (about 4.5 km deep)

_____ 29. long, high, linear regions of the ocean floor, roughly symmetrical across their crests, at depths of only 2–2.5 km

_____ 30. the crest of a mid-ocean ridge

_____ 31. long, narrow depressions that run along ridge axes

_____ 32. extremely deep (8–12 km) elongated troughs on the ocean floor

_____ 33. chains of active volcanoes that roughly parallel deep-ocean trenches

_____ 34. isolated submarine mountains that usually occur in chains

_____ 35. narrow, linear zones of rough topography on the ocean floor, broken by many fractures, paralleling each other, and occurring roughly at right angles to the mid-ocean ridge

_____ 36. the process in which old ocean floor sinks back into Earth's interior

_____ 37. the concept that continents move apart because new sea floor forms between them

_____ 38. a place where the strength of Earth's magnetic field is greater than can be accounted for solely by the flow in the liquid iron core

_____ 39. a place where the strength of Earth's magnetic field is weaker than it would be were the only influence flow in the liquid iron core

_____ 40. the component of Earth's magnetic field that is due to the flow of liquid iron in the outer core

_____ 41. areas of the sea floor at which the basalt has recorded, by means of its atomic dipoles, that north is the magnetic north pole of today (the strength of the magnetic field here is above normal)

_____ 42. areas of the sea floor at which the basalt has recorded, by means of its atomic dipoles, that north is the magnetic south pole of today (the strength of the field here is below normal)

_____ 43. a situation in which the north-seeking end of a compass needle (or an atomic dipole) would point to the current north magnetic pole

_____ 44. a situation in which the north-seeking end of a compass needle (or an atomic dipole) would point to the current south magnetic pole

_____ 45. a technique for determining the age of a radioactive element by comparing the amount of radioactive element present to the amount of its decay product

_____ 46. the history of Earth's magnetic reversals

_____ 47. the major time periods between magnetic reversals

_____ 48. subdivisions of polarity epochs, in periods no longer than 200,000 years

Matching Questions

INCLINATION AND LATITUDE

The inclination of magnetic field lines indicates the observer's latitude. Match each latitude below with the correct inclination.

a. lines are vertical
b. lines tilt at an angle to Earth's surface
c. lines lie parallel to Earth's surface

_____ 1. right at the equator
_____ 2. right at the poles
_____ 3. at mid-latitudes

INDIVIDUAL CONTRIBUTIONS

Match each person with his contribution.

a. Robert Dietz
b. Arthur Holmes
c. Lawrence Morley
d. Harry Hess
e. Drumond Mathews
f. Fred Vine
g. Alfred Wegener

_____ 1. the developer of the continental drift hypothesis between 1915 and 1930

_____ 2. the Princeton professor who, in 1962, published ideas he cautiously called *geopoetry*, about how new ocean floor forms and causes the sea floor to widen

_____ 3. the British geoscientist who, in the 1930s, believed that giant convection cells carried hot rock from deep in Earth up toward the surface

_____ 4. the scientist who, in 1961, coined the phrase "sea-floor spreading"

_____ 5. the three English and Canadian geologists
_____   who, in the 1960s, correctly interpreted the
_____   significance of marine magnetic anomalies

EVIDENCE FOR PLATE TECTONICS

Following is a list of evidence supporting plate tectonics theory. Label all the evidence offered by Wegener W; label all evidence offered later (after World War II) L.

_____ 1. analysis of bathymetric profiles
_____ 2. distribution of Late Paleozoic glaciation
_____ 3. fit of the continents
_____ 4. earthquake data from undersea and coastal areas
_____ 5. drilling of the sea floor
_____ 6. distribution of equatorial paleoclimatic belts
_____ 7. numerous discoveries about paleomagnetism
_____ 8. distribution of the same species of fossil land plants and animals on widely separated continents
_____ 9. similar assemblages of rock types and structures across vast oceans
_____ 10. age of sea-floor rocks

Short-Answer Questions

Fill in the blanks with a word or a brief phrase.

1. Earth's lithosphere is broken into about _____ (number) plates.

2. Because we can now confirm the plate tectonic model by many different observations, this model has gained the important status of a scientific _____.

3. *Glossopteris* (fern), *Cynognathus* (reptile), *Mesosaurus* (reptile), and *Lystrosaurus* (reptile) are all fossils whose distribution suggest that _____.

4. The ship on which scientists studied the sea floor by drilling holes in it was named _____.

5. The names Brunes, Matuyama, Gariss, and Gilbert—all names of geoscientists who made important contributions to the study of rock magnetism—are now used to designate _____.

6. Wegener's model had a major flaw in that it couldn't adequately explain _____ _____.

7. The flow of liquid iron in Earth's outer core generates an _____ that in turn generates a _____.

8. Magnets can either attract or repel objects that are rich in the element _____.

9. Basaltic lava contains tons of _____, an iron-rich mineral that can act as a miniature dipole (or compass).

10. Every year Earth's magnetic declination changes by about _____ degrees.

11. Earth's last magnetic reversal occurred about _____ ago.

12. The oldest sea-floor rock is _____ years old.

13. The oldest continental rock is _____ years old.

14. Two ways the age of rock can be determined are _____ and _____.

15. In seamount chains like the Hawaiian Islands, active volcanism typically occurs in only _____.

16. The magnetic field at any location on Earth is the sum of _____ and _____.

APPARENT POLAR-WANDER PATH

It is very easy to get lost in any explanation of polar-wander paths. Complete the following summary, either choosing from the items listed below or selecting one of the two choices offered in parentheses.

apparent polar-wander paths
inclination
dipole
paleomagnetic dipole
Earth's magnetic poles
polar wander
imaginary circles around Earth

A _____ is an imaginary line that connects one pole of a magnet to its other pole. A _____ is a dipole of past geologic time that has been recorded in iron-rich rock. It points to the location of the magnetic pole at the time that rock cooled. Investigation showed that projections of paleomagnetic dipoles _____ (did, did not) lie parallel to Earth's current dipole. This at first suggested that _____ wandered radically over geologic time. This idea was named _____. Projections of the paleomagnetic dipoles of rocks onto Earth's surface created _____ passing through the supposed paleopole and the rock. The paleopole's position was pinpointed by latitude determination, which was based on the _____ of the dipole. Results from polar-wander path studies on different continents sometimes showed the same north pole, which _____ (made sense, didn't make sense) and sometimes showed multiple north poles in existence at the same time, which _____ (made sense, didn't make sense). The only reasonable interpretation was that _____ (the continents were fixed and the poles wandered, the poles were fixed and the continents wandered). _____ (Pangaea existed, Pangaea had split) when the polar-wander paths showed the same north pole; _____ (Pangaea existed, Pangaea had split) when the polar-wander paths showed multiple north poles. Polar-wander paths were renamed _____.

SEA-FLOOR-SPREADING PROBLEMS

1. Assuming a constant rate of sea-floor spreading and knowing the age of the rock making up the ocean floor and the distance of that rock from the mid-ocean ridge, you can calculate the rate at which the sea floor has spread. Calculate the rate at which the sea floor would be spreading if the rock making up the sea floor were 4.5 million years old and 225 km away from the axis of the East Pacific Rise. Give your answer in centimeters per year. _____

2. Sea-floor rock has been found that is 450 km away from its point of origin. The rock is 4.5 million years old. What is the rate of sea floor spreading in this area? Give your answer in centimeters per year. _____

Making Use of Diagrams

MAGNETIC ANOMALIES

The stripes in the following illustration represent magnetic anomaly patterns. They're symmetrical across the mid-ocean ridge. Some stripes have atomic dipoles indicating that magnetic north when this portion of sea floor cooled was the magnetic north of today (normal polarity); others have dipoles indicating that magnetic north in the era in which this portion of sea floor cooled would be south in today's world (reversed polarity). Remember that basalt wells up in the mid-ocean trough.

24 | Chapter 3

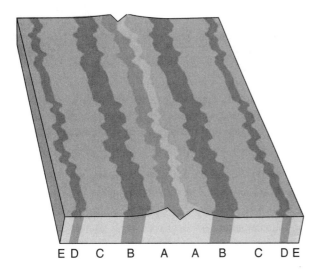

1. What area on the illustration (marked by a letter) indicates the area of maximum heat flow?

2. What letter indicates the oldest ocean floor shown?

3. The basalt in the stripes labeled *A* must be showing that the magnetic north, when this portion of sea floor cooled, was the magnetic north of today (that these stripes have normal polarity). Briefly explain why this must be true.

4. Do stripes *E* show normal or reverse polarity?

5. If the width of stripes *A* through *C* is 45 km and the age of rock *D* is 4.5 million years, what is the rate of sea-floor spreading? Give your answer in centimeters per year.

PAST (PALEOZOIC AGE) GLACIATION

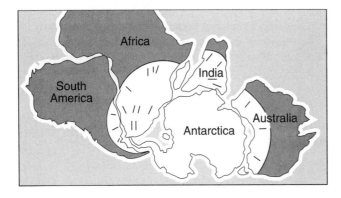

1. Color the outlined areas of southeastern South America, South Africa, southern India, Antarctica, and southern Australia light blue to represent the areas covered by glacial ice in the late Paleozoic.

2. There are lines drawn in each area to represent glacial scratches. Put arrowheads on each line to show that all ice moved outward from a starting point in southeastern Africa.

GEOLOGIC UNITS (ROCK TYPE AND STRUCTURES) ACROSS OCEANS

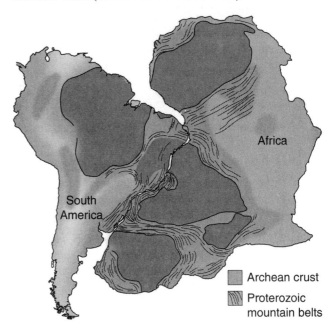

1. On the above figure, color the Precambrian rock units green. Which two continents would you have to move next to each other to make these deposits continuous?

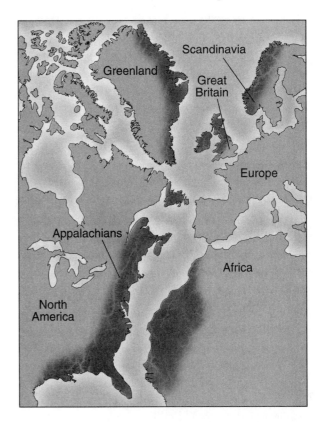

2. On the above figure, color all mountain areas red and draw a boundary line around them to show how they were once one continuous mountain belt. This suggests that the United States and Canada used to be adjacent to which four large land areas?

FOSSIL DISTRIBUTION

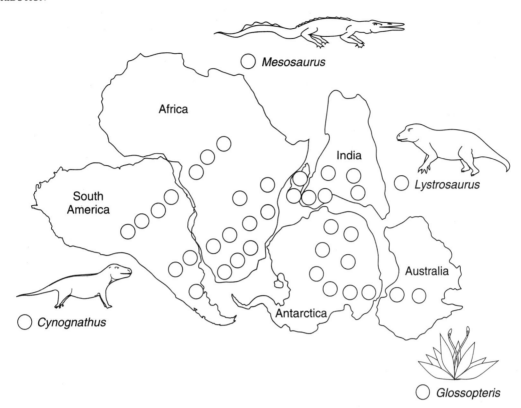

Cynognathus, *Glossopteris*, *Mesosaurus*, and *Lystrosaurus* are all land-dwelling creatures of the late Paleozoic/early Mesozoic.

1. Using a different color for each creature, mark on the figure above each location at which their fossils have been found.

2. The distributions you have highlighted suggest that which five land masses must have been joined in those times?

Short-Essay Questions

1. Your author's comments about the German meteorologist who "dared to challenge a long-held assumption" suggest some reasons why Wegener's ideas were not accepted by the scientific community of his time. In terms of the scientific method, what was good about this reluctant attitude on the part of the scientific community and what was bad about it?

2. Why does the distribution of the same land-dwelling plants and animals on land masses currently separated by vast oceans suggest that the land was once joined?

3. Why can't scientists use the radiometric method of dating rocks to date ocean-floor rocks that are older than 4.5 million years, and thus why can't they date magnetic reversals that occurred more than 4.5 million years ago?

4. What environment, climate, and geographic location are suggested by the presence of coal? sand dunes? reefs? salt? How did Wegener use this information to support his idea of shifting continents?

5. Why do the dipoles of molten basalt not point to a magnetic north pole, whereas the dipoles of cold basalt do point to a magnetic north pole?.

6. Does each atom, of any material, behave like an electromagnet (that is, like a tiny dipole)? If so, why are some, but not most, materials magnetic?

Practice Test

1. Plate tectonics theory took decades to be accepted because
   a. most of Wegener's ideas turned out to be wrong.
   b. of pure stubbornness by the scientific community.
   c. Wegener offered only some of the needed evidence.
   d. climate, fossil distributions, and land shape offered conflicting evidence.
   e. communications were bad, so only German scientists were familiar with Wegener's theory.

2. Choose the FALSE statement. Pangaea
   a. means "all land."
   b. started to break apart during the Mesozoic.
   c. broke apart due to centrifugal forces.

d. is the name for the most recent supercontinent; others have existed.
e. is Wegener's name for the supercontinent that split to produce our continents of today.

3. Which is NOT one of the ideas Wegener offered to support his theory?
   a. the good fit of the outline of the continents
   b. the matching of the distribution of similar fossils across oceans
   c. the existence of the mid-ocean ridge, where sea-floor spreading starts
   d. paleoclimatic evidence of extreme climate changes in some areas
   e. the matching of similar rock types and structures across oceans

4. Choose the one type of deposit that is not typical of a tropical/subtropical climate.
   a. coal
   b. reefs
   c. sand dunes
   d. salt
   e. till

5. *Cynognathus*, *Glossopteris*, *Mesosaurus*, and *Lystrosaurus* are
   a. the names of the most recent periods of magnetic reversal.
   b. land-dwelling species whose distribution suggested joined continents.
   c. names of supercontinents that existed before Pangaea.
   d. distinctive assemblages of rocks.
   e. scientists who supported Wegener's proposals.

6. Choose the FALSE statement. Glaciers
   a. are slow-moving sheets of ice on land.
   b. leave deposits of sediment called bathymetries.
   c. covered large areas of continents during ice ages.
   d. occurred during the late Paleozoic in places they cannot exist today.
   e. occur today at high altitudes or in polar regions.

7. Choose the FALSE statement. Magnetic anomalies are
   a. places where the magnetic field strength is either greater or less than normal.
   b. normal if the atomic dipoles of local rocks match Earth's current north pole.
   c. found both on land and in the sea floor.
   d. recorded in basalt only while it remains extremely hot.
   e. the sum of Earth's dipole field and the local rocks' atomic dipoles.

8. The rate of sea-floor spreading
   a. is the same worldwide.
   b. varies between 1 and 6 ft/year.
   c. is faster along the Mid-Atlantic Ridge than along the East Pacific Rise.
   d. has remained constant throughout Earth's history.
   e. is fast enough to account for the formation and destruction of oceans many times in Earth's history.

9. Choose the FALSE statement. The age of the sea floor
   a. has been determined by radiometric dating.
   b. is oldest at the mid-ocean ridge and youngest farthest from the ridge.
   c. is never older than 200 million years.
   d. correlates well with calculated rates of sea-floor spreading.
   e. has been confirmed by sea-floor drilling.

10. Choose the FALSE statement. Magnetic reversals
    a. have occurred many times, but not at regular intervals.
    b. reflect changes in the direction of flow of molten iron in Earth's outer core.
    c. happen slowly as the poles wander and finally cross the equator.
    d. are recorded in strips of rock parallel to the mid-ocean ridge.
    e. are a global phenomenon.

11. Earth's magnetic north pole
    a. coincides with Earth's geographic North Pole.
    b. is always within 2° of the geographic North Pole.
    c. stays constant over geologic time.
    d. is the place your compass needle points toward.
    e. All of the above are true.

12. Which of the following statements is NOT true about the ocean floor?
    a. It is covered by a layer of sediment composed of clay and plankton shells.
    b. The heat flow through it is greatest at the mid-ocean ridges.
    c. Oceanic crust contains granite and metamorphic rocks.
    d. Ocean crust contains basalt and gabbro.
    e. Oceanic crust is quite different from continental crust.

13. Basalt is
    a. cooled iron-rich lava.
    b. typical continental crust.
    c. lacking in iron.
    d. incapable of recording paleomagnetism.
    e. All of the above are true.

14. "Grains of magnetite become permanent magnets aligned along Earth's magnetic field as molten rock cools." This is a description of the way
    a. gabbro records polar wandering.
    b. magnetic reversals occur.
    c. paleopoles migrate in respect to fixed continents.
    d. continents wander in respect to fixed paleopoles.
    e. basalt creates a paleomagnetic record.

15. The relative widths of magnetic anomalies on the sea floor correspond to the
    a. intervals between magnetic reversals.
    b. relative durations of polarity epochs.
    c. magnetic strengths of the anomalies.
    d. All of the above.
    e. Only a and b are true.

16. Choose the FALSE statement. The paleomagnetic record
    a. is preserved in strips of rock parallel to the mid-ocean ridge.
    b. is preserved in basalt.
    c. is symmetrical across the mid-ocean ridge.
    d. is a puzzle; it contradicts the sea-floor spreading concept.
    e. shows magnetic reversals over geologic time.

17. Only one supercontinent has existed and split apart in all of Earth's history.
    a. True
    b. False

18. Plate tectonic theory was widely accepted by scientists in the 1930s.
    a. True
    b. False

19. No one noticed the fit of the continents until computers were available to manipulate the images.
    a. True
    b. False

20. Earth's magnetism is due to the flow of liquid iron in Earth's molten outer core.
    a. True
    b. False

21. Earth's magnetism changes through slight yearly shifting and occasional reversals of polarity.
    a. True
    b. False

22. Military needs during World War II resulted in increased study of the sea floor.
    a. True
    b. False

23. Science had no knowledge of submarine mountain ranges and oceanic deeps before World War II.
    a. True
    b. False

24. Sea-floor crust is the same age and made up of the same type rock as continental crust.
    a. True
    b. False

25. The sea floor is covered by a thick, uniform blanket of sediment.
    a. True
    b. False

26. Locations of past glaciers can be deduced by deposits of till.
    a. True
    b. False

27. Late Paleozoic glaciation can be explained by the location of the southern part of Pangaea at the North Pole.
    a. True
    b. False

28. Deposits of coal, reef, sand dunes, and salt were found where Wegener predicted; this supported his ideas.
    a. True
    b. False

29. All the mountains of a volcanic arc are active.
    a. True
    b. False

30. All mountains in a seamount chain are active volcanoes.
    a. True
    b. False

31. Earthquake epicenters in oceanic basins are found only on the abyssal plains, never by deep-ocean trenches or mid-ocean ridges.
    a. True
    b. False

32. High heat-flow values along the mid-ocean ridge are explained by molten basalt's sinking into the troughs here on its return to the mantle.
    a. True
    b. False

ANSWERS

Completion

1. Pangaea
2. continental drift hypothesis
3. Mesozoic
4. Paleozoic
5. Precambrian
6. glaciers
7. sediment
8. till
9. fossils
10. ice ages
11. electromagnetism
12. magnetic field lines
13. geographic poles
14. magnetic poles
15. paleomagnetism
16. dipole
17. magnetic declination
18. magnetic inclination
19. magnetite
20. magnetometer
21. paleopoles
22. polar-wander path
23. apparent polar-wander path
24. bathymetry
25. sonar
26. basalt
27. bathymetric profile
28. abyssal plains
29. mid-ocean ridges
30. ridge axis
31. axial troughs
32. deep-ocean trenches
33. volcanic arcs
34. seamounts
35. fracture zones
36. subduction
37. sea-floor spreading
38. positive magnetic anomaly
39. negative magnetic anomaly
40. dipole field
41. marine positive magnetic anomalies
42. marine negative magnetic anomalies
43. normal polarity
44. reversed polarity
45. radiometric dating
46. magnetic reversal chronology
47. polarity epochs
48. polarity events

Matching Questions

INCLINATION AND LATITUDE

1. c
2. a
3. b

INDIVIDUAL CONTRIBUTIONS

1. g
2. d
3. b
4. a
5. f; e; c

EVIDENCE FOR PLATE TECTONICS

1. L
2. W
3. W
4. L
5. L
6. W
7. L
8. W
9. W
10. L

Short-Answer Questions

1. 20
2. theory
3. all continents were united in the early Mesozoic
4. *Glomar Challenger*
5. recent polarity time periods
6. how and why continents move
7. electric current; magnetic field
8. iron
9. magnetite
10. 0.2–0.5
11. 700,000 years
12. 200 million
13. about 4 billion
14. by radiometric dating; by the age of the fossils present
15. the youngest end of the chain
16. the dipole field caused by iron flow in Earth's outer core; the pull of the magnetic minerals in rocks near the surface

APPARENT POLAR-WANDER PATH

dipole; paleomagnetic dipole; did not; Earth's magnetic poles; polar wander; imaginary circles around Earth; inclination; made sense; didn't make sense; the poles were fixed and the continents wandered; Pangaea existed; Pangaea had split; apparent polar-wander paths

SEA-FLOOR-SPREADING PROBLEMS

1. 225 km × 1,000 m/km × 100 cm/m = 22,500,000 cm
   22,500,000 cm /4,500,000 years = 5 cm/year
2. 450 km × 1,000 m/km × 100 cm/m = 45,000,000 cm
   45,000,000 cm /4,500,000 years = 10 cm/year

Making Use of Diagrams

MAGNETIC ANOMALIES

1. *A* and *A* (the stripes closest to the ridge axis)
2. *E* (the stripes farthest from the ridge axis)
3. The basalt stripes labeled *A* are right at the ridge axis, so they are recently cooled rock and would naturally show the north of today.
4. normal
5. 45 km × 1,000 m/km × 100 cm/m = 4,500,000 cm;
   4,500,000 cm /4,500,000 years = 1 cm/year

PAST (PALEOZOIC AGE) GLACIATION

1. See Figure 3.3b in your text.
2. See Figure 3.3b in your text.

GEOLOGIC UNITS (ROCK TYPE AND STRUCTURES) ACROSS OCEANS

1. See Figure 3.6a in your text; South America and Africa
2. See Figure 3.6b in your text; Greenland, Great Britain, Scandinavia, and Africa

FOSSIL DISTRIBUTION

1. See Figure 3.5 in your text.
2. Africa; South America; India; Antarctica; Australia

Short-Essay Questions

1. You should consider the source of any argument, scientific or otherwise, before blindly accepting the information. You should not change carefully thought-out ideas on a whim. However, scientists should be open minded. Scientific conclusions should be drawn by using the best available instruments and techniques to interpret all the relevant data available. Scientific thoughts have changed—and should continue to change—as better instruments, techniques, and data become available. Ultimately, the instruments, techniques, and data became available to scientists studying Earth, so eventually plate tectonics theory became accepted by the scientific community.

2. Land-dwelling creatures could not have swum the oceans to distribute themselves over such vast areas. Since evolutionary theory shows that separated species develop differently due to isolation from each other, it makes sense to interpret the fossil evidence showing a single species distributed over vast areas separated by oceans as indicating that those land masses were once joined.

3. There's a percentage of error involved in radiometric dating, and numbers larger than 4.5 million would allow such numerically large errors that the error might turn out to be larger than the rate of motion itself.

4. Coal forms in warm swampy conditions. Sand dune deposits suggest ancient hot, sandy deserts. Reefs form in warm shallow seas. Salt deposits suggest salty waters, heat, and evaporation. All of those environments suggest warm climates, and all except the sand dunes indicate moisture. Logical locations would be in the Tropics or subtropics, or at least at low latitudes, yet today many of these deposits are in areas where the current climate couldn't possibly have produced them. Wegener suggested that the deposits formed while the land was located in warm regions and that the land later moved to cooler latitudes.

5. In molten basalt the grains move randomly because of their high thermal energy. Only when the rock cools and loses thermal energy can the grains slow down enough to align themselves with the surrounding magnetic field.

6. All atomic dipoles in magnetic materials lock into alignment with each other, and their combined strength makes the material magnetic. All atomic dipoles in nonmagnetic materials are randomly oriented and thus cancel each other out.

Practice Test

1. c Wegener could not adequately explain how or why the continents moved. Sea-floor studies following World War II supplied the needed proof.
2. c Centrifugal forces were much too weak to cause Pangaea to break apart.
3. c Sea-floor studies didn't occur until after Wegener's death.
4. e Till is glacial sediment.
5. b
6. b Glacial deposits are called till; bathymetry is the measurement of water depth to determine the topography of the sea floor.
7. d Basalt records magnetism only when it has low thermal energy and its grains can align themselves with the surrounding magnetic field.
8. e The rate has varied over time and around the world. It ranges from less than an inch to a few inches per year, and is slowest along the Mid-Atlantic Ridge and fastest along the East Pacific Rise.
9. b It is youngest at the ridge, oldest far from the ridge.
10. c The magnetic pole does wander slowly, but it always stays relatively close to the geographic pole; reversals are a separate phenomenon.
11. d A compass needle lines up with Earth's magnetic lines of force and therefore points to Earth's magnetic north pole, not its geographic north pole; the magnetic pole shifts a fraction of a degree each

year, but always stays within 15° latitude of the geographic north pole.

12. c The only rock types found in oceanic crust are basalt and gabbro.

13. a Basalt is typical iron-rich ocean crust that shows changes in paleomagnetism.

14. e The statement describes the process by which sea-floor basalt records magnetic reversals over time.

15. e Choice b is just a more scientific way of stating choice a; choice c is wrong because the strength of the magnetic field has nothing to do with the width of the anomaly.

16. d The paleomagnetic record offers strong support for sea-floor spreading.

17. False There has been sufficient time for several cycles of splitting, and scientists are convinced that there have been several.

18. False This theory was not generally accepted until the 1960s.

19. False The fit of the continents was noticed as early as the 1500s.

20. True The flow of iron alloy in Earth's molten outer core makes it an electromagnet.

21. True Earth's magnetic poles shift by a fraction of a degree yearly and the magnetic poles reverse suddenly and sporadically on a time scale of thousands of years, but the two phenomena are not related.

22. True War-related action on and in the oceans prompted studies of the sea environment.

23. False Soundings in the 1870s hinted at their existence.

24. False Oceanic crust is basalt and is much younger than continental crust.

25. False The sediment blanket is surprisingly thin and varies in thickness depending on the distance of the sea floor from the ridge.

26. True Till is a glacial deposit so its presence indicates past glaciation.

27. False It was at the South Pole.

28. True Wegener believed certain land masses currently at northern latitudes had moved there from more southern latitudes. Discovery of these subtropical deposits supported his beliefs.

29. True The volcanic mountains are the result of a plate subducting along the entire length of the arc.

30. False Only one end—the youngest—is active.

31. False They're found by deep-ocean trenches and mid-ocean ridges, not on the abyssal plains.

32. False The molten basalt here is rising, not sinking back into the mantle.

CHAPTER 4 | The Way the Earth Works: Plate Tectonics

GUIDE TO READING

In Chapter 3 the author argued that there are good reasons to believe continents move; in this chapter he explains how and why they move. Great scientific minds found plate tectonics mindboggling at first—the acceptance of plate tectonics theory required an intellectual revolution—so try not to feel intimidated by the material. In fact, you may even find this chapter easier than the preceding one. Why? Because understanding *why* something is believed (Chapter 3) can be more demanding than understanding *what* is believed (Chapter 4). Wegener and the other pioneers of plate tectonics did the hard work: they developed a truly new idea. To do this required creativity, insight, and some moments of brilliance. Then they offered it to their peers. This took nerve. Compiling the proof needed to persuade a skeptical public took hard work and perseverance, with some anguish thrown in. But understanding plate tectonics is beautifully simple. Stripped to its barest essentials, the theory contends that Earth's outer surface is split into pieces, or plates, that slowly shift around in relation to each other. Plates can move away from each other (divergence), toward each other (convergence), or past each other (transform motion). With all these huge masses shifting around, it's not surprising that there are huge consequences, and these usually occur at plate boundaries, where Earth's surface may be deformed, built up, or destroyed—all on a grand geologic scale.

Naturally there are details to be studied as well. Some of them weren't recognized and studied until after the theory was accepted; some were part of the evidence used to develop the theory. (You'll recognize the latter because they were introduced in Chapter 3.) This chapter will introduce you to or let you take a closer look at these details.

The chapter begins with a discussion of the types of plate boundaries and the features associated with them, including:

- earthquake belts
- mid-ocean ridges
- trenches and the subduction that takes place there
- strange and numerous offset segments of the mid-ocean ridge and the transform faults that bracket them
- the far fewer transform faults that cross land, such as the infamous San Andreas Fault

Next the author calls attention to two special types of boundaries, where the action is not caused by the usual plate-against-plate motion. The first of these is the triple junction, where three plates meet in a point. The second is the hot-spot phenomenon, illustrated by many exotic places, like Yellowstone National Park and the Hawaiian Islands.

This chapter explains that plate boundaries do not remain unchanged forever. Instead, old boundaries disappear, as illustrated by India's collision with Asia to produce the Himalayas, and new ones appear, as illustrated by the rifting that produced both the East African Rift Valley and the Basin and Range Province of the U.S. West. Exactly how does all of this happen? The author concludes the chapter by discussing the probable explanations for actual mechanisms of plate motion and by explaining how we are able to determine the velocity of plates' motions.

In his closing remarks for this chapter, the author reminds us that, directly or indirectly, plate tectonics is the key to understanding just about everything geologic.

LEARNING ACTIVITIES

Special Advice for This Chapter

To understand what's going on in the world geologically, it's important to be aware of where you are. *Be sure to look at the maps provided.* Not only does the old adage "A picture is worth a thousand words" apply very nicely to this section, but your teacher will likely expect you to associate plate tectonics phenomena with real geographic locations and might even ask you to identify on a map where such phenomena occur.

Completion

Test your recall of new vocabulary terms.

_____ 1. a framework for thinking about a certain scientific topic; a pattern or mold into which all related ideas must fit to complete the picture

_____ 2. a totally new approach to explaining a scientific phenomenon

_____ 3. the outermost layer of Earth, consisting of the crust and the topmost, coolest part of the mantle

_____ 4. the layer of Earth that lies just under the lithosphere, which since it is warmer and softer than the lithosphere, is capable of flowing slowly

_____ 5. the 20–32 segments (depending on your definition) into which the lithosphere is divided

_____ 6. the divisions (breaks) between plates

_____ 7. the coasts of continents that are also plate boundaries

_____ 8. the coasts of continents that are not plate boundaries

_____ 9. the broad but shallow areas along passive continental margins consisting of thinned, broken-up continental crust topped by thick layers of sediment

_____ 10. a plate boundary at which the plates move apart

_____ 11. a plate boundary at which the plates move toward each other

_____ 12. a plate boundary at which the plates move horizontally past each other

_____ 13. the upward force on an object immersed in a fluid, which may (or may not) cause the object to float

_____ 14. the belt of earthquakes that is associated with the down-going slab of a subduction zone

_____ 15. a deep, narrow, elongated area on the sea floor that is the seaward edge of a subduction zone

_____ 16. a wedge-shaped mass of material which forms where a plate is subducting and which has been scraped off the subducting plate or fallen off the overriding plate

_____ 17. a chain of volcanoes that forms on Earth's surface after a subducting plate has reached a depth of about 150 km

_____ 18. the portion of a fracture zone between two segments of the mid-ocean ridge, where slippage occurs

_____ 19. the intersection of three plate boundaries

_____ 20. locations of volcanoes (or volcanic-type activity), either on land or on the sea floor, not close to plate boundaries

_____ 21. columns of hot rock rising from deep in the mantle that serve as heat sources for hot spots

_____ 22. an underwater, extinct (dead) volcano that began as a volcanic island formed above a hot spot

_____ 23. the process by which a continent is split and pulled apart into two continents (i.e., the process by which a divergent boundary is created on land)

_____ 24. the situation where two buoyant pieces of lithosphere (continental or island-arc material) smash into each other after the oceanic lithosphere between them has subducted

_____ 25. the pushing away of adjacent lithosphere laterally from the ridge axis by oceanic lithosphere, high on the ridge axis

_____ 26. the sinking of oceanic lithosphere older than 10 million years into the less dense asthenosphere and the pulling of the entire plate along with it

_____ 27. a system that uses a series of satellites orbiting Earth to determine positions on the Earth very accurately

Matching Questions

MAPPING

Match locations a through l to their corresponding features 1 to 10. A map is provided for your reference.

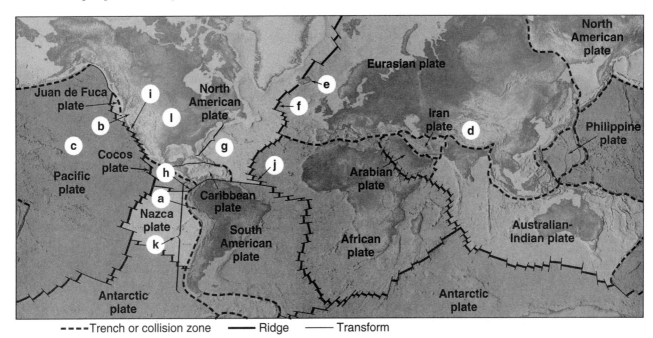

- - - - Trench or collision zone ——— Ridge ——— Transform

a. Andes Mountains
b. the Cascades
c. Hawaii
d. Himalayas
e. Iceland
f. Mid-Atlantic Ridge
g. North American east coast
h. Peru-Chile trench
i. San Andreas Fault
j. transform fault on the Mid-Atlantic Ridge
k. west coast of South America
l. Yellowstone

_____ 1. a divergent plate boundary

_____ 2. a convergent plate boundary

_____ 3. two transform plate boundaries

_____

_____ 4. a continental hot spot

_____ 5. a hot spot on a divergent plate boundary

_____ 6. an oceanic hot spot not on a plate boundary

_____ 7. a passive continental margin and wide continental shelf

_____ 8. two volcanic arcs on land

_____

_____ 9. the result of a continent-continent collision

_____ 10. a subduction zone along the South American coast

RECOGNIZING PLATE BOUNDARIES

Plate boundaries are real places, not just theoretical constructs. Some are easily recognized, even if it is often difficult to get to them; others can only be inferred by recognizing associated features like volcanoes. Match each plate boundary below with the numbered descriptions of what you would see there.

a. Iceland
b. Juan de Fuca Plate–North American Plate boundary
c. Mid-Atlantic Ridge
d. San Andreas Fault

_____ 1. a linear depressed area (valley) filled with ponds and marshes, offset curbs in towns, offset tree rows in orchards, and contorted rock layers in road cuts

_____ 2. the Cascade Range (e.g., Mt. Ranier, Mt. St. Helens)

_____ 3. molten rock oozing up through cracks, steep cliffs, and fountains of dark, boiling water

_____ 4. fault-bounded troughs and fissures sometimes with lava emerging

Chapter 4

CRUST VERSUS LITHOSPHERE

Indicate whether the statements below are true of oceanic crust (OC), continental crust (CC), or both (BC).

_____ 1. The average thickness is 40 km.
_____ 2. The average thickness is 10 km.
_____ 3. It consists of low-density granite.
_____ 4. It consists of high-density basalt.
_____ 5. It is a component of the lithosphere.

Indicate whether the statements below are true of oceanic lithosphere (OL), continental lithosphere (CL), or both (BL).

_____ 6. It is composed of crust and the uppermost portion of the mantle.
_____ 7. The average thickness if old is 100 km.
_____ 8. The average thickness if new is 10 km.
_____ 9. The average thickness is 150 km.
_____ 10. It floats higher than the other type of lithosphere and therefore sticks up higher.

Short-Answer Questions

Fill in the blank with a word, a number, or a brief phrase.

1. Earth's interior is not homogenous (all the same). Starting on the surface and moving inward, list the five different layers of Earth's interior: _____, _____, _____, _____, and _____.

2. Scientists define the boundaries between the different layers in question 1 by _____.

3. Earth's outermost layer (consisting of the crust and the uppermost portion of the mantle), which bends, flexes, or breaks when stressed, is called the _____. The next layer down, which flows slowly when stressed, is called the _____.

4. The lowest temperature at which the mantle is warm and soft enough to flow (and thus, by definition, is asthenosphere) is _____.

5. An excellent way to determine plate boundaries is to plot locations of _____.

6. "Spreading boundary," "mid-ocean ridge," or simply "ridge" are all terms that describe two oceanic plates _____ each other.

7. "Subduction zone," "consuming boundary," and "trench" are all terms that describe two oceanic plates or one oceanic plate and one continental plate _____ each other.

8. The Mid-Atlantic Ridge, the East Pacific Rise, and the Southeast Indian Ocean Ridge are all examples of a _____ plate boundary.

9. The portion of the North American Plate–Pacific Plate boundary that cuts across California is called the _____.

10. Los Angeles sits on the _____; San Francisco sits on the _____.

11. Los Angeles is moving in a _____ direction, at a rate of _____ cm/year.

12. Slippage along a transform fault destroyed _____ in 1906 and caused extensive damage to this same California city in 1989.

13. Iceland, which sits astride the Mid-Atlantic Ridge, sticks up above the ocean's surface because it is located _____.

14. The Yellowstone area has had a violent volcanic past. Although it is no longer volcanically active, the presence of _____ suggests that there is still magma close to the surface.

15. Yellowstone lies at the northeast end of the Snake River Plain, which is covered with basalt that grows progressively older to the southwest. This information indicates that the mantle plume is currently located _____ and that the North American Plate has been moving in a _____ direction.

16. The following three mountain ranges are all the result of collision: _____, _____, and _____.

17. India crashed into Asia _____ million years ago.

18. The collision of continents results most obviously in mountains; it also makes the crust in the area _____–_____ km thick, which is _____ the normal thickness.

19. The East African Rift Valley and the Basin and Range Province of the western United States result from _____.

20. Use two well-chosen words to explain the origin and distribution of earthquakes, mid-ocean ridges, deep-sea trenches, fracture zones, volcanoes, atmospheric and oceanic gases, mountain belts, erosion, land shifts, and changes in life forms: _____ and _____.

21. Scientists use the _____ to detect land displacements as small as a few millimeters per year and to accumulate further evidence that plates move.

22. The three other terrestrial (Earth-like) planets are _____, _____, and _____.

23. Scientists _____ (do, do not) believe that the other terrestrial planets and/or our Moon show plate tectonic action.

24. Because Earth's interior is _____, and the interiors of the other terrestrial planets and our Moon are not, plate tectonics can occur.

25. _____ is a Canadian geologist who correctly interpreted oceanic fracture zones; he is also the first person to explain hot spots correctly.

TAKING A CLOSER LOOK AT SOME PLATE TECTONICS FEATURES

Fill in the blanks with a word, number, or brief phrase, or select one of the choices offered in parentheses.

Plates

1. There are _____ (number) major plates.
2. There are _____ (number) to _____ (number) microplates.
3. The _____ Plate is all ocean floor.
4. The _____ Plate includes a major continent plus half of a major ocean.
5. The _____ Plate is mostly ocean, with just a little land (California and the Baja) along one edge.
6. The slowest and fastest possible rates of motion for any single plate are _____ (slowest rate) and _____ (fastest rate).

Mid-Atlantic Ridge

1. The Mid-Atlantic Ridge is part of a larger system called the _____.
2. The Mid-Atlantic Ridge sits in the middle of the _____.
3. The Mid-Atlantic Ridge begins in the far north, at _____ (latitude), in line with the northern edges of _____ and _____.
4. The Mid-Atlantic Ridge extends as far south as _____ (latitude), in line with the southern tip of _____.
5. The Mid-Atlantic Ridge is generally _____ km (or _____ mi) wide.
6. The plate boundary is located in a _____ on the Mid-Atlantic Ridge, approximately _____ m deep, which runs along the _____.
7. The _____ rises through the trough of the Mid-Atlantic Ridge as _____, which may solidify at depth in the trough as _____ or closer to the surface as _____.

East Pacific Rise

The East Pacific Rise, like the Mid-Atlantic Ridge, is part of the mid-ocean ridge, but it differs from the Mid-Atlantic Ridge in three significant ways:

1. It _____ (is, is not) in the middle of its ocean basin.
2. It _____ (does, does not) have an axial trough.
3. It is _____ (narrower, wider) than the Mid-Atlantic Ridge.

Oceanic Lithosphere

Oceanic lithosphere gets _____ (thicker, thinner) as it gets farther from the ridge axis because its _____ (crust, mantle) component increases as the rock cools. Oceanic lithosphere is at maximum thickness when it is about _____ years old and located _____–_____ km away from the ridge axis. A mid-ocean ridge is higher than the adjoining sea-floor basin because the hot, young lithosphere of the ridge is _____ (more, less) dense than the cooler, older lithosphere farther out, making the ridge lithosphere _____ (more, less) buoyant than the lithosphere farther out. When oceanic lithosphere is about _____ years old, it is cool enough to be denser than the asthenosphere. From then on, if one end gets pushed into the mantle, it can subduct. During subduction, the lithosphere of the downgoing slab _____ (is always oceanic, is always continental, can be either oceanic or continental), and the lithosphere of the overriding plate _____ (is always oceanic, is always continental, can be either oceanic or continental).

Triple Junctions

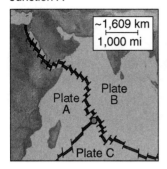
Junction A

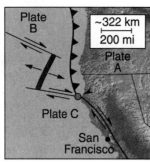
Junction B

Compare the two triple junctions in the illustration above to answer the following questions.

1. Junction _____ is a ridge-ridge-ridge triple junction.

2. Junction _____ is a trench-transform-transform triple junction.

3. The infamous U.S. fault called the _____ is involved in junction type B.

Hot Spots and Mantle Plumes

1. Plumes originate _____ (location) in the Earth.

2. Plumes stream upward because _____.

3. When the plume reaches the base of the overlying lithosphere, it _____.

4. A _____ forms on Earth's surface above the mantle plume (i.e., at the hot spot).

5. When a volcanic mountain moves off its hot spot, it _____.

Rate of Motion

1. It is jokingly said that in 100 million years Los Angeles, California, will have moved far enough north to become a suburb of Anchorage, Alaska. On the basis of this observation and knowing that the Pacific Plate is moving north at the rate of 6 cm/year, indicate approximately how far Los Angeles is from Anchorage: _____ cm; _____ km; _____ mi. (Note: There are 100 cm/m, 1,000 m/km, and 0.62 mi/km.)

2. The diagram below shows South Island, New Zealand.

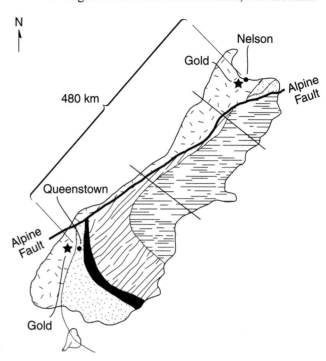

The Alpine Fault cuts across the land in a northeasterly-southwesterly direction. Gold was discovered in 1861 near Queenstown on the east side of the fault. The same gold-bearing rock masses are found on the west side of the fault, but they are 480 km to the northeast. Geologists believe movement along the fault has been occurring for 23 million years. The average rate of motion along the fault is _____ cm/year.

Making Use of Diagrams

FRACTURE ZONES OF THE MID-OCEAN RIDGE

In the 1960s geologists got very excited about the information conveyed by one of the diagrams at the top of the next page, because it offered additional proof of Hess's sea-floor-spreading idea.

1. Which diagram shows the ridge starting out continuous and getting offset along a fault? _____

2. Which diagram shows the ridge starting out offset and linked by the fault line? _____

3. Which diagram is correct? _____

4. On diagram B color the sections of the fault lines where earthquakes occur.

5. Which is older, the rock at point 1 or the rock at point 2? _____

Diagram A

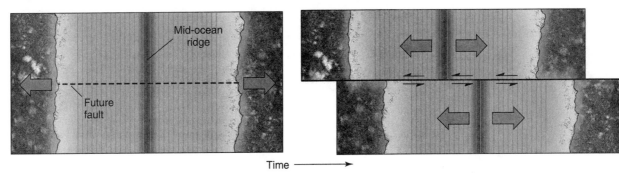

Time →

Diagram B

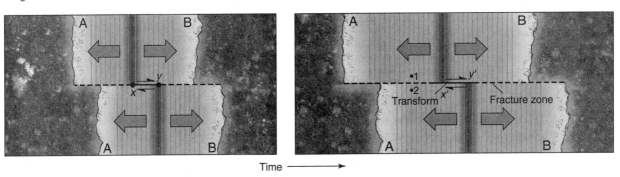

Time →

HAWAIIAN ISLAND CHAIN

1. On the figure below, label the only island with currently active volcanoes *VA* (volcanically active); label the oldest island of the chain *O*; and label the location of the heat source (mantle plume) *HS*.

2. Mark the location of Loihi, the newest Hawaiian Island, which is still below the surface, with an *X*.

3. The distance between point *A* (on Kauai) and point *B* (on Hawaii) is 550.5 km. The ages of rocks at these points are shown on the diagram. Given the age of these rocks, what has the average rate of plate motion been during the last 4.7 million years (in cm/year)?

HOT-SPOT TRACKS

The Hawaiian Islands, Easter Island, and Macdonald Seamount are all hot spots today. Referring to the diagram below and selecting from the options below, complete the following statement:

The direction of motion of the Pacific Plate changed

1. from north northwest to northwest.
2. from northwest to north northwest.
3. from south southeast to southeast.
4. from southeast to south southeast.

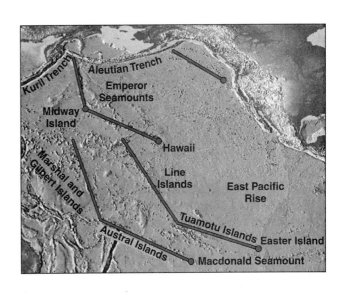

PLATE VELOCITIES

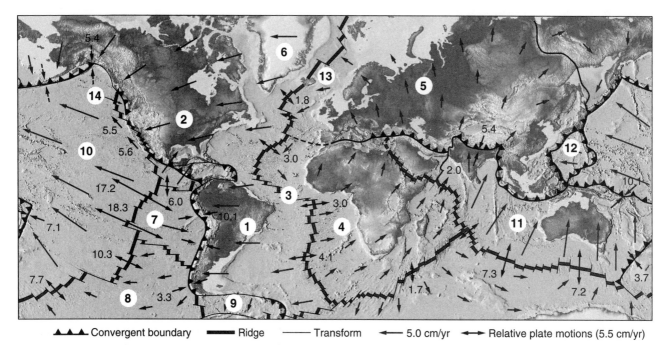

▲▲▲ Convergent boundary   ▬▬ Ridge   ── Transform   ←── 5.0 cm/yr   ←─→ Relative plate motions (5.5 cm/yr)

1. South American Plate
2. North American Plate
3. Mid-Atlantic Ridge
4. African Plate
5. Eurasian Plate
6. Greenland
7. Nazca Plate
8. Antarctic Plate
9. Scotia Plate
10. Pacific Plate
11. Australian-Indian Plate
12. Philippine Plate
13. Iceland
14. Juan de Fuca Plate

1. What is the absolute velocity of the North American Plate in its extreme northern area along the Mid-Atlantic Ridge near Iceland and Greenland?

2. What is the velocity of the North American Plate in relation to the Eurasian Plate?

3. Which plates show the fastest absolute motion? Where is the boundary between these plates? What type of boundary is it? What is the velocity of these plates relative to each other?

Short-Essay Questions

1. Why are some volcanic arcs continental and others oceanic?

2. What occurs when new sea floor forms? (Be sure to include the words "magma," "magma chamber," "gabbro," "basalt," "dikes," "submarine volcanoes," and "*Alvin*.")

3. What does tension have to do with the numerous earthquakes that occur along the mid-ocean-ridge axis?

4. Oceanic lithosphere gets no older than about 200 million years. Why? (Include in your answer what happens to it and where it goes.)

5. Why are there no earthquakes along subducting plates below about 670 km?

6. Because of its unique location, what will eventually happen to Iceland?

7. Explain how studies of the Appalachian Mountains suggest that there have been more supercontinents in Earth's history than just Pangaea.

8. Explain why Kilimanjaro in Africa is volcanic.

Practice Test

1. Continents
   a. plow their way through the sea floor.
   b. passively ride along as the sea floor spreads.
   c. may subduct if they are old and dense enough.
   d. consist of rock that is younger than sea-floor rock.
   e. have retained the same size and shape throughout Earth's history.

2. Pick out the FALSE statement.
   a. As the sea floor spreads, the asthenosphere rises, melts to become magma, and fills the space between plates.
   b. Some magma generated during sea-floor spreading spills out to produce a new layer of sea floor called gabbro.

c. Some magma generated during sea-floor spreading erupts from submarine volcanoes.
d. Observers in *Alvin* have seen submarine volcanoes.
e. The sea floor is made of basalt.

3. The chain of Hawaiian Islands extends northwest across the Pacific as shown in the figure.

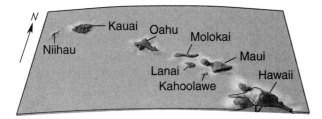

The island of Hawaii is the youngest, and the islands get progressively older to the northwest. Plate tectonic theory explains this as
a. a hot spot moves under the Pacific plate toward the southeast.
b. a fixed hot spot is currently situated under Kauai across which the Pacific Plate is moving in a southeasterly direction.
c. microplates break loose from the Pacific Plate as it moves, leaving this chain of islands in its wake.
d. the rifting of the Pacific Plate, starting at the northwest end of the chain and extending southeast to Hawaii.
e. a fixed hot spot currently situated under Hawaii across which the Pacific plate is moving in a northwesterly direction.

4. Major earthquakes and volcanic eruptions
a. do not occur at similar locations.
b. usually occur on the boundaries of plates or at hot spots.
c. usually occur at the center of plates.
d. prove through their locations that Earth is splitting apart (rifting) at all plate boundaries.
e. occur at divergent plate boundaries and hot spots only.

5. Pick out the FALSE statement.
a. The direction of movement of the Pacific Plate changed about 40 million years ago, so that hot-spot tracks bend.
b. Plates move in response to two forces: the ridge-push force and the slab-pull force.
c. The North American Plate is currently moving toward the northeast at about 10 ft/year.
d. India crashed into Asia to create the Himalayas about 40 million years ago.
e. Southwestern California and the Baja Peninsula are moving toward the northwest along the section of the Pacific Plate–North American Plate boundary called the San Andreas Fault.

6. The chain of volcanoes along the west coast of South America (the Andes Mountains) exist because
a. the Pacific Plate is sliding past the South American Plate and heading toward the northwest.
b. the Pacific Plate is rifting apart from the South American Plate.
c. an oceanic plate is subducting under the western edge of the South American Plate.
d. a continental plate is colliding with the South American Plate and causing it to buckle.
e. basalt is pushing up through the trench that parallels the west coast of South America.

7. Pick out the FALSE statement. According to present plate tectonics theory,
a. plates move on the asthenosphere.
b. plates move at speeds of a few centimeters per year.
c. a continental plate will subduct under an oceanic plate when the two converge.
d. plates are composed of the crust and the uppermost part of the mantle.
e. all continental material started out as one supercontinent called Pangaea.

8. The huge trough down the center of the mid-ocean ridge is where
a. molten basalt subducts back into Earth.
b. the actual plate boundary is located.
c. the Hawaiian Island chain formed.
d. two continental plates converge.
e. you find the oldest, densest, hottest oceanic crust.

9. Both earthquakes and volcanoes occur in Alaska because
a. two plates are moving past each other there.
b. a continental plate is colliding with another continental plate there.
c. an oceanic plate is converging with a continental plate and subducting there.
d. two plates are diverging there.

10. Earthquakes but not volcanoes occur in San Francisco because
a. two plates are moving past each other there.
b. a continental plate is colliding with another continental plate there.
c. an oceanic plate is converging with a continental plate and subducting there.
d. two plates are diverging there.

11. The Himalayas are growing because
    a. two plates are moving past each other there.
    b. a continental plate is colliding with another continental plate there.
    c. an oceanic plate is converging with a continental plate and subducting there.
    d. two plates are diverging there.

12. Pick out the FALSE statement. A subducting plate
    a. can be either continental or oceanic lithosphere.
    b. has a trench along its seaward edge.
    c. can be charted by noting its Wadati-Benioff zone.
    d. has a volcanic arc associated with it.
    e. has an accretionary prism associated with it.

13. Asthenosphere
    a. is warm enough to flow slowly.
    b. subducts when it collides with continental lithosphere.
    c. is the lower layer of both continental and oceanic lithosphere.
    d. is the uppermost layer of Earth's core.
    e. contains the Wadati-Benioff zone where submarine quakes originate.

14. The rate of plate motion
    a. can be determined to within millimeters by using the Global Positioning System (GPS).
    b. is faster along the northern Mid-Atlantic Ridge than along the East Pacific Rise.
    c. was first correctly determined by J. Tuzo Wilson in the 1960s.
    d. changes about every 1,000 years.
    e. All of the above are true.

15. Pick out the FALSE statement.
    a. When an oceanic plate subducts under another oceanic plate, a volcanic island arc is formed.
    b. Oceanic lithosphere gets no older than 200 million years.
    c. All oceanic lithosphere eventually subducts into the mantle and probably melts there.
    d. Iceland will eventually split apart because it sits astride a diverging plate boundary.
    e. Kilimanjaro is a volcano because it is located on a hot spot.

16. Only one supercontinent has broken up during Earth's history.
    a. True
    b. False

17. The Alps, Himalayas, and Appalachians are all the results of collision.
    a. True
    b. False

18. Crust beneath mountains can be twice the thickness of normal continental crust.
    a. True
    b. False

19. There is convective flow (with hotter rock rising, cooler sinking) within the asthenosphere.
    a. True
    b. False

20. Convective flow within the asthenosphere, which drags plates along, is the basic cause of plate motion.
    a. True
    b. False

21. The asthenosphere rising from depth along the mid-ocean ridge causes the plates there to move apart.
    a. True
    b. False

22. Movement along plate boundaries is slow but steady; this explains why earthquakes don't occur there.
    a. True
    b. False

23. Earthquakes typically occur along passive continental margins.
    a. True
    b. False

24. Hot spots result when columns of especially hot material move slowly beneath plates.
    a. True
    b. False

25. Hot spots may occur under land, as illustrated by Yellowstone, or under the sea floor, as illustrated by Hawaii.
    a. True
    b. False

26. The mid-ocean ridge is under water for most of its length.
    a. True
    b. False

27. All islands along a hot-spot chain remain volcanically active for the lifetime of the mantle plume involved.
    a. True
    b. False

28. The San Andreas Fault is a long, narrow valley from which lava often oozes onto the surface.
    a. True
    b. False

29. The Mid-Atlantic Ridge began as a continuous mountain chain and has been broken into segments along its transform faults.
    a. True
    b. False

30. The East Pacific Rise is narrower than the Mid-Atlantic Ridge but has a deeper trough down its center.
    a. True
    b. False

31. Slab-pull and ridge-push are the two processes that cause sea-floor spreading.
    a. True
    b. False

ANSWERS

Completion

1. scientific paradigm
2. scientific revolution
3. lithosphere
4. asthenosphere
5. plates
6. plate boundaries
7. active continental margins
8. passive continental margins
9. continental shelves
10. divergent plate boundary
11. convergent plate boundary
12. transform plate boundary
13. buoyancy
14. Wadati-Benioff zone
15. ocean trench
16. accretionary prism
17. volcanic arc
18. transform fault
19. triple junction
20. hot spots
21. mantle plumes
22. seamount
23. continental rifting
24. collision
25. ridge-push force
26. slab-pull force
27. global positioning system (GPS)

Matching Questions

MAPPING

1. f
2. k
3. i, j
4. l
5. e
6. c
7. g
8. a, b
9. d
10. h

RECOGNIZING PLATE BOUNDARIES

1. d
2. b
3. c
4. a

CRUST VERSUS LITHOSPHERE

1. CC
2. OC
3. CC
4. OC
5. BC
6. BL
7. OL
8. OL
9. CL
10. CL

Short-Answer Questions

1. crust; upper mantle; lower mantle; outer core; inner core
2. abrupt changes in the velocity of earthquake waves
3. lithosphere; asthenosphere
4. 1,280°C
5. earthquakes
6. diverging from
7. converging toward
8. divergent
9. San Andreas Fault
10. North American Plate; Pacific Plate
11. northerly (relative to North America); 6
12. San Francisco
13. over a hot spot
14. geysers
15. at Yellowstone National Park; southwesterly
16. Himalayas; Alps; Appalachians
17. about 40
18. 60; 70; about twice
19. continental rifting
20. plate tectonics
21. global positioning system (GPS)
22. Mercury; Venus; Mars
23. do not
24. warm enough to allow the slow flow of mantle rock
25. J. Tuzo Wilson

TAKING A CLOSER LOOK AT SOME PLATE TECTONIC FEATURES

Plates

1. 12
2. 8; 20
3. Nazca
4. North American
5. Pacific
6. 1 cm/year; 15 cm/year

Mid-Atlantic Ridge

1. mid-ocean ridge
2. Atlantic Ocean
3. 80° north; Greenland; Scandinavia
4. 55° south; South America
5. 1,500; 930 mi (1 km equals 0.62 mi.)
6. narrow trough; 500; axis of the ridge
7. asthenosphere; magma; gabbro; basalt

East Pacific Rise

1. is not
2. does not
3. wider

Oceanic Lithosphere

thicker; mantle; 80 million; 800; 1,500; less; more; 10 million; is always oceanic; can be either oceanic or continental

Triple Junctions

1. A
2. B
3. San Andreas Fault

Hot Spots and Mantle Plumes

1. just above the core-mantle boundary
2. the hot rock there is less dense than the overlying mantle
3. partially melts and the magma thus formed rises through the lithosphere
4. volcano
5. becomes extinct

Rate of Motion

1. 100,000,000 years × 6 cm/year = 600,000,000 cm; 600,000,000 cm/100,000 cm/km = 6,000 km; 6,000 km × 0.62 = 3,720 mi
2. 480 km × 1,000 m/km × 100 cm/m = 48,000,000 cm; 48,000,000 cm/23,000,000 years = 2.09 cm/year

Making Use of Diagrams

FRACTURE ZONES OF THE MID-OCEAN RIDGE

1. Diagram A
2. Diagram B
3. Diagram B
4. Should be colored along sections XY and X′Y′.
5. Point 2 rock is older. Despite the fact points 1 and 2 are directly across the fault line from each other, point 1 is closer to the ridge/rift, and was therefore formed more recently by upwelling basalt.

HAWAIIAN ISLAND CHAIN

1. *VA* on Hawaii, *O* on Kauai, and *HS* on Hawaii
2.

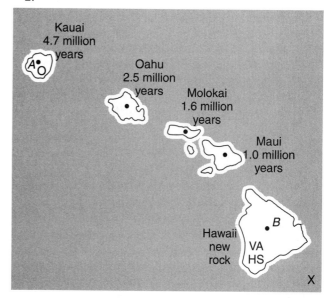

3. 550.5 km × 1,000 m/km × 100 cm/m = 55,050,000 cm; 55,050,000 cm/4,700,000 years = 11.7 cm/year

HOT-SPOT TRACKS

1. from north northwest to northwest.

PLATE VELOCITIES

1. 1.8 cm/year
2. 3.6 cm/year

3. Pacific and Nazca Plates; at the East Pacific Rise; divergent; 2 ×18.3 cm/year = 36.6 cm/ year

Short-Essay Questions

1. A continental volcanic arc forms when ocean plate subducts beneath continental lithosphere, so the resulting volcanoes are a chain on land. An oceanic arc forms when one ocean plate subducts under another ocean plate, so the resulting volcanoes form a chain of volcanic islands.

2. Hot asthenosphere rises below the trough of a mid-ocean ridge. It melts; the molten rock is called magma. Some of the magma solidifies in a space (magma chamber) below the sea-floor surface of the trough as the coarse-grained mafic rock called gabbro. Some of the magma continues to rise higher and solidifies, as the fine-grained mafic rock called basalt, in vertical cracks, still within the sea floor. The vertical sheet-like structures formed this way are called dikes. Some rising magma makes it all the way to the sea floor, spills out through small submarine volcanoes, and solidifies as a new layer of basaltic sea floor. The preceding is not just all theoretical; scientists on the research submarine *Alvin* have seen submarine volcanoes in action.

3. Newly formed sea floor is literally pulled apart by tensional forces that result from the spreading motion; this creates numerous faults that parallel the mid-ocean ridge. Slippage along these faults results in earthquakes along the divergent boundary.

4. By the time oceanic lithosphere is 200 million years old, it has become denser than asthenosphere and has subducted into it. It probably goes deep enough into the mantle to become sufficiently warm and soft to lose its identity and mix and flow with the rest of the mantle.

5. Probably the rock below 670 km is too warm and soft to break and quake; it just flows.

6. Iceland is a hot-spot volcano that straddles the diverging Mid-Atlantic Ridge. It will eventually be pulled apart by the diverging process and become two islands. They will cease to be volcanic as they leave the ridge, and eventually will become members of a symmetrical seamount chain extending on both sides of the ridge.

7. The study of rocks of the Appalachian Mountains show that these mountains are the result of an ancient continental collision that was much older than any events associated with either the formation or splitting apart of Pangaea.

8. Kilimanjaro lies along the East African Rift, where the stretched continental crust has become thinner than normal. Hot asthenosphere rises below the rift area, and eventually volcanoes result at the surface.

Practice Test

1. b Continental lithosphere (continental plate) is always older and less dense than oceanic lithosphere, so continents never subduct under the ocean; continents' shapes have changed as they have merged or split; continents don't move *through* the sea floor but are passively moved *by* the sea floor.

2. b Choice b is false; the sea floor is basalt, as stated in e, not gabbro.

3. e Rifting and microplates have nothing to do with the Hawaiian Islands. Since Hawaii is the youngest island, the stationary hot spot must be located there. The older islands to the northwest were created when their portions of plate sat over the hot spot.

4. b The shifting of huge plates in any direction (past, into, or away from each other) creates major disruptions like earthquakes and volcanoes.

5. c The Pacific plate is moving toward the southwest; even if you didn't know the direction of the plate's movement, you should have realized that answer c was false, because no plate moves faster than a few inches per year.

6. c Subduction at this location causes the plate to sink deep and melt. Neither transform motion nor continental collision produces melted rock and volcanics. Basalt does not push up through trenches.

7. c Continental plates never subduct; any subducting plate is oceanic.

8. b Molten basalt is rising, not sinking, in the trough. The ridge has nothing to do with the Hawaiian hot spot or with continental convergence. It's associated with an oceanic divergent boundary.

9. c Transform motion and continental collision cause only earthquakes; subduction of ocean plate causes earthquakes and volcanic activity. The Pacific Plate is subducting under the North American Plate and producing Alaska's earthquakes and volcanoes.

10. a San Francisco is located along the San Andreas Fault, which is part of a transform plate boundary. Plates moving past each other generate earthquakes, but not molten rock and volcanics.

11. b India crashed into Asia 40 million years ago, and this continental collision produced the Himalayas.

12. a Any subducting plate is oceanic lithosphere.

13. a Asthenosphere is part of the mantle, not the core. It supports the lithosphere, rather than colliding with or subducting under it.

14. a The rate of plate motion is slowest along the northern Mid-Atlantic Ridge. Rates remain constant for millions of years. Wilson worked on transform faults and hot spots, not rates of plate motion.

15. e Kilimanjaro is a volcano because it is in a continental rifting zone.

16. False Relicts of ancient collision mountain ranges, like the Appalachians, predate the existence of Pangaea and demonstrate there were previous supercontinents.

17. True The Alps and Appalachians were formed by continental collisions in a previous cycle of plate tectonics. The Himalayas continue to grow as the Australian-Indian Plate collides with the Eurasian Plate.

18. True Due to the principle of buoyancy, the low-density continental crust that floats the highest must be balanced by the continental crust that reaches the deepest; thus the overall thickness of mountain crust exceeds the thickness of normal continental crust.

19. True The asthenosphere is hot enough to flow; the lithosphere above it is not.

20. False There is convective flow, but it cannot explain the motion of plates.

21. False The asthenosphere passively fills the space created by the sea-floor plates' moving apart; it doesn't push them apart.

22. False Plate movement is not slow and steady, and there are numerous earthquakes along plate boundaries.

23. False They occur along active continental margins.

24. False Hot spots don't move; the plate moves across the stationary hot spot.

25. True Geothermal activity of Yellowstone today and its volcanic activity in the past happen because the region overlies a hot spot; the Hawaiian islands all formed as the Pacific Plate passed over the hot spot.

26. True

27. False Only the island currently over the hot spot (plume) is volcanically active.

28. False The San Andreas Fault is a valley, but lava does not emerge from it.

29. False The Mid-Atlantic Ridge began as offset segments linked by transform faults.

30. False The East Pacific Rise is wider and doesn't have a trough down its center.

31. True Upwelling basalt does not push the sea floor apart; slab-pull and ridge-push are the mechanisms of sea-floor spreading.

CHAPTER 5 | Patterns in Nature: Minerals

GUIDE TO READING

Although there are over 3,000 recognized minerals, only a few dozen are common on Earth. The author begins by presenting a table of fewer than thirty of these. If you're fortunate enough to be taking this as a lab class or if you have access to a mineral collection, you'll probably find samples of all of these in your collection. Do look at them as you read along. Nothing beats hands on for mineral identification and appreciation.

Most of Chapter 5 is devoted to the scientific approach to mineral study, but woven throughout the chapter are consumer-type topics:

- the asbestos concerns
- the questionable magical powers of crystals
- timekeeping with quartz watches
- the beauty and legends of gemstones (diamonds and more)
- even the fabled "dilithium crystals" of *Star Trek* fame (which don't actually exist)

The author begins with a very detailed definition of minerals. This leads quite naturally into a discussion of crystal formation and crystals' unique internal structures, which determine their external shapes and symmetry.

Mineral identification is the next topic. How do you tell one kind from another when there are so many? Fortunately there are few enough common ones that examination of a few physical properties generally allows identification of specimens in either the field or the lab.

Keeping things organized is a challenge when dealing with large numbers of anything. Since minerals are simply chemical elements or compounds formed naturally on Earth, it's logical and useful to study them as chemical groups. Minerals that are single *elements* are part of the group native metals. Most minerals that are *compounds* fit into one of the following chemical groups:

- Silicates
- Oxides
- Sulfides
- Sulfates
- Halides
- Carbonates

The chapter leaves many practical questions unanswered, including questions dealing with mankind's use of minerals as resources for manufacturing products and producing energy. These issues are dealt with in Part 5, Earth Resources, Chapters 14 and 15. For now the author pursues the fact that minerals are the building blocks of rocks, and the next three chapters deal with the very geologic topics of igneous, sedimentary, and metamorphic rocks.

LEARNING ACTIVITIES

Completion

Test your recall of new vocabulary terms.

_____ 1. minerals that serve as the raw materials for manufacturing chemicals, concrete, and wallboard

_____ 2. minerals that serve as either the source of valuable metals (like copper and gold) or as energy resources (like uranium)

_____ 3. the study of minerals

_____ 4. a naturally occurring, homogenous, inorganic solid with a definable chemical composition and an internal structure characterized by the orderly arrangement of atoms, ions, or molecules in a lattice (a crystalline arrangement)

_____ 5. the same throughout

_____ 6. materials manufactured in laboratories that are identical to naturally formed minerals but technically don't fit the mineral definition, which requires minerals to be naturally occurring

_____ 7. compounds composed of carbon and hydrogen, and often additional elements like oxygen and nitrogen, that were initially (but incorrectly) believed to occur only in living organisms

_____ 8. materials, such as minerals, that don't contain organic compounds

_____ 9. a specific, repeating pattern, at the atomic or molecular level, that serves as an orderly framework to accept the atoms, ions, or molecules that make up minerals

_____ 10. the quality of materials that contain a crystal lattice

_____ 11. a single continuous piece of crystalline material, bounded by flat surfaces (crystal faces)

_____ 12. the splitting apart of a beam of electromagnetic radiation as it passes through a material composed of regularly spaced obstacles (if the spacing is comparable to the wavelength of radiation involved), which can be observed using X-rays and minerals

_____ 13. a type of visual balance, in which one side of a mineral is the mirror image of the other side

_____ 14. two different minerals that have the same chemical composition but different crystal lattice structures and therefore exhibit different physical properties

_____ 15. the freezing of a liquid

_____ 16. the slow movement of atoms or ions through a solid to arrange into a new crystal lattice

_____ 17. the process in which ions dissolved in water chemically bond together and come out of solution as a solid compound

_____ 18. an extremely small crystal that serves as an attachment center for more mineral material and thus the growth of a larger crystal

_____ 19. an adjective describing a liquid in which ions, atoms, or molecules dissolved in it are so numerous they get close enough to chemically bond

_____ 20. a roughly spherical cavity in rock that is lined with crystals, which precipitated out of the water solution that percolated through the rock

_____ 21. a crystal that was unconfined in its growing area and therefore has developed well-formed faces and angles

_____ 22. crystalline minerals that grew in confined spaces and thus did not develop well-formed faces and angles

_____ 23. the temperature at which there's enough thermal energy to break the chemical bonds holding atoms or ions in place in a crystal lattice and thus change the material from its solid to its liquid state

_____ 24. characteristics of a mineral (like color, hardness, streak, luster, specific gravity, crystal form, cleavage, and cleavage habit) that help in mineral identification

_____ 25. the physical characteristic of a mineral that's the result of wavelengths of light that are not absorbed but instead are reflected back to the observer

_____ 26. color of the powder of a mineral, which is generally produced by rubbing the mineral on an unglazed ceramic plate

_____ 27. the way a mineral handles light, the possibilities of which include metallic, nonmetallic, silky, glassy, satiny, resinous, pearly and earthy

_____ 28. the ability of a mineral to resist scratching, which depends on the strength of the chemical bonds involved

_____ 29. the classic scale, from number 1 (softest) to number 10 (hardest), used to designate the ability of a mineral to resist scratching

_____ 30. the density of a mineral compared to the density of water

_____ 31. a geometry term, based on the angular relations between crystal faces and used to classify different types of well-formed (euhedral) crystals

_____ 32. the general shape or character of a crystal or cluster of crystals that grew unimpeded (examples: fibrous, cubic, grape-like)

_____ 33. the characteristic breaking of a mineral along planes and angles of weak chemical bonding

_____ 34. planar surfaces created when a mineral breaks along weak chemical bonds

_____ 35. smooth, curving, shell-shaped surfaces produced by the breakage of quartz and some other minerals

_____ 36. an assortment of rather unique characteristics that pertain to few minerals and help identify them

_____ 37. a stone that has been cut and finished for use in jewelry

_____ 38. a mineral that is prized for its rarity and beauty

_____ 39. a jeweler's term to designate stones that are very rare and expensive (diamonds, rubies, sapphire, emeralds)

_____ 40. a jeweler's term to designate stones that are considered valuable but are not particularly rare or expensive (topaz, tourmaline, aquamarine, garnet)

_____ 41. ground and polished surfaces on stones, created by a jeweler's machine, not the crystal faces or cleavage faces on stones, which are naturally occurring features of a mineral

_____ 42. extremely coarse-grained igneous rock formed by hardening of water-rich magma

Matching Questions

MINERALS CONSIDERED AS CHEMICAL GROUPS

Match the appropriate chemical group with the mineral names.

a. carbonates
b. halides
c. native metals
d. oxides
e. silicates
f. sulfates
g. sulfides

_____ 1. quartz and 95% of all deep continental crustal rock and almost all mantle rock

_____ 2. magnetite and hematite (iron ores)

_____ 3. pyrite (fool's gold), galena, and many other metallic ore minerals

_____ 4. gypsum, formed by precipitation from water at or near Earth's surface

_____ 5. table salt and fluorite

_____ 6. calcite ("lime") and dolomite

_____ 7. a gold nugget or copper

GEMS

Match the gem name with its description.

a. amber
b. amethyst
c. aquamarine
d. diamond
e. emerald
f. garnet
g. jade
h. opal
i. pearl
j. ruby
k. sapphire
l. topaz
m. tourmaline
n. turquoise

_____ 1. a very valuable gem variety of beryl, noted for its rich green color

_____ 2. microscopic spheres of hydrated silica for which Australia is noted

_____ 3. the gem name for the mineral aragonite when it's been produced by irritated oysters

_____ 4. a copper mineral with a distinctive blue-green color

_____ 5. a mineral brought to Earth's surface from 150 km deep in igneous bodies called kimberlite pipes

_____ 6. purple-colored quartz

_____ 7. fossilized tree sap

_____ 8. a blue and very valuable version of corundum

_____ 9. a bluish version of an emerald

_____ 10. the precious stone produced by impurities of chromium that color corundum red

_____ 11. a family of aluminum silicate minerals (made up of calcium, iron, magnesium, and manganese versions) which is so common its members are often used as industrial abrasives and whose beauty is frequently marred by inclusions, bubbles, or fractures

_____ 12. the gem name for two different minerals of metamorphic rocks: pyroxene and amphibole

_____ 13. a hydrated aluminum silicate, formed when hot water reacts with igneous rocks

_____ 14. a semi-precious stone which is found in both igneous and metamorphic rocks and which is a complex borate-silicate compound that illustrates how elements (like sodium, magnesium, and iron) can substitute for each other in a crystalline lattice

Short-Answer Questions

Answer with a few words, numbers, or short sentences, or choose from the words in parentheses.

1. What is an asbestos mineral? _____

2. According to the scientific definition of a mineral, why aren't the following materials minerals?
   a. ivory _____
   b. pearls _____
   c. coral from reefs _____
   d. glass _____
   e. oil _____

3. Ice cubes are not minerals. Glacial ice is a mineral. Explain. _____

4. What raw material might be used to manufacture synthetic diamonds? _____

5. Do chemical formulas of minerals have to be totally unvariable? Give an example. _____

6. Sodium and chlorine fit together in a cubic crystal lattice. What shape is a grain of table salt? _____

7. What mineral does the starship *Enterprise* need in order to achieve "warp speed"? _____

8. Mineral crystals cause X-ray diffraction. What does this say about the internal arrangement of the atoms or ions that make up mineral crystals? _____

9. Name the polymorph of diamonds. _____

10. State three ways mineral crystals can form. _____

11. The youngest part of a crystal is always on its outer edge because crystals grow _____ (inward, outward) from a starting point called a _____.

12. Are the crystals of a geode euhedral or anhedral? _____

13. Is color a useful physical property for identifying quartz? Explain. _____

14. Why can the same mineral have so many different colors? _____

15. Why can a streak color be more useful in mineral identification than the color of the solid mineral? _____

16. If you can scratch a mineral with a copper penny but can't scratch it with your fingernail, what is its possible range of hardness? _____

17. a. What are the three hardest minerals on Mohs hardness scale, and what do they have in common? _____

    b. If these same three minerals had hardnesses at the other end of the scale, would it be likely they'd have the same value to society as they do now? _____

18. Galena has a specific gravity of 7.6.
    a. From this information alone you can deduce it is probably a _____ (metallic, nonmetallic) mineral.
    b. A cubic foot of water weighs 62.4 lb. What would a cubic foot of galena weigh? _____

19. The mineral mica can be pulled apart with your fingers, sheet by sheet, like a stack of papers. This must mean mica molecules have _____ (strong, weak) bonding within a sheet and _____ (strong, weak) bonding between sheets. How many directions of cleavage must mica have to separate this way? _____

20. Galena breaks into cubic pieces. How many directions of cleavage must it have? _____
    What size angles are between its cleavage faces? _____

21. Give the special properties of the following:
    a. halite _____   b. magnetite _____
    c. graphite _____   d. plagioclase _____
    e. calcium carbonate _____

22. a. What is the chemical composition of all silicate minerals? _____

b. What is the geometric form of their fundamental building block? _____

23. How are the different groups of silicate minerals distinguished from each other? _____ _____

24. a. Name two very special and valuable varieties of corundum. _____
    b. Name a very special and valuable variety of beryl. _____.

25. a. Why aren't pearls and amber considered true minerals? _____ _____
    b. What is an additional reason *cultured* pearls are not considered true minerals? _____ _____

26. It's theoretically possible to make diamonds out of anything with high carbon content if you expose the material to extreme _____ and extreme _____

27. How does the carbon necessary for diamond formation get deep into the mantle? _____ _____

28. What are nongem quality diamonds used for? _____ _____

29. Diamonds are brought to Earth's surface in kimberlite pipes. How do they get from the pipes into river gravels? _____ _____ _____

30. A young man spent his last cent (or more likely used his credit card to its limit) and bought his fiancée a 1-carat diamond engagement ring. Nice as it was, how much larger a ring would she have if he'd been able to afford the uncut Cullinan diamond? _____ _____

31. What would be the karat rating of a gold ring that was 50% gold and 50% silver? _____

32. The gold band of a typical high school ring is marked 18K (karat). How pure (in percent) is the gold in the ring? _____

APPLIED CHEMICAL TERMS

2 NaCl, (sodium chloride, halite, salt)
$CaCO_3$, (calcium carbonate, calcite)
Na^+ Cl^- Ca^{2+} $(CO_3)^{2-}$

Answer the following questions. Some of them require direct reference to the chemical formulas and symbols above. Some simply require knowledge of chemical terms.

1. What are the two elements in salt? _____ _____

2. a. How many different elements are there in calcium carbonate? _____
   b. Name them. _____ _____

3. a. How many atoms of oxygen are in one molecule of calcite? _____
   b. How many total atoms are there in one molecule of calcite? _____

4. a. The atomic number of sodium (Na) is 11. How many protons does it have? _____
   b. The most common variety of sodium has an atomic weight of 23. How many neutrons does an atom of sodium have? _____

5. How many molecules of NaCl are in the chemical expression above? _____

6. How many molecules of $CaCO_3$ are in the chemical expression above? _____

7. An ion is a charged atom. A negatively charged ion is called a(n) _____; a positively charged ion is called a(n) _____.

8. List two cations in the chemical expression above: _____

9. List two anions in the chemical expression above: _____

10. a. When cations attract anions and form a molecule, the chemical bond formed is called a(n) _____ bond.
    b. If instead a bond is formed because of mutual sharing, this bond is called a(n) _____ bond.

11. Rocks are aggregates (mixtures) of minerals.
    a. Can you write formulas for rocks? _____
    b. Can you write formulas for minerals? _____

12. a. Is a chemical reaction necessary to separate elements in a mineral? _____
    b. Is a chemical reaction necessary to separate mineral grains in a rock? _____

13. When salt is put in water it dissolves. The salt is called the _____ (solute, solvent); the water is called the _____ (solute, solvent).

14. If the sodium and chlorine ions of salt in a solution reunite and settle out as a solid, the process is termed _____ and the solid salt is called a _____.

Short-Essay Questions

1. Use the text for ideas and look around you with a discerning eye to write a paragraph about several ways minerals serve as practical resources for human needs.

2. You've probably heard about the asbestos problem; maybe you've had personal experience with the intensive asbestos-removal efforts in the United States during the 1990s. Write a paragraph about this that addresses these three issues:
   a. What is asbestos?
   b. How valid is its designation as a health hazard?
   c. Your personal opinion: Should asbestos removal continue with the vigor of the past decade or is more study needed?

3. Are all garnets valuable and expensive? Explain your answer.

4. It is fact that a diamond has four cleavage directions, yet a typical wedding ring diamond with a brilliant cut has 57 surfaces. Explain.

Practice Test

1. Which of the following statements is FALSE? Formation of table salt
   a. is, in mineralogic terms, formation of the halide halite.
   b. takes place when a solution has become saturated.
   c. is an example of solidification of a melt.
   d. needs a seed crystal to begin.
   e. results in cubic crystals.

2. Which of the following statements about asbestos is FALSE?
   a. It's a generic name for fibrous minerals with strong, nonflammable fibers.
   b. Chrysotile (white asbestos) is the form used most often.
   c. It's most dangerous when it occurs as dust in the air.
   d. Current regulations make distinctions between the different kinds of asbestos and regulate accordingly.
   e. Asbestos can cause asbestosis, which clogs lung tissue and causes lungs to stop functioning.

3. Silicates
   a. have the silicon tetrahedron as their structural unit.
   b. are a major component of continental crust.
   c. are classified on the basis of how the tetrahedrons join and share oxygen atoms.
   d. are a huge category of minerals and include the very common mineral quartz.
   e. All of the above.

4. Which of the following statements is FALSE? Crystals
   a. have an orderly internal arrangement of atoms attached to a crystal lattice framework.
   b. grow outward from a seed.
   c. are formed by the high-temperature, extremely rapid process known as solid-state diffusion.
   d. display symmetry.
   e. cause X-ray diffraction.

5. Which of the following is a mineral?
   a. amber (tree sap)
   b. icicles
   c. sugar (formula $C_6H_{12}O_6$)
   d. ivory (from animal tusks)
   e. coral (from reefs)

6. Which of the following is NOT a mineral?
   a. glacial ice, because ice just isn't a rocky material
   b. salt, because it's not crystalline
   c. sulfur, because it's an element
   d. pyrite, because it's a chemical compound
   e. oil, because it's a liquid, not a solid

7. The tendency of a mineral to break and produce smooth, curving shell-shaped surfaces is termed
   a. luster.
   b. fibrous fracture.
   c. conchoidal fracture.
   d. streak.
   e. perfect cleavage.

8. There are two physical properties of minerals that both result in smooth, flat surfaces with specific angles between them. The first property is the result of how the mineral forms; the second is the result of breaking a sample of the mineral. These properties are called
   a. hardness and cleavage.
   b. cleavage and hardness.
   c. crystal form and cleavage.
   d. cleavage and crystal form.
   e. crystal form and hardness.

9. Graphite is
   a. the polymorph of galena.
   b. harder than glass.
   c. pure silicon.
   d. the lead in the pencil you write with.
   e. the proper name for fool's gold.

10. Which is not a physical property commonly used in the lab or field to identify minerals?
    a. color
    b. streak
    c. luster
    d. diffraction
    e. specific gravity

11. An unknown mineral scratches glass, has only average specific gravity, and shows no cleavage but does show conchoidal fracture. Which of the following could it be?
    a. quartz
    b. talc
    c. asbestos
    d. mica
    e. galena

12. Which of the following statements is FALSE? Diamonds
    a. are brought from the mantle to the surface in magma that hardens into kimberlite.
    b. are found in carrot-shaped structures called pegmatites.
    c. of industrial quality are used as abrasives.
    d. which have weathered out of ore bodies can end up in stream gravels.
    e. have been found in South Africa, India, northwest Canada, and the United States.

13. The expression "24 karat" must mean that
    a. you're talking about the biggest diamond ever found.
    b. the substance is flawed by impurities.
    c. there are 24 cleavage planes present.
    d. the substance is 100% pure.
    e. the metal's composition is 24% gold and 76% silver.

14. Glass is not a mineral because it doesn't have crystal lattice structure.
    a. True
    b. False

15. A geode is filled with anhedral crystals on its walls.
    a. True
    b. False

16. Galena and halite both have one direction of cleavage.
    a. True
    b. False

17. Chrysotile, amosite, and crocidolite are all asbestos minerals.
    a. True
    b. False

18. Mica peels apart in parallel sheets because the chemical bonds between sheets are weak and the chemical bonds within sheets are strong.
    a. True
    b. False

19. The energy of various crystals has been shown to cure human ailments.
    a. True
    b. False

20. Ice crystals form by solidification of a melt.
    a. True
    b. False

21. Bronze, an alloy produced by melting together copper and tin, is a mineral.
    a. True
    b. False

22. Minerals can be destroyed, sometimes by heating them, sometimes by dissolving them.
    a. True
    b. False

23. Pegmatites are rocks composed of extremely fine grained mineral crystals so it's easy to see the gemstones in them.
    a. True
    b. False

24. Color and cleavage are two very useful physical properties used to identify quartz.
    a. True
    b. False

25. Gold and copper can both be found as native metals.
    a. True
    b. False

26. Ore minerals are often sulfides and oxides of metallic elements.
    a. True
    b. False

27. Facets are simply a jeweler's encouragement of crystal faces.
    a. True
    b. False

28. The Hope Diamond, now in the Smithsonian Institution in Washington, D.C., has exceptional value because in addition to being the subject of legends and curses, it's a rare blue diamond.
    a. True
    b. False

29. Examples of precious stones are diamonds, rubies, sapphires, and emeralds; examples of semi-precious stones are aquamarines and garnets.
    a. True
    b. False

30. All precious and semi-precious stones must be transparent.
    a. True
    b. False

ANSWERS

Completion

1. industrial minerals
2. ore minerals
3. mineralogy
4. mineral
5. homogenous
6. synthetic minerals
7. organic compounds
8. inorganic substances
9. crystal lattice
10. crystalline
11. crystal
12. diffraction
13. symmetry
14. polymorphs
15. solidification of a melt
16. solid-state diffusion
17. precipitation
18. seed
19. saturated
20. geode
21. euhedral crystal
22. anhedral grains
23. melting temperature
24. physical properties
25. color
26. streak
27. luster
28. hardness
29. Mohs hardness scale
30. specific gravity
31. crystal form
32. crystal habit
33. cleavage
34. cleavage planes
35. conchoidal fracture
36. special properties
37. gem
38. gemstone
39. precious stones
40. semi-precious stones
41. facets
42. pegmatites

Matching Questions

MINERALS CONSIDERED AS CHEMICAL GROUPS

1. e
2. d
3. g
4. f
5. b
6. a
7. c

GEMS

1. e
2. h
3. i
4. n
5. d
6. b
7. a
8. k
9. c
10. j
11. f
12. g
13. l
14. m

Short-Answer Questions

1. a fibrous mineral whose fibers are strong and nonflammable

2. a. organic origin   b. organic origin   c. organic origin   d. no ordered structure, i.e., not crystalline   e. organic origin and not a solid

3. The material ice fits the mineral definition, but ice cubes are not naturally occurring; glacial ice is naturally occurring.

4. Anything that can supply carbon is a possibility; diamonds literally have been made from peanuts.

5. No, sometimes atoms of two different elements can substitute for each other in the same crystal lattice, for example, biotite (black mica), in which iron and magnesium may substitute for each other.

6. Cubic; you can easily check this out under a magnifying lens.

7. a dilithium crystal (Note: no such crystal exists, but the author thought you *Star Trek* fans would appreciate this one.)

8. It's an orderly internal arrangement, with regularly spaced particles, as required to cause diffraction.

9. graphite

10. a. solidification of a melt; precipitation from solution; solid-state diffusion

11. outward; seed

12. euhedral

13. No, quartz can be many colors, including clear, white, purple (amethyst), gray (smoky), and red (rose).

14. Trace amounts of impurities influence the mineral's color.

15. The mineral powder color is less variable than the color of a whole crystal, which is easily influenced by the presence of even a trace amount of impurity.

16. greater than $2^{1/2}$, less than or equal to $3^{1/2}$

17. a. Diamond (10), corundum (ruby, sapphire 9), topaz (8); they're either precious or semi-precious stones.
    b. No, because they could easily get scratched and ruined, or even fall apart, which would not be desirable in an expensive piece of jewelry; also they wouldn't last long enough to become legendary, like the Hope Diamond.

18. a. metallic   b. $62.4 \times 7.6 = 474$ lb

19. strong; weak; one

20. three; right angles (90°)

21. a. salty taste   b. magnetic   c. leaves a dark gray mark on paper (If you're writing with pencil, you're using it now.)   d. striations (i.e., thin parallel "scratches" on cleavage surfaces)   e. effervesces (fizzes) with dilute hydrochloric acid (It's releasing carbon dioxide gas, which bubbles through the liquid.)

22. a. one silicon atom and four oxygen atoms
    b. put together as a tetrahedron

23. by the way their tetrahedrons are linked

24. a. rubies and sapphires   b. emeralds

25. a. Both have organic origins; pearls come from activity of oysters and amber is fossilized tree sap.
    b. They are not naturally occurring; the irritating particle that causes growth of the pearl was put there by a person.
26. heat; pressure
27. Carbon-containing rocks and sediment on the ocean floor are carried downward as plates subduct at convergent plate boundaries.
28. abrasives, because any diamond is harder than any other substance
29. Weathering causes most minerals in kimberlite to disintegrate, leaving the diamonds exposed to be eventually washed away into stream gravels.
30. 3,106 times as large
31. 12 karat
32. 18/24 = 75%

APPLIED CHEMICAL TERMS

1. sodium and chlorine
2. a. three   b. calcium, carbon, and oxygen
3. a. three   b. five: one calcium, one carbon, three oxygen
4. a. 11 (The atomic number is the number of protons.)
   b. 12 (Atomic weight = number of protons plus number of neutrons.)
5. two ($\underline{2}$ NaCl)
6. one
7. an anion, a cation
8. Na^+, Ca^{2+}
9. Cl^-, CO_3^{2-}
10. a. ionic bond   b. covalent bond
11. a. no   b. yes
12. a. yes   b. no
13. solute, solvent
14. precipitation, precipitate

Short-Essay Questions

1. Many answers are possible, on the basis of your powers of observation and life experience. As a minimum, your answer should include the following information from the text: Minerals are the building blocks of the planet; we literally live on minerals. We also use them to make just about every product of modern society. They are the raw materials for industrial chemicals, concrete, wallboard, and both decorative and practical metal products (metals like gold, silver, platinum, iron, aluminum, lead, zinc, copper, and tin). Uranium is an energy resource. Asbestos, now in disfavor, is still found in many insulating materials.

2. a. Asbestos can be one of three minerals, all of which have strong, nonflammable fibers: chrysotile (white asbestos); amosite, (brown), or crocidolite (blue). About 95% of all asbestos products are made of chrysotile. This is fortunate because it's thought to be the least harmful, due to the fact its fibers are curly and can't pierce lung tissue.
   b. There are serious health concerns about asbestos use. It probably is the cause of mesothelioma (a rare cancer of the chest and abdomen) and asbestosis (a disease in which lungs get clogged with asbestos and cease to function). Uncertainties exist about exactly how asbestos causes disease, and it's even uncertain the risk is great unless you're exposed to large quantities of asbestos dust in the air (as in asbestos factories, mines, or buildings where asbestos is currently being removed). Presently regulations are based on the idea all asbestos is *equally* dangerous and *extremely* dangerous under all circumstances.
   c. Your personal opinion is your opinion. You may agree with current regulations, on the premise it's best to err on the side of caution. You may say asbestos removal is more dangerous than just leaving it alone. You may say more study is needed before an intelligent decision can be made.

3. No, not all garnets are valuable. Except for gem-quality specimens, they're not rare. Many have inclusions that detract from their beauty, and they get used simply as industrial abrasives.

4. Diamonds, like most gems used in jewelry, are cut stones. This means their flat surfaces are facets that have been cut and polished on a faceting machine. A jeweler may encourage cleavage breakage to make a large stone smaller and may produce facets parallel to cleavage directions, but facets are neither naturally occurring cleavage planes nor crystal faces.

Practice Test

1. c Salt is an example of precipitation from a solution.
2. d It is still debatable, but evidence supports the viewpoint that different kinds of asbestos pose different health hazards and should be regulated accordingly.

3. e Statements a to d are all true.

4. c Solid-state diffusion is the very slow migration of atoms through solid material as they rearrange themselves on a crystal lattice.

5. b Icicles fit the mineral definition criteria; the others have organic origins.

6. e Minerals by definition must be solid; all of the rest are either invalid statements about the material or not disqualifying reasons.

7. c Only conchoidal fracture matches the definition given.

8. c Crystal form is the geometry of a euhedral crystal; cleavage describes the way a mineral breaks.

9. d Graphite is very soft, gray, and metallic; consists of carbon; and is used for pencil lead (see Table 5.1 in the text).

10. d These physical properties are: color, streak, luster, hardness, specific gravity, crystal form, crystal habit, cleavage, and special properties; choice d, X-ray diffraction, is a procedure used to examine crystal lattice patterns.

11. a Only quartz meets these specifications; talc is too soft, asbestos breaks in fibers, mica has excellent cleavage, and galena has a high specific gravity.

12. b The carrot-shaped structures in which diamonds are found are called kimberlite pipes.

13. d Karat is a measure of purity; "24 karat" means a gold item is 100% gold.

14. True By definition a mineral has a crystal lattice structure; glass doesn't have this structure.

15. False The crystals had room to form and therefore are euhedral.

16. False Both are excellent examples of cubic cleavage, with three cleavage directions.

17. True Asbestos is a generic name for many fibrous, nonflammable minerals, including chrysotile, amosite, and crocidolite.

18. True Mica has perfect cleavage in one direction because chemical bonds within each layer are strong but bonds between layers are weak.

19. False Although crystal therapists claim to cure ailments, a series of experiments has shown the proximity of crystals to have no demonstrable effect on health.

20. True Ice is the solid crystal form of the liquid (or melt) form of water.

21. False A mineral must be naturally occurring.

22. True High temperature and solvent molecules both can destroy chemical bonds that attach atoms to their crystal lattices.

23. False Pegmatites are extremely coarse-grained rocks.

24. False Quartz can come in various colors and shows no cleavage.

25. True Metals that exist as elements, not compounds, are called native metals; gold and copper are two examples.

26. True Many metallic elements bond with sulfur and oxygen in the environment to produce sulfides and oxides.

27. False Facets are produced by a jeweler's tool; though they may be in agreement with crystal faces and cleavage planes, they are more numerous than either of the others.

28. True Pale color of a diamond diminishes its value; strong color, such as the blue color of the Hope Diamond, increases its value.

29. True These are all good examples of very valuable precious gemstones and less valuable semi-precious gemstones.

30. False Lapis, malachite, and opal are semi-precious stones that are opaque or translucent.

CHAPTER 6 | Up from the Inferno: Magma and Igneous Rocks

GUIDE TO READING

Sometime in your previous school years you probably learned there are three basic kinds of rocks: igneous, sedimentary and metamorphic. The classification is based on the origin of the rock. *Igneous* rocks, the fire-formed ones, are the logical group to start with for two reasons: (1) they make up the greatest part of planet Earth, and (2) they were the first rocks to exist on Earth.

The igneous category is subdivided into two main branches: *intrusive igneous rocks* and *extrusive igneous rocks*. In the first, molten rock below Earth's surface (*magma*) hardens and creates various igneous rocks (including granite, diorite, and gabbro) and structures (irregular shapes called plutons and tabular intrusions). In the second, molten rock spills out on the surface as *lava* that produces rocks (including rhyolite, andesite, basalt, obsidian, pumice, and pyroclastic rocks), structures (volcanoes and lava flows), and some wild and exciting events, for "extrusive igneous activity" is synonymous with the word "volcanics." This chapter concentrates on magmas and rocks and touches only briefly on volcanic activity. You'll have to wait until Chapter 9 for the wild and exciting events; that's the chapter devoted to volcanic eruptions and their importance to humans throughout history.

Any study of igneous activity begins with a discussion of magma. As you learned in Chapter 2, Earth's interior is *not* composed chiefly of molten rock. Instead there are relatively few interior areas that are liquid, and there has to be some special reason for the solid rock of these areas to have become molten rock, that is, magma. Your author organizes his discussion of magma formation as follows:

- the conditions that cause melting of mantle and crustal rocks (decreased pressure, addition of volatiles, and heat transfer)
- the chemical composition of magmas
- the four major types of magmas (silicic, intermediate, mafic, and ultramafic)
- movement of magma, why it goes where it does, and why it has different viscosities

Once formed, magma doesn't always stay molten. Why and how does it harden? Your author discusses:

- the sequence of hardening of a melt (fractional crystallization, Bowen's Reaction Series)
- factors that control the cooling rate
- structures that result when magma solidifies within the Earth (plutons, tabular intrusions, laccoliths, batholiths, xenoliths and the stoping process, sills, and dikes)
- classification of igneous rocks (based on chemical compositions and textures)

What determines where igneous activity occurs? Those same two words crop up again: "plate tectonics." Once again you'll read about volcanic arcs, mid-ocean ridges, subducting plates, continental rifting, and hot spots.

A word of advice: A discussion of magma is not conceptually difficult, but it can be confusing. After all, the intrusive activity is happening in the unseen and unfamiliar world of Earth's interior, and extrusive lava is not something the average person has had experience dealing with. Keep in mind the following two sets of word associations to help you follow the discussions of solid rock melting and liquid rock freezing.

Set one: A chemical composition high in silica, low in iron and magnesium (silicic); light-colored, light-weight rocks; low-temperature but high-viscosity melts

Set two: A chemical composition high in iron and magnesium, low in silica (mafic); dark-colored, heavy rocks; high-temperature but low-viscosity melts

Earth is a dynamic place; given enough geologic time, nothing on it remains unchanged. Once igneous rocks exist, they become part of that rock cycle you no doubt learned about long ago, at the same time you learned to recite "igneous, sedimentary, and metamorphic." Chapters 7 and 8 continue the story, as natural forces change igneous rock into sedimentary and metamorphic rocks.

Completion

Test your recall of new vocabulary terms.

_____ 1. the generic term for any molten rock, either above or below Earth's surface

_____ 2. a vent from which melt issues onto Earth's surface

_____ 3. a stream of hot molten rock moving across Earth's surface and the sheet it forms when it solidifies

_____ 4. rock that is formed by the freezing of a melt, either below or on top of Earth's surface

_____ 5. molten rock that exists below Earth's surface

_____ 6. molten rock that has erupted onto Earth's surface

_____ 7. the category of rock that has formed by the freezing of a melt within the Earth after it has pushed its way into preexisting rock

_____ 8. the category of rock that has formed by the freezing of a melt after it's been extruded onto Earth's surface

_____ 9. the line on a graph that defines the temperature as a function of depth

_____ 10. the melting of hot mantle rock as it rises to shallower depths and is subjected to lower pressures

_____ 11. the rate of increase in temperature as you go deeper into the Earth

_____ 12. the line on a graph, determined by laboratory measurements, that defines the conditions of pressure and temperature at which mantle rock begins to melt

_____ 13. elements or compounds, like water and carbon dioxide, that can exist in the gaseous state under ordinary Earth surface conditions

_____ 14. the melting of crustal rock when its temperature is raised by contact with hot rising mantle magma

_____ 15. magmas that are rich in silica (70%) but have little magnesium and iron

_____ 16. magmas whose composition is between silicic and mafic and which contain about 55% silica

_____ 17. magmas that are rich in magnesium and iron but have less than 50% silica

_____ 18. magmas that are the richest in magnesium and iron but have the least silica (40%)

_____ 19. the melting process in which the source rock doesn't all become molten

_____ 20. the situation in which a magma's composition gets altered as it incorporates some of the rock it's passing through into itself

_____ 21. the process in which blocks of crustal rock fall into magma that's migrating through and are melted into the magma ("digested") (Crustal material thus becomes magma material.)

_____ 22. the progressive crystallization of minerals at their respective freezing temperatures from magma as its temperature lowers, so the magma hardens a little at a time, not all at once

_____ 23. holes in rock that began as gas bubbles and were preserved as the rock froze

_____ 24. a measure of the resistance of any material to flow

_____ 25. tiny glass shards formed when a fine spray of exploded lava freezes instantly on contact with air

_____ 26. the structure formed when volcanic ash billows upward and is temporarily suspended in the atmosphere

_____ 27. the event in which volcanic ash falls to Earth like snow

_____ 28. volcanic ash that has fallen to earth, tumbled across its surface, settled and welded itself together with its own heat

_____ 29. preexisting rock into which magma intrudes

_____ 30. the boundary between preexisting rock and intruding rock

_____ 31. the process by which magma makes room for itself as it intrudes existing rock and breaks off blocks of wall rock

_____ 32. blocks of country rock that have been broken off by intruding magma and do not melt, but instead end up as "foreign bodies" surrounded by frozen magma

_____ 33. magma intrusions, that have planar shape and roughly uniform thickness

_____ 34. tabular intrusions, often horizontal, table-top-shaped structures, that slipped between existing rock layers and therefore are parallel to rock layering in the area

_____ 35. wall-like tabular intrusions that cut across existing rock layers in an approximately vertical direction

_____ 36. a blister-like structure formed when magma that has intruded between horizontal rock layers domes the overlying layers upward

_____ 37. any irregular or blob-shaped structure resulting from the intrusion of magma

_____ 38. huge masses of irregularly shaped intrusions, which can cover several hundred square kilometers

_____ 39. any igneous rock composed of mineral grains of any size

_____ 40. a texture of all crystalline igneous rocks in which mineral grains lock into each other like puzzle pieces

_____ 41. the quality of igneous rocks that the mineral grains are so small they are not visible to the naked eye (not a specialized scientific term)

_____ 42. the quality of igneous rocks that the mineral grains are large enough to be identified with the naked eye (not a specialized scientific term)

_____ 43. the large mineral grains of an igneous rock with a porphyritic texture, that were formed when the rock was cooling slowly

_____ 44. the "background," small mineral grains of an igneous rock with a porphyritic texture, that were formed when the rock was cooling quickly

_____ 45. coarse-grained igneous rock (a scientific term)

_____ 46. fine-grained igneous rock (a scientific term)

_____ 47. the texture of an igneous rock that cooled so quickly no crystals formed

_____ 48. the texture of igneous rocks composed of phenocrysts surrounded by matrix

_____ 49. a very coarse-grained igneous rock with crystals that can be several tens of centimeters long

_____ 50. a thin, tabular intrusive structure composed of very coarse-grained igneous rock

_____ 51. rocks composed of fire pieces, volcanic debris of all sizes that were exploded out of a volcano, that have been naturally cemented together

_____ 52. rock that is a solid mass of volcanic glass

_____ 53. glassy volcanic rock that results from the cooling of lava which is frothy with escaping gas bubbles and which is often light enough to float on water

_____ 54. volcanic rock which is full of air-filled holes but which is not light enough to float on water because the voids make up less than 50% of the rock

_____ 55. pyroclastic igneous rock composed mainly of volcanic ash and pumice

_____ 56. tuff that settles from the air as an ash fall and eventually gets cemented together

_____ 57. tuff in which the volcanic fragments have welded themselves together with their own heat

_____ 58. any happenings involving molten rock—its formation, movement, and possibly eruption

_____ 59. volcanoes that are on land, "under air," not underwater

_____ 60. curving chains of volcanoes on land, which are adjacent to deep-ocean trenches that mark convergent plate boundaries

_____ 61. curving chains of volcanic islands, which are adjacent to deep-ocean trenches that mark convergent plate boundaries

58 | Chapter 6

_____ 62. volcanoes which exist because they're above hot mantle plumes and which are usually far from plate boundaries and exist independently of plate boundary interactions

_____ 63. a chain of hot-spot-created volcanoes, all of which are dead except the end one

_____ 64. huge flows of low-viscosity basalt associated with rifting and hot spots

_____ 65. a form of lava resulting from the extrusion of basalt into seawater and consisting of blobs of lava encrusted with glass rinds

Matching Questions

MAGMA TYPES

Match the type of magma with the appropriate descriptive phrase.

a. intermediate   b. mafic   c. silicic   d. ultramafic

_____ 1. coolest of the magmas, with temperatures as low as 700°C

_____ 2. hottest of the magmas, with temperatures over 1,200°C

_____ 3. magma with 70% silica and little magnesium and iron

_____ 4. magma with 55% silica and little magnesium and iron

_____ 5. magma with 40–50% silica and quite rich in iron and magnesium

_____ 6. magma with 40% or less silica and extremely rich in iron and magnesium

PUTTING SCIENCE IN THE SCENERY

Match the location or geologic feature with its description.

a. Andes Mountains, South America
b. batholith
c. glaciers
d. Juan de Fuca
e. Kilauea
f. Lake Superior, Michigan
g. Mono Lakes
h. Mt. Fujiyama, Japan
i. sill
j. Yellowstone

_____ 1. a feature like the Palisades (cliffs along the Hudson River opposite New York City)

_____ 2. a volcanic area on the western edge of the Basin and Range rift of the western United States, which is currently being monitored for signs of volcanic activity

_____ 3. a type of pluton like the Sierra Nevada Mountains

_____ 4. the objects that carved out the valley at the heart of Yosemite National Park

_____ 5. according to legend, the hot-spot volcano in which Pele lives

_____ 6. the plate whose subduction under the North American Plate caused the continental volcanic arc that is the Cascade Range (Mt. Ranier, Mt. St. Helens, Mt. Baker, etc.)

_____ 7. two locations on the Ring of Fire that are not
_____   in North America

_____ 8. location that marks the end of a continental hot-spot track

_____ 9. location that marks the beginning of an ancient, narrow, deep rift that started to split apart a continent about 1 billion years ago, filled with basalt, and is totally inactive today

Short-Answer Questions

Answer with a few words, numbers, or short sentences, or choose from the words in parentheses.

1. Strange as it may sound liquid rock "freezes" to form solid rock. What is freezing temperature of liquid rock? _____

2. Igneous rock, which is formed from extremely hot molten rock, is appropriately named because ignis is Latin for what word? _____

3. By sheer volume, which of the three major rock types (igneous, sedimentary, and metamorphic) is the most abundant on Earth and why? _____
_____
_____

4. What were the four heat sources of the youthful Earth?
_____
_____
_____
_____

5. a. Which of the three main rock types was the first to form on Earth? _____
   b. Did all existing igneous rocks originate during Earth's youth? _____

6. a. What is the temperature of mantle-derived magma? _____
   b. What is the minimum temperature needed to melt crustal rock? _____

7. List three conditions that can cause melting of solid rock. _____
   _____
   _____

8. List eight common chemical elements found in magma, excluding those which are volatile components. ____
   _____
   _____
   _____

9. What is a wet magma? _____
   _____

10. Name the two most abundant volcanic gases, and give a typical percentage for each. _____
    _____

11. List two reasons magma rises from depth. _____
    _____
    _____

12. Hotter magma is _____ (more, less) viscous than cooler magma because _____

13. Magmas that contain more volatiles are _____ (more, less) viscous than dry, volatile-free magmas because _____
    _____

14. Magmas that contain more silica are _____ (more, less) viscous than those with less silica because _____
    _____

15. Summarize your answers for questions 12, 13, and 14 by listing the three factors that control the viscosity of a magma. _____
    _____
    _____

16. Which type of lava is associated with the production of large amounts of volcanic ash, silicic or mafic, and why? _____
    _____

17. Color is a useful but not infallible clue to the chemical composition of igneous rocks. Choose the most likely composition (silicic, intermediate, or mafic) for the following:
    a. black or dark-gray rocks _____
    b. light-tan, pink, or maroon rocks _____
    c. light- or greenish-gray rocks _____

18. What instrument do geologists use to actually see the mineral ingredients of a rock and thereby identify it with certainty? _____

19. Usually the faster a melt cools, the smaller the crystal formed. Pegmatites cool quickly in thin, dike-shaped intrusions, yet they have extremely large crystals (are very coarse grained). Why? _____
    _____
    _____
    _____

20. Name the igneous rock textures that would result from the following situations:
    a. a melt that cools so fast atoms don't have time to arrange into crystal lattices _____
    b. slightly slower cooling which allows many small mineral seeds to grow and develop at once _____
    c. very slow cooling which results in the formation of few mineral seeds _____

21. a. Basalt is the fine-grained rock that's the equivalent of what coarse-grained rock? _____
    b. Diorite is the coarse-grained rock that's the equivalent of what fine-grained rock? _____
    c. Rhyolite is the fine-grained rock that's the equivalent of what coarse-grained rock? _____

22. The Juan de Fuca Plate that exists today is the remnant of a much larger plate, almost all of which has already subducted under North America. Name this larger plate. _____

23. Mantle plumes that underlie rifts may create extremely hot, low-viscosity basaltic lava flows like those found on the Columbia River Plateau of Oregon and Washington, the Parana Plateau of Brazil, the Karoo region of Africa, and the Deccan region of India. What's the name for this type of lava flow? _____

24. Where is most of Earth's igneous rock formed? (It's a significant plate tectonics feature.) _____
    _____

Making Use of Diagrams

GEOTHERM AND MELTING CURVE TEMPERATURE

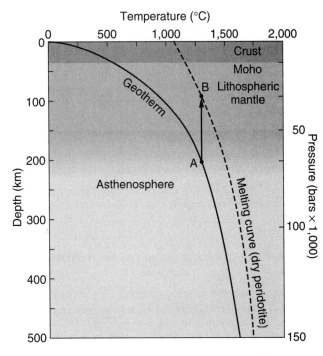

Questions 1 to 5 refer to the geotherm and melting curve figure above.

1. What does the solid line labeled Geotherm represent?
   _____

2. Does the area to the right of the melting curve (the dashed line) represent solid rock or melted rock?
   _____

3. Does the area to the left of the melting curve represent solid rock or melted rock? _____

4. Fill in the blanks with the temperatures and pressures found at the following depths:

Depth	Temperature	Pressure
a. 50 km		
b. 150 km		
c. 300 km		
d. 500 km		

5. Complete the following statements:
   a. The *rate* of increase in Earth's temperature (the *geothermal gradient*) _____ (increases, decreases) with increasing depth.
   b. Mantle rock (dry peridotite) located at point *A* would have a temperature of ____°C, be at a depth of ____ km, and have a pressure of ____ kbar. Its physical state would be _____ (entirely solid, just beginning to melt, partially melted).
   c. If the mantle rock above rose without cooling (as in a rising plume) to point *B*, its temperature would still be ____°C, its depth would be ____ km, its pressure would be ____ kbar, and it would be _____ (entirely solid, just beginning to melt, partially melted).
   d. On the graph plot point *C* at a depth of 50 km and a temperature of 1,300°C. The pressure here would be ____ kbar, and the rock would be _____ (entirely solid, just beginning to melt, partially melted).

ADDITION OF VOLATILES

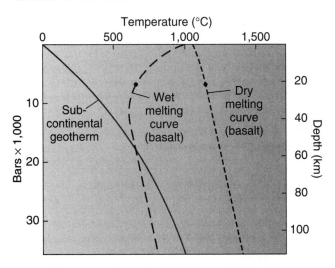

Questions 1 to 6 refer to the volatiles graph above.

1. What is the temperature of the Earth at a depth of 80 km? _____

2. At a depth of 80 km, at what temperature does dry basalt begin to melt? _____

3. At a depth of 80 km, at what temperature does wet basalt (that is, basalt containing volatiles) begin to melt? _____

4. As you go from Earth's surface to a depth of about 30 km, the *difference* between the melting point of dry basalt and of wet basalt _____ (increases, decreases, stays about the same).

5. As you go from a depth of about 30 km to a depth of 110 km, the *difference* between the two temperatures _____ (increases, decreases, stays about the same).

6. At Earth's surface and at all depths, wet basalt melts at a _____ (higher, lower) temperature than dry basalt does.

BOWEN'S REACTION SERIES

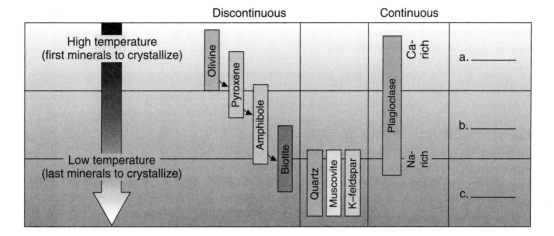

Questions 1 to 3 refer to the figure above. As you answer these questions, remember this figure does *not* represent a big vat filled with molten rock and chunks of minerals; it represents a sequence of crystallization that begins at the top, where the temperatures are hotter, and works downward as time passes and temperatures cool.

1. Fill in the blanks (a to c) on the figure with the terms that describe the silica and/or iron/magnesium content of the minerals found in each level.
2. What material did Bowen use to quickly freeze the melt and thus be able to look at various "moments in time" in the crystallization process?
3. Describe the actual product Bowen's freezing method produced.

Short-Essay Questions

1. In the 1700s geologists were either Neptunists of Plutonists. What do these terms mean? Which viewpoint was correct? Who was the geologist whose work helped decide the issue?
2. Use the term "fractional crystallization" to explain why magmas become progressively more silicic as they cool.
3. List four reasons why there are so many different types of magmas and briefly discuss each.
4. List and briefly explain the three factors that control the cooling rate of magma.
5. Hawaii produces basaltic lava; the Andes produce intermediate-composition lava, and Yellowstone produced silicic lava. Each happens because magma rises up from the mantle. Briefly explain why the lavas are different.

Practice Test

1. Which of the following statements is FALSE? Igneous rocks
   a. form from the freezing of either lava or magma.
   b. form in great quantity along the mid-ocean ridge.
   c. were the first rocks to exist on Earth.
   d. are termed "phaneritic" if they're fine grained.
   e. are coarse grained if they're intrusive.

2. The dramatic scenery of Yosemite National Park
   a. owes its existence to glacial carving of a batholith.
   b. is the result of a continental hot spot.
   c. is especially impressive because it's dark-colored basalt rock.
   d. is composed of fine-grained volcanic rock.
   e. All of the above are true.

3. A rock has a medium-gray, fine-grained matrix with large crystals of plagioclase. Which of the following statements about it is FALSE?
   a. It could logically be porphyritic andesite.
   b. The lava flow that produced it got quenched in ocean water.
   c. The plagioclase crystals are called phenocrysts.
   d. It began to crystallize at depth from an intermediate-composition magma, then rose to the surface and became a lava flow.
   e. Its chemical composition is intermediate, not silicic or mafic.

4. A rock is porphyritic. Which of the following words COULD NOT be used to describe it?
   a. matrix
   b. phenocryst
   c. plagioclase
   d. andesite
   e. pyroclastic

5. Granite
   a. is a phaneritic rock.
   b. is silicic in composition.
   c. logically could be found in a batholith.
   d. is the coarse-grained equivalent of rhyolite.
   e. All of the above.

6. Which of the following igneous rocks does NOT have a glassy texture?
   a. air-fall tuff
   b. obsidian
   c. welded tuff (ignimbrite)
   d. pegmatite
   e. pumice

7. After an explosive volcanic eruption on an island, the surrounding sea is full of floating rock. What must it be?
   a. rhyolite
   b. pumice
   c. obsidian
   d. basalt
   e. scoria

8. Which is NOT a setting for igneous activity?
   a. hot spots
   b. continental rifts
   c. continental transform fault zones
   d. areas bordering ocean trenches
   e. mid-ocean ridges

9. Which of the following is not a location on the Ring of Fire?
   a. Kilauea, Hawaii
   b. Cascade Range of Washington and Oregon (Mt. St. Helens, Mt. Ranier, etc.)
   c. Andes Mountains, South America
   d. Mt. Fujiyama, Japan
   e. Aleutian Islands, Alaska

10. Which of the following statements is FALSE? Volcanic arcs
    a. form where subduction takes place.
    b. are long, curving mountain chains adjacent to deep-ocean trenches.
    c. can be continental or island.
    d. make up the Ring of Fire circling the Atlantic Ocean.
    e. occur at convergent plate boundaries.

11. Solid blocks of country rock (or wall rock) may
    a. be broken off by intruding magma.
    b. be broken off during a process called stoping.
    c. melt entirely and thus change the chemical composition of the intruding magma.
    d. not melt but instead remain as recognizable blocks called xenoliths.
    e. All of the above.

12. Which statement is TRUE?
    a. James Hutton was the geologist who in the late 1700s proved the Neptunists were correct in their views about the origin of igneous intrusions.
    b. Liquid rock doesn't freeze until its temperature gets down to 200°C.
    c. Hot solid mantle rock may experience decompression melting as it rises in the Earth.
    d. Bowen's reaction series was based on field observation of Hawaiian volcanoes.
    e. All of the above are true statements.

13. The two most abundant gases produced by volcanoes are
    a. sulfur and water vapor.
    b. sulfur and chlorine.
    c. water vapor and carbon monoxide.
    d. water vapor and carbon dioxide.
    e. carbon monoxide and carbon dioxide.

14. Intrusive igneous rocks
    a. are volcanic rocks.
    b. are fine grained because they cooled slowly.
    c. are never seen by humans because they form deep in the Earth and are never exposed at the surface.
    d. are rocks like basalt, andesite, and rhyolite.
    e. cool slowly and are coarse grained.

15. Bowen's reaction series
    a. shows that minerals crystallize in a random order, with no particular pattern involved.
    b. allows a geologist to predict what minerals will be together in igneous rocks.
    c. is an attempt to explain the logic of formation of sedimentary rocks.
    d. explains why some compounds use ionic bonds and others have covalent bonding.
    e. All of the above are true statements.

16. Bowen's reaction series
    a. shows the sequence in which different silicate minerals form during the progressive cooling of a mafic melt.
    b. has a discontinuous track in which each step yields a different class of silicate mineral.
    c. has a continuous track in which there's a progressive change from calcium-rich to sodium-rich plagioclase.
    d. was established by laboratory experiments in which mafic melt was quenched in mercury.
    e. All of the above are true statements.

17. A black, fine-grained tabular intrusion between two layers of horizontal sedimentary rock must logically be a
    a. basaltic sill.
    b. granitic sill.

c. basaltic dike.
   d. granitic dike.
   e. batholith.
18. A light-tan tabular intrusion that cuts vertically across layered country rock is most logically a
   a. basaltic sill.
   b. granitic sill.
   c. basaltic dike.
   d. granitic dike.
   e. batholith.
19. Which of the following statements is FALSE? Silicic magma
   a. is the coolest of magmas.
   b. has about 70% silica and little magnesium and iron.
   c. is associated with explosive volcanic eruptions.
   d. is less viscous than mafic magma.
   e. is likely to form light-tan-, pink-, or maroon-colored rocks.
20. Which of the following statements is FALSE? Mafic minerals
   a. freeze at higher temperature than silicic minerals do.
   b. are those rich in iron and magnesium.
   c. would be at the top of the Bowen's reaction series, crystallizing first out of the melt.
   d. form rhyolite and granite.
   e. form black or dark-gray rocks.
21. Pillow basalts
   a. occur in tabular intrusions called dikes.
   b. are composed of silicic lava.
   c. are formed in a submarine environment.
   d. contain large bulbous crystals called pillows.
   e. All of the above are true statements.
22. The lava of Pele's hair and Pele's tears cools faster than the lava of a sill.
   a. True
   b. False
23. The yellow stone of Yellowstone is pillow lava that has been stained by iron and sulfur.
   a. True
   b. False
24. The plume of a hot spot consists of molten magma rising slowly from the core-mantle boundary.
   a. True
   b. False
25. Volatile-rich magma develops gas bubbles as it rises, and this creates even more buoyant force to move the magma upward more aggressively.
   a. True
   b. False
26. Vesicles are pieces of country rock broken off by intruding magma.
   a. True
   b. False
27. The deeper an igneous intrusion, the slower it cools.
   a. True
   b. False
28. Igneous rocks formed of igneous debris blown out of a volcano are called pegmatites.
   a. True
   b. False
29. The Farallon Plate is currently subducting under the North American Plate and producing the Cascade Range of mountains.
   a. True
   b. False
30. Yellowstone National Park marks the end of a continental hot-spot track.
   a. True
   b. False
31. The Palisades (cliffs along the Hudson River across from New York City) are dikes radiating out from an old volcanic neck.
   a. True
   b. False
32. The bedrocks of the Columbia River Plateau in Oregon and Washington are flood basalts formed by hot spots associated with continental rifts.
   a. True
   b. False

ANSWERS

Completion

1. melt
2. volcano
3. lava flow
4. igneous rock
5. magma
6. lava
7. intrusive igneous rock
8. extrusive igneous rock
9. geotherm
10. decompression melting
11. geothermal gradient
12. melting curve
13. volatiles
14. heat-transfer melting
15. silicic magmas
16. intermediate magmas
17. mafic magmas
18. ultramafic magmas
19. partial melting
20. magma contamination
21. assimilation
22. fractional crystallization
23. vesicles
24. viscosity
25. volcanic ash
26. ash cloud
27. ash fall
28. ash flow
29. country rock (or wall rock)
30. intrusive contact

31. stoping
32. xenoliths
33. tabular intrusions (or sheet intrusions)
34. sills
35. dikes
36. laccolith
37. pluton
38. batholiths
39. crystalline igneous rock
40. interlocking texture
41. fine grained
42. coarse grained
43. phenocrysts
44. matrix
45. phaneritic
46. aphanitic
47. glassy
48. porphyritic
49. pegmatite
50. pegmatite dike
51. pyroclastic rocks
52. obsidian
53. pumice
54. scoria
55. tuff
56. air-fall tuff
57. welded tuff (or ignimbrite)
58. igneous activity
59. subaerial volcanoes
60. continental volcanic arcs
61. island volcanic arcs
62. hot-spot volcanoes
63. hot-spot track
64. flood basalts
65. pillow basalt

Matching Questions

MAGMA TYPES

1. c
2. d
3. c
4. a
5. b
6. d

PUTTING SCIENCE IN THE SCENERY

1. i
2. g
3. b
4. c
5. e
6. d
7. a, h
8. j
9. f

Short-Answer Questions

1. between 700 and 1,200°C

2. "fire"

3. igneous, because it makes up all of the mantle, all ocean crust, and much continental crust

4. a. Gravity caused material to compress, and the compression caused heat.
   b. The kinetic energy of meteors changed to heat energy on impact.
   c. Earth's radioactive elements generated heat as they decayed.
   d. The potential energy of heavy elements like iron alloy changed to heat energy as the material sank.

5. a. igneous b. No, igneous rock forms both in and on the Earth today.

6. a. 1,200°C b. between 700 and 850°C

7. decrease in pressure, called decompression melting; addition of volatiles; transfer of heat by contact with rising magma (heat-transfer melting)

8. silicon, oxygen, aluminum, calcium, sodium, potassium, iron, magnesium

9. a magma that includes up to 15% dissolved volatiles, like water, carbon dioxide, nitrogen, hydrogen, and sulfur dioxide

10. water vapor, 50%; carbon dioxide, 20%

11. Magma is less dense than surrounding rock so buoyancy drives it upward; the weight of overlying rock creates pressure at depth, which squeezes the molten magma upward.

12. Less; thermal energy breaks restricting chemical bonds and allows more freedom of motion

13. Less; circulating atoms of the volatile materials break restricting chemical bonds and allow more freedom of motion

14. More; silica tetrahedrons link together and create chains that, at the molecular level, tangle and restrict flow

15. temperature, content of volatiles, silica content

16. Silicic; these are more viscous melts, less flowable and more explosive, and ash results from exploded lava.

17. a. mafic b. silicic c. intermediate

18. a petrographic microscope

19. They crystallized from very water rich magma in which atoms and ions could move quickly, reach each other, and chemically bond. Therefore individual crystals grew quickly and could achieve large size before they bumped into others.

20. a. glassy b. fine grained c. coarse grained

21. a. gabbro b. andesite c. granite

22. the Farallon Plate

23. flood basalt

24. at mid-ocean ridges

Making Use of Diagrams

GEOTHERM AND MELTING CURVE TEMPERATURE

1. the temperature of the Earth as a function of depth

2. melted rock

3. solid rock

4.
   a. 625°C, 17 kbar
   b. 1,200°C, 50 kbar
   c. 1,400°C, 95 kbar
   d. 1,550°C, 150 kbar

5.
   a. decreases
   b. 1,300°C, 200 km, 65 kbar, entirely solid
   c. 1,300°C, 100 km, 35 kbar, just beginning to melt
   d. 15 kbar, partially melted

ADDITION OF VOLATILES

1. 850°C (read from the subcontinental geotherm curve)
2. about 1,300°C
3. about 800°C
4. increases
5. stays about the same
6. lower

BOWEN'S REACTION SERIES

1. a. mafic   b. intermediate   c. silicic
2. mercury
3. The remaining melt formed a glassy case around the crystals produced at that moment in time.

Short-Essay Questions

1. Neptunists (from Neptune, god of the sea) believed igneous intrusive rock formed by the settling of sediments in, or precipitation of ions from, seawater. Plutonists (from Pluto, god of the underworld) believed these rocks formed when molten rock cooled and hardened (in today's terminology, freezing of a melt). James Hutton's observations of crystallized rock squeezed between sandstone layers showed that the Neptunist's viewpoint was wrong.

2. Most rocks consist of a variety of different minerals which freeze (solidify) at different temperatures. Therefore a melt turns solid in a gradual fashion, a fraction at a time, thus the term "fractional crystallization." The more mafic the mineral, the higher the temperature at which it freezes and removes from the melt. As magma cools, at first much iron and magnesium and very little silica are removed, so the amount of silica remaining in the melt becomes a larger percentage of the magma composition.

3. The reasons there are different types of magma are:
   a. Different source rock composition: The composition of a melt reflects the composition of the solid it came from. Different rocks melt to produce different magmas.
   b. Partial melting: Magma forms as solid rock melts, but the rock doesn't melt all at once. Often the magma moves away from the source rock before it's all melted, so the magma composition is not the same as the rock it came from.
   c. Contaminating itself: Magma melts country rock and "digests it" into itself—a process called assimilation. Thus the magma mixes the new material with itself, or contaminates itself.
   d. Fractional crystallization: The chemical composition of a magma at any particular point in time is not the same as it was originally or the same as it would be at a different point in time (see previous question).

4. The cooling rate of magma depends on:
   a. Depth of intrusion: Rock is a good heat insulator, not a good conductor, so the deeper the magma is the more insulation it has and the slower it cools.
   b. Shape and size of the magma body: Anything, including a magma body, loses heat from its surface. The greater the surface area, the faster it cools. Therefore one must consider the volume/surface area relationship when talking about rate of cooling. Plutons cool more slowly than dikes, large plutons cool more slowly than small plutons, droplets of lava cool more quickly than a lava flow, and a thin lava flow cools more quickly than a thick lava flow.
   c. Presence of circulating groundwater: Flowing water carries away heat.

5. Hawaii: The heat from the plume of the Hawaiian hot spot melts the ocean crust it comes in contact with. Ocean crust is basalt; it's turned into basaltic lava.
   The Andes: The mafic subducting plate melts mafic mantle rock, and this all rises up through silicic continental crust. Basalt plus silicic rock result in intermediate composition lava.
   Yellowstone: The Yellowstone hot-spot plume melts continental crust where it comes in contact with it. Continental crust is silicic, so the lava it becomes is also silicic.

Practice Test

1. d Igneous rocks are aphanitic if they're fine grained, phaneritic if they're coarse grained.

2. a The batholith is light-colored granite, an intrusive, coarse-grained, nonvolcanic rock.

3. b Not only would water quenching create a glassy texture, but this explanation doesn't touch at all on the origin of the large crystals.

4. e "Pyroclastic" refers to fire pieces, fragments that have been exploded out of a volcano; it has nothing to do with a porphyritic situation.

5. e Statements a to d are true.

6. d Pegmatite is a very coarse-grained rock (which means it has very large crystals); "glassy" means no crystalline structure.

7. b Although it's true that scoria is vesicular, it doesn't have enough vesicles (air spaces) to be less dense than water.

8. c Transform motion is plates' sliding past each other; the heat from friction is not enough to melt rock; other choices involve molten rock's welling up from depth or plates' subducting and melting at depth.

9. a Hawaii sits on a hot spot, far from any plate boundary; the Ring of Fire marks converging plate boundaries where subduction is taking place.

10. d Volcanic arcs do make up the Ring of Fire, but that circles the Pacific Ocean, not the Atlantic Ocean.

11. e Statements a to d are true.

12. c Hutton proved the Neptunists were wrong; freezing temperature need be only as low as 700–1,200°C, and Bowen's reaction series was based on laboratory experiments.

13. d Water vapor constitutes 50% of volcanic gas, carbon dioxide 20%; the remaining gases are largely nitrogen, hydrogen, and sulfur dioxide.

14. e Volcanic rocks, like basalt, andesite, and rhyolite, are extrusive igneous rocks; intrusive igneous rocks solidify in the earth but eventually are exposed by erosion.

15. b Bowen's reaction series was deduced through laboratory procedures that cooled molten rock; the crystal formation in the igneous rocks produced always followed the same order.

16. e Statements a to d are true.

17. a Only choice a fits the description; granite would be light colored, a dike would cut across existing layers, and a batholith is a massive, not tabular intrusion.

18. d Only choice d fits the description; basalt would be dark colored, a sill would fit between the layers, and a batholith is a massive, not tabular intrusion.

19. d Silicic magma is more viscous because the silicon tetrahedrons link up in chains which, at the microscopic level, tangle and impede smooth flow.

20. d The rocks formed of mafic minerals would be basalt and gabbro.

21. c Pillow basalts are bulbous flows of molten basalt, low in silica, that cool and harden as they are expelled into seawater.

22. True With such a large surface area and so small a volume, Pele's hair and tears cool almost immediately on expulsion.

23. False The yellow stone is stained volcanic ash.

24. False It's not molten material; it's solid rock rising slowly because it's hot and buoyant.

25. True The development of gas decreases the density of the magma and thus increases its buoyancy.

26. False Vesicles are gas bubbles frozen into the rock; pieces of broken off country rock are xenoliths.

27. True Earth's temperature increases with depth, so the deeper an intrusion is, the hotter its environment and the longer it takes for it to cool.

28. False They're called pyroclastic rocks ("fire pieces").

29. False The Juan de Fuca Plate—a mere remnant of the Farallon Plate—is doing the subduction.

30. True Yellowstone is at the northeast end of a hot-spot track that is marked by older volcanics extending toward the southwest.

31. False The Palisades are a horizontal sill.

32. True The extensive basalt deposits of the Columbia River Plateau are a flood basalt region created by continental rifting associated with hot-spot activity.

CHAPTER 7 A Surface Veneer: Sediments and Sedimentary Rocks

GUIDE TO READING

In this chapter we encounter the second basic rock type: sedimentary rocks. Formed from detritus, mineral crystals, and shells, sediments and sedimentary rocks cover 80% of Earth's surface and are part of a multitude of different environments. Past environments have influenced the types of sedimentary rocks created, and current sedimentary rocks influence the looks, characteristics, and resources of current environments.

The chapter starts by explaining that sedimentary rocks are created at or near Earth's surface in one of three general ways: (1) cementing together loose grains of rock, (2) precipitating ions from water solution, or (3) concentrating skeletal material of aquatic organisms.

The rock grains needed to create sedimentary rocks are the result of the breakdown (disintegration) and chemical change (decomposition) of existing rock by physical (mechanical) weathering and chemical weathering. Several types of physical weathering are discussed, including jointing, frost wedging, root wedging, salt wedging, thermal expansion, and animal attack. Common categories of chemical weathering are offered next. These are:

- dissolution, which is just the plain dissolving of a solid in water
- hydrolysis, in which water facilitates the chemical change of minerals
- oxidation, in which an element looses some electrons and which may or may not directly involve oxygen
- hydration, in which water absorbed into the crystal structure may cause the mineral to expand

A word of advice to the readers who are not chemists: don't worry. Words are supplied for every formula and equation so you can understand what's going on, looking at the formulas and equations and using simple arithmetic will let you keep track of all the ingredients, and some of the reactions are as simple as iron rusting (iron oxidizing).

A discussion of soils comes next. Soil science is complex and can be the subject matter for numerous courses. Here the author offers the simple basics: (1) why soil is more than just broken down rock, (2) the physical structure of typical soils (zones and horizons), and (3) soils' relations to environments, using the relations of pedalfer, pedocal, and laterite as examples.

A large part of the chapter is devoted to classifying and describing common sedimentary rocks. There are four main classifications:

- clastic sedimentary rocks (examples: breccia, conglomerate, arkose, sandstone, shale, siltstone, mudstone, and graywacke)
- biochemical sedimentary rocks (examples: limestone, including fossiliferous limestone, micrite and chalk, and chert)
- chemical sedimentary rocks (examples: the evaporites gypsum and halite, travertine, dolostone, and several varieties of chert)
- organic sedimentary rocks (examples: coal and oil shale)

Sedimentary rocks occur in layers called beds or strata, which may display special features such as cross beds, graded beds, ripple marks, mud cracks, and fossils.

The very existence of a certain type of sedimentary rock is a clue to its past environment. It may have been a terrestrial environment (possibly glacial valley, mountain stream, mountain front, sand dune, lake, or river), or it may have been a marine environment (a delta, shallow-marine clastic area, shallow-marine carbonate area, or deep-ocean water.) The sequence of sedimentary beds can even tell the geologist whether the sea was encroaching on the land (transgression) or receding (regression) during the time of the sediment deposition.

The chapter ends by relating sedimentary rock formation and distribution to that grand unifying concept, plate tectonics. Once again you read about rifts, passive continental margins, intracontinental areas, and foreland basins.

A word of advice: though the content covered in this chapter is not conceptually difficult, you will encounter considerable new vocabulary. The matching sections that follow offer one means of mastering new terms; you may also wish to make up flash cards or develop mnemonic aids to memory.

By chapter's end you have covered two of the three major rock types, igneous and sedimentary. What is characteristic of the third type? The more you learn about geology the more you'll realize Earth is a very dynamic place. Even solid rock doesn't stay the same forever. And that's what Chapter 8 is all about, changed rocks—metamorphic rocks.

Completion

Test your recall of new vocabulary terms.

_____ 1. the wearing away and removal, by wind or runoff, of rock debris or soil from its place of origin

_____ 2. loose, small pieces of rock, mineral, or shell material derived from once-intact rock, water by precipitation, or dead organisms

_____ 3. rock that's formed on Earth's surface by the cementing together of small pieces of preexisting rock, by the precipitation of minerals from water solutions, or by the growth of skeletal material of organisms

_____ 4. rock whose mineral grains have their original composition and shape because they have not been reworked by surface air or water

_____ 5. rock whose minerals have been altered by reaction with surface air and water

_____ 6. the process that takes intact rock and by air and water action physically breaks it into smaller pieces which are chemically the same as the larger rock piece

_____ 7. smaller pieces of rock produced by physical (mechanical) weathering of larger rocks

_____ 8. cracks that result when erosion reduces the pressure and temperature on rock that has been deeply buried in Earth

_____ 9. a wedge of rock rubble at the base of a slope

_____ 10. the process in which water trapped in joints freezes, expands, and thus breaks blocks of rock free from intact bedrock

_____ 11. the process in which plant roots enlarge existing cracks and pores in rocks

_____ 12. the process in which dissolved salts in groundwater crystallize and grow in pore spaces in rock, and thus enlarge the openings

_____ 13. verb that means the outer part of a rock breaks off in sheet-like pieces

_____ 14. the process in which various types of animals, from earthworms and gophers to humans, break up and remove rock from its original site

_____ 15. the process that takes intact rock and by reactions with air or water solutions chemically alters or destroys its minerals

_____ 16. a layer of rotten rock produced in warm, wet climates by chemical weathering

_____ 17. the dissolving of minerals in water

_____ 18. the quality of a material that has large amounts of hydrogen ions (H^+)

_____ 19. the reaction of minerals with water that causes them to break down chemically, not merely go into solution

_____ 20. a chemical process in which an element loses electrons, so that its positive electric charge (valence) increases and which often but not always involves combination with oxygen

_____ 21. the absorption of water into the crystal structure of a mineral which may cause it to expand

_____ 22. the situation when different rocks in an outcrop weather at different rates

_____ 23. the sediment that has undergone changes on Earth's surface that enable it to support plant growth

_____ 24. the unconsolidated debris (rock fragments, various sizes of sediment, and soil) that covers bedrock

_____ 25. a horizontal layer in soil, fairly close to the surface, in which downward-percolating rainwater dissolves ions and picks up fine clay particles

_____ 26. the horizontal layer of soil in which downward-percolating water leaves behind the fine clay particles and precipitates out the ions it picked up in the zone of leaching just above

_____ 27. distinct horizontal zones of soil that change with depth

_____ 28. a vertical sequence of different soil horizons

_____ 29. the typical soil of temperate climates, which forms on granite and possesses an organic-rich A-horizon

_____ 30. the typical soil of desert climates, which is thin, with little organic matter, lots of unweathered rock fragments, and a high concentration of soluble minerals like calcite

_____ 31. an accumulation of solid calcite in the B-horizon of pedocal soils

_____ 32. the typical soil of tropical environments, which has no B-horizon and from which most mineral components have been leached out, leaving only dark-red insoluble iron and/or aluminum oxide

_____ 33. the laterite soil that contains an exceptionally large amount of aluminum

_____ 34. the major class of sedimentary rock that forms by the cementing together of detritus from previously existing rocks

_____ 35. the major class of sedimentary rock that consists of pieces of shells of organisms

_____ 36. the major class of sedimentary rock that consists of carbon-rich, partially decayed plant remains

_____ 37. the major class of sedimentary rock that forms from minerals precipitated from water solution

_____ 38. rocks that contain quartz

_____ 39. rocks that contain clay

_____ 40. rocks that contain calcite or dolomite

_____ 41. the process in which sediment settles out of its transporting medium (air, water, or ice)

_____ 42. the transformation of unconsolidated pieces (sediment) into solid rock

_____ 43. the process in which the pressure of overlying rock squeezes out water and air between sediment pieces and allows them to pack together more tightly

_____ 44. the process in which compacted sediment pieces get stuck together, by minerals precipitated from groundwater, to form solid rock

_____ 45. the rock formed of *angular* fragments cemented together

_____ 46. the rock formed of *rounded* pebbles, cobbles, and sand cemented together

_____ 47. the smaller pieces (clasts) of a breccia or conglomerate rock, the "background" mass of material surrounding the bigger pieces

_____ 48. a clastic sedimentary rock consisting of quartz and feldspar grains

_____ 49. a clastic sedimentary rock consisting of sand-sized grains of a mineral (often, but not always quartz)

_____ 50. a clastic sedimentary rock consisting of silt (particles smaller than sand and larger than mud)

_____ 51. a clastic sedimentary rock consisting of mud (clay and water) that breaks into thin sheets

_____ 52. a clastic sedimentary rock consisting of mud (clay and water) that breaks into chunks, not thin sheets

_____ 53. a clastic sedimentary rock composed of pieces ranging in size from mud to coarse sand or pebbles, deposited not in streams or lakes but on continental margins by submarine avalanches

_____ 54. the chips or fragments of rocks that are not single mineral grains (The term means literally "stone pieces.")

_____ 55. the range of clast (grain) sizes in a collection of sediment, which varies from all the same size to a wide range of grain sizes

_____ 56. the degree of rounding of clast, how close it gets to being a true sphere

_____ 57. the degree of sharpness of the edges and corners of a clast

_____ 58. the degree of development of sediment, how much it has changed from being raw, broken-up rock to being sorted, rounded, and chemically altered rock

_____ 59. the sedimentary rock which is composed of calcite or aragonite and whose origin may be chemical or biochemical

_____ 60. the microscopic animals with calcite shells that make up part of the ocean plankton (microscopic floating life forms)

_____ 61. the limestone composed of shells and shell fragments visible to the naked eye

_____ 62. the limestone composed of lime mud

_____ 63. the limestone composed of plankton shells

_____ 64. the process in which ions in existing crystals slowly rearrange themselves into new crystals

_____ 65. a biochemical sedimentary rock composed of the silica shells of plankton, or a chemical sedimentary rock of silica precipitated from water solution, the crystal structure of either of which is too small to be seen except under an electron microscope

_____ 66. an organic sedimentary rock composed of plant remains from forests and swamps

_____ 67. an organic sedimentary rock consisting of shale that's rich (50% or more) in organic matter, typically from partially decayed plankton

_____ 68. a category of chemical sedimentary rock resulting from the evaporation of salt water

_____ 69. the limestone formed on cave walls or around hot springs

_____ 70. the recrystallization of already formed sedimentary rock as a result of presure and temperature changes and reaction with water

_____ 71. the chert that forms by gradual replacement of calcite crystals in an existing mass of limestone

_____ 72. black chert

_____ 73. red chert

_____ 74. the chert formed by the penetration of volcanic ash into tree wood

_____ 75. the chert with a pattern of varicolored, concentric rings

_____ 76. a layer of sedimentary rock with a recognizable top and bottom

_____ 77. the boundary between two sedimentary beds

_____ 78. a geologist who specializes in studying strata

_____ 79. the study of strata

_____ 80. a sequence of strata, distinctive enough to be traced across a fairly large region, that suggests the depositional environment that was there during a long period of past geologic time

_____ 81. a type of bedding in which thin bands tilt at an angle to the boundary of the overall bedding

_____ 82. a cloud of sediment, triggered by storm or earthquake, that avalanches down an underwater slope

_____ 83. the bedding situation that results from the settling out of the sediment of a turbidity current, with sediments which are coarse on the bottom and which grade upward into finer sediments

_____ 84. the features that appear on the surface of beds, during or soon after deposition, as a consequence of environmental factors

_____ 85. the elongate ridges in a layer of sediment, evenly spaced and perpendicular to the direction of the wind or water currents that created them

_____ 86. the cracks between roughly hexagonal plates of dried mud, filled with sediment and preserved over geologic time

_____ 87. a set of sediments and sedimentary structures that indicate a particular type of environment, such as a beach or deep-ocean waters

_____ 88. dry land environments in which sediments were deposited

_____ 89. coastal and submarine environments in which sediments were deposited

_____ 90. unsorted, unstratified sediment, from clay size to boulder size, carried on glacial ice and deposited when it melts

_____ 91. a large, wedge-shaped apron of sediment deposited by a stream when it leaves the mountains and abruptly empties onto a flat plain

_____ 92. sediments that are rusty to bright red because the iron in them has been oxidized in the air

_____ 93. large piles of well-sorted sand, carried and deposited by wind

_____ 94. fine clay sediments that settle out in a lake

_____ 95. flat regions on both sides of a river that are covered by water only during floods

_____ 96. sediments carried and deposited by streams

_____ 97. a triangular-shaped structure formed where a river empties out into the sea

_____ 98. horizontal layers of mudstone that build up on the tops of deltas

_____ 99. tilted layers of silts and clays deposited along the front of a delta

_____ 100. horizontal layers of fine mudstone deposited in deep water at the foot of a delta

_____ 101. snowball-like spheres of calcite that were formed as sand grains and roll back and forth in carbonate-rich water

_____ 102. deposits of graywacke placed by turbidity currents in the transition zone between coastal waters and deep ocean

_____ 103. a fan-shaped underwater deposit of marine turbidities

_____ 104. the process in which the surface of Earth's lithosphere sinks relative to sea level

_____ 105. any depression caused by subsidence, that fills with a thick accumulation of sediment

_____ 106. sedimentary basins that form over continental rifts where stretching and thinning have occurred

_____ 107. sedimentary basins that form along edges of continents where the lithosphere has been stretched

_____ 108. sedimentary basins that form in continental interiors, probably over failed rifts

_____ 109. sedimentary basins that form on the continent side of a growing mountain belt

_____ 110. the encroachment of the sea on land, in which the coastline moves inward as the sea level rises

_____ 111. the withdrawal of ocean waters in which the coast migrates seaward as the sea level falls

_____ 112. the succession of strata deposited during a cycle of transgression and regression

Matching Questions

SOIL TYPES

Match the soil type with a geographic location.

a. laterite   b. pedalfer   c. pedocal

_____ 1. high plains of Colorado (semi-arid region)

_____ 2. jungles of Indochina

_____ 3. forests of New York State (temperate climate, moderate rainfall)

_____ 4. tropical rainforests of Brazil

_____ 5. Arizona deserts

_____ 6. the Swiss Alps (granite mountains, temperate climate, moderate rainfall)

Answer the following questions, again using the soil choices supplied above.

_____ 7. In which soil type are you likely to find caliche?

_____ 8. Which of the soil's names hints at its high calcite content?

_____ 9. Which of the soil's names hints at its high aluminum and iron concentrations?

SEDIMENT TRANSPORT

Match the maximum-sized sediment it can move with the transporting agent.

a. sand and gravel   b. huge boulders   c. silt and mud
d. cobbles and small boulders

_____ 1. glacier

_____ 2. mid-sized stream in rolling hills

_____ 3. turbulent stream high in the mountains during spring runoff

_____ 4. small creek meandering across a valley meadow

_____ 5. large river in flood stage

Chapter 7

ENVIRONMENTS OF DEPOSITION

Remember: Sedimentary rocks and structures are clues to the past environment. Match the sedimentary rock or structure with the environmental situation it suggests.

a. alluvial fan
b. coal
c. chalk, mudstone, bedded chert
d. conglomerate
e. cross bedding
f. delta
g. graded beds of graywacke in a fan-shaped deposit
h. mud cracks
i. mudstone
j. ooids in white sand
k. salt
l. silt
m. till (unsorted, unstratified deposits with mixed clast sizes)
n. travertine

_____ 1. large stream, high-velocity flow
_____ 2. small, slow-moving stream
_____ 3. lake with no outlet
_____ 4. hot springs and geysers
_____ 5. swamp
_____ 6. turbidity current in a submarine canyon
_____ 7. sand dunes
_____ 8. any region that periodically dries out
_____ 9. glacier
_____ 10. mountain stream coming out onto plains
_____ 11. lake bed
_____ 12. foreset, topset, and bottomset beds
_____ 13. tropical beach
_____ 14. deep ocean

Short-Answer Questions

Answer with a few words, numbers, or short sentences, or choose from the words in parentheses.

1. What percentage of Earth's surface is covered by sedimentary rocks? _____

2. In what type of environment is the thickest sedimentary rock veneer on Earth, and just how thick does it get?
_____

3. Starting with the smallest, list the following types of detritus in order of increasing size:

boulders   cobbles   mud   pebbles   sand   silt
_____

Label each term on your list as C if it's considered coarse grained, M if medium grained, or F if fine grained.

4. Sand can be different colors and different chemical compositions. The term simply means a particle with grain size ranging from _____ mm to _____ mm.

5. In what types of environment is frost wedging the most active? _____

6. In respect to the weathering of rocks, what do earthworms, gophers, and humans have in common?
_____

7. Name four categories of chemical weathering. _____
_____

8. Name two rocks that are particularly vulnerable to dissolution because their chief ingredient is calcite.
_____

9. Analyze the following chemical-weathering event:

$2KAlSi_3O_8 + 2H^+ + 2HCO_3^- + H_2O$
potassium feldspar + carbonic acid + water
$\rightarrow Al_2Si_2O_5(OH)_4 + 2K^+ + 2HCO_3^- + 4SiO_2$
$\rightarrow$ kaolinite clay + potassium ions + bicarbonate ions + dissolved silica

a. Which chemical weathering process is occurring and which reactant tells you this?
Process: _____ (dissolution, hydrolysis, oxidation, hydration)
Reactant: _____
b. The presence of which reactant makes water more efficient in decomposing rock? _____
c. What new mineral is produced? _____

10. Analyze the following chemical-weathering event:

$4Fe_2^+ + 3O_2 \rightarrow 2(Fe_3^+)2O_3$
ferrous iron + oxygen $\rightarrow$ hematite
and
$4Fe_2^+ + O_2 + 2H_2O \rightarrow 4(Fe_3^+)O(OH)$
ferrous iron + oxygen + water $\rightarrow$ goethite

a. Which chemical weathering process is occurring and which reactant tells you this?
Process: _____ (dissolution, hydrolysis, oxidation, hydration)
Reactant: _____

b. Under the influence of oxygen, each iron ion releases an electron and changes from its +2 state, termed (ferrous, ferric) _____, to its +3 state, termed _____ (ferrous, ferric).

c. Name the two minerals produced in the reaction (both of them have reddish-brown or yellow-brown color). _____

d. A very simple one-word way of saying "iron combines with oxygen" is "iron _____."

11. Why do beaches in granitic rock areas typically consist of quartz sand and nothing else? _____

12. The presence of iron, magnesium, sodium, potassium, and aluminum makes minerals _____ (more, less) vulnerable to chemical weathering.

13. How can plant roots, fungi, and lichens accomplish *chemical* weathering of rock? _____

14. Why do the corners of broken blocks of rock weather faster than its planar surfaces or its edges? _____

15. Why did the soil of the high plains of the United States blow away and create the Dust Bowl of the 1930s? _____

16. Name three transporting agents of sediment. _____

17. What two factors determine the strength of a transporting agent? _____

18. Which rock logically is formed farther from its sediment source, a conglomerate or a breccia? _____ Why do you conclude this? _____

19. What three factors must be considered when classifying sedimentary rocks? _____

20. a. Conglomerate, breccia, and graywacke are all examples of _____ (well-, poorly) sorted rock.
    b. Sandstone, shale, and siltstone are all examples of _____ (well-, poorly) sorted rock.

21. Polymorphs are minerals with the same chemical composition but different crystalline form and therefore different physical properties. From Chapter 5, the polymorph of diamond is _____ From this chapter, the polymorph of calcite is _____.

22. Ancient limestone, whose origin *is* shell material, even under a microscope doesn't look like it's composed of pieces of shell. Why? _____

23. a. Which silicious sedimentary rock, sometimes biochemical and sometimes just chemical in origin, breaks with conchoidal fracture? _____
    b. How have humans made use of this characteristic? _____

24. Chemical sedimentary rocks differ from clastic sedimentary rocks in these two ways:
    a. They do not contain _____.
    b. They can become solid rock without undergoing _____.

25. Chemical sedimentary rocks differ from biochemical sedimentary rocks because their minerals precipitate, at least in part, without _____.

26. What are two possible times during which the crystalline texture of chemical sedimentary rocks is established? _____

27. a. What element must be added to calcite to change it into dolomite? _____
    b. In natural situations, where does this element come from? _____

28. Complete with "bed/strata" words:
    a. A layer of sediment or sedimentary rock with a recognizable top and bottom is called _____.
    b. The boundary between two beds is a _____.
    c. Several beds together are called _____
    d. The arrangement of sediment into a sequence of beds is called _____.

    (Don't be too worried about the many terms; in practice, the slight differences generally don't cause misunderstanding.)

29. Name three factors that control the type of sediment deposited in a region. _____

Making Use of Diagrams

DIFFERENTIAL WEATHERING OF ROCKS

Land surface features are often determined by differential weathering.

Name each lettered rock layer in the diagrams below with a logical rock type. Choose from the list in the column on the right.

Name	Degree of Resistance to Weathering and Erosion
Conglomerate	Resistant
Granite	Resistant (an igneous rock)
Limestone	In dry climates resistant to weathering and erosion; in wet climates vulnerable to dissolution
Sandstone	Resistant
Shale	Very nonresistant
Travertine	A variety of limestone with resistance equal to regular limestone

Bad Lands Topography

1.

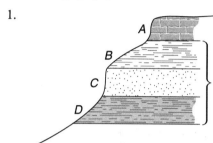

A — A chemical sedimentary rock

B, C, D — All are clastic, well-sorted sedimentary rocks.

A _____

B _____

C _____

D _____

Hogbacks of the Semi-arid Southwest

2.

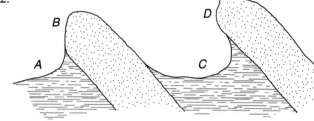

All strata are clastic, well-sorted sedimentary rocks.

A _____

B _____

C _____

D _____

Rock Overhang

3.

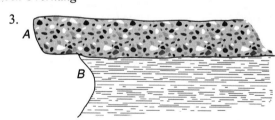

A — Clastic, poorly sorted sedimentary rock

B — Clastic, well-sorted sedimentary rock

A _____

B _____

Kentucky Cave

4.

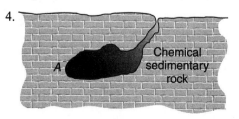

A — Chemical sedimentary rock

A _____

Cemetery Geology

5. Bob and Tom died and were buried the same day one hundred years ago. Who chose a marble stone? _____

Plateau Country of the Southwest

6. The sketch below shows the profile of the landscape 200,000 years ago.

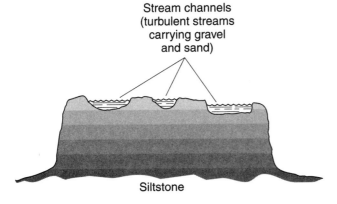

Based on the figure above, sketch a logical profile of this same area today.

Geyser Basin Scenery

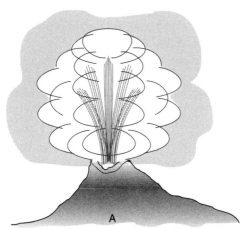

A = a chemical sedimentary rock deposited around the geyser

name of rock A _____

7. A is a chemical sedimentary rock deposited around the geyser. What is the name of rock A? _____

Short-Essay Questions

1. Hieroglyphics carved on a granite obelisk, Cleopatra's needle, stood in Egypt and were still readable after more than 3,000 years. Then the needle was taken down, stored, moved out of Egypt, contaminated by groundwater, and erected in Central Park, New York City, in the late 1800s. The hieroglyphics were destroyed in less than 100 years. Explain why this happened.

2. Briefly explain how mineral-eating bacteria function and what they have to do with the question of whether there's life on Mars.

3. Physical (mechanical) and chemical weathering happen to rocks simultaneously. How does each process help the other process?

4. Discuss three processes that take place at Earth's surface and contribute to soil formation.

5. List and explain five factors that affect the quantity and quality of the soil produced.

6. Rocks are defined as mixtures of minerals. Minerals are defined as having inorganic origin. Technically speaking, what's the problem in classifying coal as a rock?

7. Bonneville Salt Flats of car-racing fame owes its existence to an ancient lake. Briefly explain why the lake formed and what happened to it.

Practice Test

1. Sedimentary rocks can form
   a. by the precipitation of minerals from water solution.
   b. by the cementing together of loose grains of preexisting rock.
   c. from shell fragments.
   d. All of the above are true.
   e. Only a and b are true.

2. Choose the proper listing of detritus terms, going from smallest to largest of the words offered.
   a. "sand," "silt," "pebbles"
   b. "mud," "silt," "sand"
   c. "silt," "mud," "sand"
   d. "sand," "cobbles," "pebbles"
   e. "mud," "sand," "silt"

3. Where would you find the most chemical weathering?
   a. Sahara desert, because it's so hot
   b. Antarctica, because it's so cold
   c. tropical rain forests, because of the warm, wet conditions
   d. high plains states, because of the high elevation, heat, and dryness
   e. high mountains, because of the frequent alternate freezing and thawing

4. Where would you find the most physical (mechanical) weathering?
   a. Sahara desert, because it's so hot
   b. Antarctica, because it's so cold
   c. tropical rain forests, because of the warm, wet conditions
   d. high plains states, because of the high elevation, heat, and dryness
   e. high mountains, because of the frequent alternate freezing and thawing

5. Which statement is true?
   a. Animal attack, whether by gopher or man, is termed spalling.
   b. Pieces of broken rock produced by physical weathering are collectively called detritus.
   c. A saprolite layer is typically found in cold, dry climates.
   d. Hydrolysis is the absorption of water into crystal structure and its subsequent expansion.
   e. All of the above are true.

6. Which is NOT a "soil term"?
   a. "bauxite"
   b. "laterite"
   c. "caliche"
   d. "regolith"
   e. "arkose"

7. Which statement is true?
   a. The downward percolation of rainwater removes ions from the zone of accumulation and deposits them in the zone of leaching.
   b. The B- and C-horizons make up the zone of leaching.
   c. The B-horizon is the subsoil.
   d. The C-horizon contains the most humus.
   e. Farmers till and plant in the C-horizon, which is topsoil.

8. A nonmarine clastic sedimentary rock composed of angular fragments surrounded by matrix is
   a. breccia.
   b. conglomerate.
   c. turbidite.
   d. fossiliferous limestone.
   e. arkose.

9. Which of the following words is associated with the formation of limestone?
   a. "deposition"
   b. "lithification"
   c. "compaction"
   d. "cementation"
   e. "crystallization"

10. If you find graywacke in the place where it formed, you know you are looking at an ancient
    a. submarine continental slope.
    b. sand dune.
    c. swamp.
    d. alluvial fan.
    e. lake bed.

11. A clastic sedimentary rock with mud-sized grains that breaks in thin sheets is called
    a. mudstone.
    b. shale.
    c. siltstone.
    d. greywacke.
    e. micrite.

12. Which term has nothing to do with the mineral calcite?
    a. "chert"
    b. "fossiliferous limestone"
    c. "chalk"
    d. "micrite"
    e. "forams"

13. The organic sedimentary rock that is composed of mud-sized particles and more than 50% partially decayed organic matter (plankton remains) is
    a. coal.
    b. fossiliferous limestone.
    c. chert.
    d. oil shale.
    e. ooid rock.

14. Which statement is true?
    a. Both chert and limestone may have either chemical or biochemical origin.
    b. Flint, jasper, agate, and petrified wood are all varieties of the siliceous rock chert.
    c. Gypsum and halite are evaporite minerals.
    d. Chemical sedimentary rocks are crystalline in texture.
    e. All of the above are true statements.

15. Flint and jasper were prized in early human cultures because
    a. large outcrops of them often contained caves to live in.
    b. they are translucent minerals and were useful for windows.
    c. they were rare and pretty and used as trade items.

d. they dissolved in water readily, contributing minerals that made healthy drinking water.
e. they broke with conchoidal fracture and thus made good cutting tools.

16. Ripple marks on a bedding plane
    a. suggest you're looking at an old delta.
    b. are parallel to the current direction that created them.
    c. could suggest an ancient stream bed or a seashore.
    d. are a sure sign the area had periodic drought.
    e. All of the above are true.

17. The diagram shown above is a profile of
    a. ripple marks created by water flowing from A to B.
    b. ripple marks created by water flowing from B to A.
    c. sand dunes created by wind blowing from A to B.
    d. sand dunes created by wind blowing from B to A.
    e. graded bedding in a submarine trench.

18. What is the most logical past environment to have produced a deposit consisting of layers of well-sorted sandstone with cross beds several meters high?
    a. alluvial fan along a mountain front
    b. sand dunes
    c. shallow-water carbonate area
    d. small-lake environment
    e. glacial valley

19. What is the most logical past environment to have produced a deposit consisting of layers of unsorted sedimentary clasts, clay sized to boulder sized?
    a. alluvial fan along a mountain front
    b. sand dunes
    c. shallow-water carbonate area
    d. small-lake environment
    e. glacial valley

20. What is the most logical past environment to have produced a deposit consisting of very little sand and mud, but instead lots of broken-up carbonate shells of marine organisms?
    a. alluvial fan along a mountain front
    b. sand dunes
    c. shallow-water carbonate area
    d. small-lake environment
    e. glacial valley

21. Sedimentary rocks differ from igneous rocks in that sedimentary rocks can never be crystalline in texture and igneous rocks always are crystalline.
    a. True
    b. False

22. Sedimentary rocks form only at or near Earth's surface, never at great depths.
    a. True
    b. False

23. The words "weathering" and "erosion" are synonyms; they can be used interchangeably.
    a. True
    b. False

24. Naturally formed cracks in rock are termed talus.
    a. True
    b. False

25. Salt wedging can happen only along the sea coast.
    a. True
    b. False

26. In the chemical-weathering process called oxidation, oxygen must be present to chemically combine with the ions present.
    a. True
    b. False

27. The purer the water, the greater its powers of dissolution and hydrolysis.
    a. True
    b. False

28. All life forms need oxygen, water, and organic (carbon) material to survive.
    a. True
    b. False

29. Rectangular blocks of granite weather to spheroidal shape because corners are most vulnerable to weathering, followed by edges and finally flat surfaces.
    a. True
    b. False

30. Only granite and sandstone tombstones hold up well to weathering; slate and marble tombstones quickly disintegrate and dissolve.
    a. True
    b. False

31. A pedalfer soil that's extremely rich in aluminum is called bauxite.
    a. True
    b. False

32. Caliche is a solid mass of calcite that collects in the C-horizon of pedalfer soils.
    a. True
    b. False

33. Graywacke and conglomerate are examples of well-sorted rocks; shales and fine-grained sandstone are examples of poorly sorted rocks.

a. True
b. False

34. A rock with large, angular fragments and lots of feldspar is classified as immature; a rock with smaller, rounded grains and no feldspar is classified as mature.
    a. True
    b. False

35. The formation of dolostone, due to the introduction of magnesium into limestone, is an example of diagenesis.
    a. True
    b. False

36. The presence of redbeds indicates the sediments lithified in water and this caused the iron to rust.
    a. True
    b. False

37. Obvious sandstone beach deposits overlain by marine mudstone indicate transgression of the sea.
    a. True
    b. False

ANSWERS

Completion

1. erosion
2. sediment
3. sedimentary rock
4. fresh rock
5. weathered rock
6. physical weathering (or mechanical weathering)
7. detritus
8. joints
9. talus
10. frost wedging
11. root wedging
12. salt wedging
13. spall
14. animal attack
15. chemical weathering
16. saprolite
17. dissolution
18. acidic
19. hydrolysis
20. oxidation reaction
21. hydration
22. differential weathering
23. soil
24. regolith
25. zone of leaching
26. zone of accumulation
27. horizons
28. soil profile
29. pedalfer soil
30. pedocal soil
31. caliche
32. laterite soil
33. bauxite
34. clastic (or detrital) sedimentary rock
35. biochemical sedimentary rock
36. organic sedimentary rock
37. chemical sedimentary rock
38. siliceous rocks
39. argillaceous rocks
40. carbonate rocks
41. deposition
42. lithification
43. compaction
44. cementation
45. breccia
46. conglomerate
47. matrix
48. arkose
49. sandstone
50. siltstone
51. shale
52. mudstone
53. graywacke
54. lithic clasts
55. sorting
56. sphericity
57. angularity
58. sediment maturity
59. limestone
60. forams
61. fossiliferous limestone
62. micrite
63. chalk
64. recrystallization
65. chert
66. coal
67. oil shale
68. evaporite
69. travertine
70. diagenesis
71. replacement chert
72. flint
73. jasper
74. petrified wood
75. agate
76. bed (or stratum)
77. bedding plane
78. stratigrapher
79. stratigraphy
80. formation
81. cross beds
82. turbidity current
83. graded beds
84. bed surface markings
85. ripple marks
86. mud cracks
87. sedimentary facies
88. terrestrial sedimentary environments
89. marine sedimentary environments
90. glacial till
91. alluvial fan
92. redbeds
93. sand dunes
94. lacustrine sediments
95. floodplains
96. fluvial sediments
97. delta
98. topset beds
99. foreset beds
100. bottomset beds
101. ooids
102. turbidites
103. submarine fan
104. subsidence
105. sedimentary basin
106. rift basins
107. passive-margin basins
108. intracontinental basins
109. foreland basins
110. transgression
111. regression
112. sedimentary sequence

Matching Questions

SOIL TYPES

1. pedocal
2. laterite
3. pedalfer
4. laterite
5. pedocal
6. pedalfer
7. pedocal
8. pedo-*cal* (*cal*cite)
9. ped-*al-fer* (*al*uminum, *fer*rous/*fer*ric = iron)

SEDIMENT TRANSPORT

1. b
2. a
3. d
4. c
5. b

ENVIRONMENTS OF DEPOSITION

1. d
2. l
3. k
4. n
5. b
6. g
7. e
8. h
9. m
10. a
11. i
12. f
13. j
14. c

Short-Answer Questions

1. more than 80%
2. on the continental shelves; almost 20 km
3. mud, F; silt, F; sand, M; pebbles, C; cobbles, C; boulders, C
4. 1/16; 2
5. in types where there's lots of alternate freezing and thawing, such as in the mountains at high altitude
6. They're all agents of physical erosion of rocks, in the category called animal attack.
7. dissolution, hydrolysis, oxidation, and hydration
8. limestone and marble
9. a. hydrolysis; water   b. carbonic acid   c. kaolinite clay
10. a. oxidation; oxygen   b. ferrous; ferric   c. hematite and goethite   d. rusts
11. Because granite rock contains the minerals quartz, feldspar, and mica. The feldspar and mica, given time, weather to clay and wash away; only the quartz is left.
12. more
13. They secrete organic acids that help dissolve minerals.
14. Weathering attacks a flat surface from only one direction and an edge from two directions, but a corner from three directions.
15. Drought killed the vegetation cover and wind stripped away the exposed soil.
16. wind, water, and ice
17. its viscosity and its velocity
18. a conglomerate; because sediment particles loose their angularity as they are transported and the longer the distance, the more rounded they become
19. mineral composition, grain size, and grain shape
20. a. poorly   b. well-
21. graphite; aragonite
22. Over time the calcite and aragonite of the shells recrystallizes, eventually becoming stable calcite crystals that show no evidence of their shell origin.
23. a. chert, in all its varieties (flint, jasper, etc.)
    b. Humans have made and still continue to make excellent cutting tools from broken-off chert, like knives, scrapers, and arrowheads.
24. a. clasts of preexisting rocks   b. burial, compaction, and cementation
25. the involvement of living organisms
26. during original precipitation or, possibly later, during recrystallization
27. a. magnesium   b. groundwater
28. a. a bed or stratum   b. bedding plane   c. strata   d. bedding; stratification
29. source of sediment, climate, and water depth

Making Use of Diagrams

DIFFERENTIAL WEATHERING OF ROCKS

1. A limestone
   B shale
   C sandstone
   D shale
2. A shale
   B sandstone
   C shale
   D sandstone
3. A conglomerate
   B shale
4. A limestone
5. Bob
6.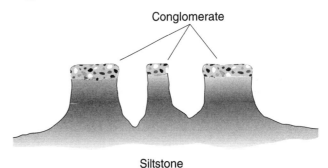

GEYSER BASIN SCENERY

7. travertine

Short-Essay Questions

1. The dry Egyptian climate did not encourage chemical weathering, so the obelisk and its inscription were well preserved for over 3,000 years. When the needle was taken down and temporarily stored, salty groundwater impregnated it. The growth of salt crystals began the process of salt wedging, and when it was finally erected in New York City, the salt wedging and exposure to the elements finished the job of destroying the inscription.

2. Microscopic bacteria found in many different environments can "eat" (metabolize) many different chemical compounds. They break down chemical bonds in minerals and use the released energy for their own life processes. On Earth these bacteria can be found up to a few kilometers deep in the crust. It is possible, but as yet unproven, that similar bacteria could live below the surface of Mars. If so, it would be true to say there was life on Mars, but life with a different physiology and different life-sustaining requirements than the majority of Earth life.

3. Chemical reactions need a surface to occur on. The greater the surface area, the more chemical activity can be taking place and the faster a rock can be chemically weathered. Physical weathering speeds up chemical weathering because it creates smaller pieces, which mean greater surface area per volume. Chemical weathering helps physical weathering because it can dissolve away grains and cement and it can change hard minerals to softer ones.

4. Surface processes that contribute to soil formation include:
   a. animals', plants', microbes', and fungi's interacting with sediment, absorbing nutrients, and leaving behind their wastes and dead remains
   b. rainwater's percolating through rock, sometimes removing ions by leaching, sometimes adding ions by precipitation
   c. the burrowing of ants, worms, and gophers, churning the soil and thus facilitating physical and chemical changes

5. Factors that affect soil include:
   Substrate: Soil formed on basalt is more iron-rich than soil formed on granite, and soils develop faster on sediment than on bedrock.
   Climate: Heavy rainfall speeds chemical weathering and increases leaching.
   Slope steepness: Thick soil accumulates on level land; soil thickness decreases as slope increases.
   Duration of soil formation: Young soils are thin; old soils are thick.
   Vegetation type: Plants add or extract different nutrients, and plant root systems prevent soil erosion.

6. Coal has an organic origin, which technically means it's not a mineral and therefore not a rock. The problem is solved by pointing out it is made of detritus (admittedly organic detritus) deposited in layers, which fits the clastic sedimentary rock description, and creating its own special rock category, organic sedimentary rock.

7. In a past wetter climate, today's Great Salt Lake was much larger, encompassing all of the area now known as Bonneville Salt Flats. The ancient lake had no outlet, so stream waters coming into it could leave only by evaporation, depositing their dissolved salts in the lake. It was a huge lake, huge volumes of stream waters emptied into it, much evaporation occurred, and over time the lake got very salty. Eventually the climate got drier, much water was lost by evaporation and not replaced, the lake diminished to its present size, and all that's left of the rest of the ancient lake is its salt.

Practice Test

1. d Despite the fact organisms are involved, shell material does constitute some sedimentary rocks, so choices a to c are all true.

2. b

3. c Heat, moisture, and increased surface area are the three factors that speed the rate of chemical weathering.

4. e Fluctuations in temperature that cause alternate freezing and thawing speed physical weathering.

5. b Spalling is rock breaking off in sheets; saprolite is rotten rock produced in warm, wet regions; hydration, not hydrolysis is the phenomenon described here (hydrolysis produces chemical change and new minerals).

6. e Arkose is clastic sedimentary rock consisting of quartz and feldspar.

7. c Ions are removed in the zone of leaching and deposited in the zone of accumulation; O- and A-horizons make up the zone of leaching; the O-horizon contains the most humus; farmers till and plant in the topsoil, horizons O and A.

8. a Fossiliferous limestone and turbidities are marine rocks. Arkose doesn't have fragments in its matrix. Conglomerate fragments are rounded. Only choice a fits the description.

9. e Choices a to d are processes associated with clastic sedimentary rocks; limestone is a chemical (or biochemical) rock that crystallizes out of water solution, with or without organic aid.

10. a Only choice a offers an environment where sand is mixed with mud, as in a greywacke.

11. b Only choices a and b are rock made of silt and mud-sized pieces, and a (mudstone) doesn't break in thin sheets.

12. a Chert is composed of the mineral quartz (i.e., it's siliceous).

13. d Coal is partially decayed particles, larger than mud-sized, and derived from swamp plants, not plankton; the other rocks are not composed of decayed organic matter; only choice d fits the description.

14. e Choices a to d are all true.

15. e Flint and jasper are highly insoluble, don't contain large cavities, and are neither translucent nor rare.

16. c Ripple marks require a constant current in permanent shallow water and form at right angles to current direction.

17. a Ripple marks are gentle on the up-current side, steep on the down-current side; sand dune cross sections would look similar, but they would by many feet high.

18. b Alluvial fans would produce conglomerates; carbonate areas would be limestone; lakes would produce mudstones; and glacial valleys would contain unsorted till. Only sand dunes would produce well-sorted sandstone with crossbedding.

19. e Glaciers are powerful enough agents of transportation to carry along unsorted large-sized clasts and drop them all simultaneously when the ice melts.

20. c Marine organisms of shallow water zones contribute their calcite (lime) shells to form the bedrock of the area.

21. False Igneous rocks need not be crystalline (example: obsidian isn't); some sedimentary rocks are crystalline (example: limestone).

22. True Sedimentary rocks result from sediment deposition by wind or water or from chemical precipitation in still bodies of water, all of which are surface occurrences.

23. False They're similar, but "erosion" means the rock has been carried away from its place of origin, not just broken down and/or decomposed.

24. False The cracks are called joints; talus is a wedge-shaped pile of rock rubble at the base of a slope.

25. False Groundwater can carry salts that penetrate rock and do salt wedging (example: the fate of Cleopatra's needle).

26. False "Oxidation" simply means an element gives up electrons and thus raises its electric charge (valence); oxygen may or may not be involved.

27. False Acidic water is more efficient than pure water for both processes.

28. False Many bacteria metabolize minerals.

29. True Rectangular rock pieces get rounded because their corners get weathered faster than their edges and flat surfaces do.

30. False Granite and slate are resistant to weathering; sandstone and marble are not as resistant. As every geologist knows, when the time comes to go, go with granite (or slate).

31. True Bauxite, the ore of aluminum, is pedalfer soil which has an extremely high concentration of aluminum.

32. False It collects in the B-horizon of pedocal soils.

33. False Just the reverse is true.

34. True The presence of angular fragments and large amounts of feldspar implies the rock has not undergone weathering for a geologically long period of time and therefore is young, or immature.

35. True Dolostone is a good example of the reworking of an existing rock, which is the process called diagenesis.

36. False Redbeds indicate subaerial lithification; the iron present is oxidized by the oxygen in the air.

37. True Shallow water deposited the sand; deeper water moved in (transgression) and deposited finer-grained mud.

CHAPTER 8 | Change in the Solid State: Metamorphic Rocks

GUIDE TO READING

In any study of rocks, metamorphic rocks always come last. This is logical; some other rock (igneous, sedimentary, or even a different metamorphic rock) must exist first to get changed into a metamorphic rock. There are limits to the changes that may occur and still yield metamorphic rock. Note the wording in the title of this chapter. It begins "Change in the Solid State," words chosen to emphasize an important limit that is sometimes forgotten. No matter how extreme the temperatures or pressures involved, the rock undergoing change must remain essentially a solid. If it were to be broken down into sediment or changed to a melt, you'd be back in the realms of sedimentary or igneous rocks.

Once the above is explained and a basic definition of metamorphic rock is established, the chapter examines the causes of metamorphism and the features associated with different types of metamorphism. The causes, known as agents of metamorphism, are heat, hot groundwater, pressure, and differential stress. Under the general topics of heat and hot groundwater (hydrothermal solutions) you read about recrystallization, compositional banding, metasomatism, and veins. Learning about the roles of pressures and stresses in metamorphism brings up discussions of shock metamorphism, differential stress, normal stress (compression and tension), and shear stress. Stresses produce plastic deformation and pressure solution, both of which can cause mineral grains to change from equant to inequant in shape and to align with preferred orientation. Temperature and pressure conditions together determine whether a particular mineral is in a stable or unstable condition, metamorphicly speaking.

Classification of metamorphic rock comes next. It's not very complex; there are only two fundamental divisions, foliated and nonfoliated. Common foliated rocks (those which exhibit a layered look) that the author describes are slate, metasandstone, metaconglomerate, phyllite, schist, amphibolite, gneiss, and the "hybrid rock" (part igneous, part metamorphic) migmatite. Common nonfoliated rocks described (which don't have the layered look because they have neither preferred mineral orientation nor compositional banding) are hornfels, quartzite, marble, and dolomitic marble. Nature often manages to defy rigid classification. As an example of this, the author points out the existence of two rocks that exhibit contradicting characteristics, foliated quartzite and foliated marble.

The significance of the existence of one kind of metamorphic rock instead of another kind is addressed in index minerals, metamorphic zones, grade of metamorphism (low, intermediate, or high grade), and metamorphic facies, with its seven main subdivisions.

Just when you start to feel you have a firm grip on what metamorphism is all about, you learn that a seemingly backward version of everything you've just been taught can occur. Retrograde metamorphism can happen to rocks under conditions of decreasing temperatures and pressures, in direct contrast to the usual prograde metamorphism associated with increasing temperatures and pressures.

The chapter concludes with a discussion of locations where you can find metamorphic rocks (environments of metamorphism). These include areas:

- adjacent to plutons (contact or thermal metamorphism)
- in fault zones (dynamic metamorphism)
- beneath mountains adjacent to subducting plates or between colliding plates (dynamothermal or regional metamorphism)

- in continental areas where rifting or transform faulting is occurring
- at mid-ocean ridges
- in subduction zones
- on continental shields

If many of the above environments remind you of the plate tectonic chapters, that's good! As you were told early in the text, in any geologic discussion today it's difficult to avoid plate tectonics. It so often provides the basic answer to the question "Why does that happen?"

The first eight chapters presented the infrastructure of Earth, what it is and how it got established. If you don't have a firm grasp of the concepts presented so far, you may want to review these fundamental chapters, as future chapters build on this material. In the remaining chapters you will learn about activities that occur within this infrastructure, beginning with some very dramatic action, volcanic activity.

LEARNING ACTIVITIES

Completion

Test your recall of new vocabulary terms.

_____ 1. rock that changed from its original form into a new one without first becoming a melt or a sediment, that is, that changed in the solid state

_____ 2. the process in which an existing rock changes, in the solid state, into a different rock

_____ 3. the parent rock (original rock) that is metamorphosed into a different rock

_____ 4. a group of minerals that forms in a rock as a result of metamorphism within a certain range of temperature and pressure

_____ 5. the phenomena that cause metamorphism: heat, hot groundwater, pressure, and differential stress

_____ 6. the formation of new crystals from existing crystals

_____ 7. layers of different composition that may exist in a metamorphic rock

_____ 8. the process in which hot water solutions pass through a rock, pick up some ions, drop off other ions, and thus change the chemical composition of the rock

_____ 9. joints (seams) in rocks that get filled with precipitated minerals (often quartz)

_____ 10. the range of temperatures and pressures at which metamorphism can occur

_____ 11. quality of a mineral while it remains within a limited range of temperature and pressure conditions and therefore does not change

_____ 12. quality of a mineral when it is no longer in a temperature-pressure environment that allows it to remain unchanged, so that its atoms can redistribute and form a new mineral.

_____ 13. the squeezing (pushing) or stretching (pulling) on a material by forces that are not equal from all directions

_____ 14. a push or pull that is perpendicular to the surface it's affecting

_____ 15. a pushing force that squashes or flattens the material it's acting on

_____ 16. a pulling force that stretches the material it's acting on

_____ 17. a force that spreads the material it's acting on in the manner of a deck of cards that is spread apart by forces on top and bottom acting in opposite directions

_____ 18. tiny pieces of mineral that are equally long in all directions

_____ 19. tiny pieces of mineral that have changed shape under stress, ending up as flattened circles

_____ 20. tiny pieces of mineral that have changed shape under stress, ending up as cigar-shaped pieces

_____ 21. the process in which a rock does not break but instead changes shape by flowing like soft plastic

_____ 22. the process in which pressure on a soluble mineral causes it to dissolve on the crystal surfaces undergoing the most compression and reprecipitate on the surfaces undergoing the least compression

_____ 23. the situation in which all inequant grains of a rock are aligned parallel to each other

_____ 24. the layered look in a metamorphic rock due to the occurrence of repeated layers, which may be paper-thin or meters thick

25. the fine-grained, foliated metamorphic rock that results from the metamorphism of shale

26. the foliation in slate along which it tends to break (proper term)

27. the metamorphic rock which results from slight metamorphism of sandstone and whose foliation is created by pancake- or cigar-shaped quartz grains

28. the metamorphic rock which results from slight metamorphism of conglomerate and whose foliation is created by the alignment of pancake- or cigar-shaped conglomerate clasts

29. a fine-grained metamorphic rock with a distinctive silky sheen caused by the preferred orientation of extremely small grains of mica

30. the way phyllite handles light and displays a silky sheen (proper term)

31. a medium- to coarse-grained metamorphic rock whose look is determined by the preferred orientation of large mica flakes and of which shale is a common protolith, although other rocks that can supply atoms to make mica are also possible protoliths (This is the rock name you pronounce very carefully.)

32. the type of foliation found in schist

33. a dark-colored metamorphic rock that results from the metamorphism of mafic rocks like basalt, which are rich in amphibole, plagioclase, and garnet

34. a high-grade metamorphic rock that displays alternating dark and light layers, ranging in thickness from millimeters to meters (compositional banding)

35. a mixed (metamorphic and igneous) rock, which is formed at temperatures even higher than those needed to produce gneiss and which consists of gneiss and some igneous rock that formed from the melting of some of the gneiss

36. a group of metamorphic rocks which form as a result of changes in temperature only and which are not foliated because the lack of differential stresses allows the crystals to remain randomly oriented

37. a nonfoliated metamorphic rock formed by the metamorphism of quartz sandstone

38. an unusual form of quartzite that has undergone deformation in response to differential stress and therefore shows parallel alignment of pancake-shaped quartz grains

39. the rock that is metamorphosed limestone

40. the rock that is metamorphosed dolostone

41. an unusual form of marble that displays color bands formed as the limestone/marble material slowly flowed during metamorphism

42. the category (grade) of metamorphic rock produced by relatively mild conditions of metamorphism, the most important of which is low temperature (less than 320°C)

43. the category (grade) of metamorphic rock produced by midlevel metamorphic conditions, the most important of which are temperatures between 320 and 500°C

44. the category (grade) of metamorphic rock produced by intense metamorphic conditions, the most important of which are temperatures above 500°C

45. minerals that serve as good indicators of metamorphic grade because they form only under relatively limited conditions of temperature and pressure

46. lines on maps which delineate the locations where an index mineral first appears and along which it can be inferred are of the same metamorphic grade

47. the region between two isograds, named after the index mineral found there

48. the metamorphism that occurs because of increasing temperature and pressure

49. the metamorphism that occurs because of decreasing temperature and pressure

50. the geologic setting in which metamorphism occurs

51. the region of "cooked rock" that extends away for a distance of tens to hundreds of meters from a pluton that supplied enough heat to recrystallize the existing rock but not to melt it

52. the type of metamorphism caused by heat from an igneous intrusion (two different names)

53. the type of metamorphism found in fault zones that is caused by shearing alone, with no changes in temperature or pressure

_____ 54. a rock formed by dynamic metamorphism, which displays pronounced foliation that roughly parallels the direction of the fault that formed it

_____ 55. the type of metamorphism, typically found in mountains along convergent plate boundaries, that results from both heat and differential stress (compression and shearing) (two different names)

_____ 56. very old regions of continents where the current ground surface is metamorphic rock that formed when it was buried much deeper in Earth and was then part of the middle and lower crust

_____ 57. the metamorphic rock that forms the lower layer of surface rocks, on which younger sedimentary rocks have been deposited

Matching Questions

ENVIRONMENTS OF METAMORPHISM

Match the type of metamorphism with the environmental setting or specific geographic location that it's found in.

a. dynamic metamorphism
b. dynamothermal (or regional) metamorphism
c. retrograde metamorphism
d. rift and transform fault metamorphism
e. subduction zone metamorphism
f. thermal (or contact) metamorphism

_____ 1. Alpine Fault, New Zealand
_____ 2. Palisades of New Jersey (a huge sill along the Hudson River)
_____ 3. Cascade Mountains of Oregon and Washington (a convergent plate boundary)
_____ 4. mountains in the Basin and Range Province, Nevada (rifting area)
_____ 5. basalts of the sea floor
_____ 6. the west coast of Peru (an accretionary prism)

Short-Answer Questions

Answer with a few words, numbers, or short sentences, or choose from the words in parentheses.

1. Any rock may change over time to become a different rock, which may be igneous, sedimentary, or metamorphic. State which of these three major rock types would form in each of the following situations:
   a. The original rock broke down into small pieces which became compacted and cemented. _____
   b. The original rock got hot enough to melt, then froze. _____
   c. The original rock underwent extreme heat and pressure changes but never became hot enough to melt. _____

2. State two ways you can visually determine a rock is metamorphic. _____

3. Name four agents of metamorphism. _____

4. Why does recrystallization occur more rapidly at high temperatures than at low temperatures? _____

5. Compositional banding, such as you find in gneiss, is a sign of extremely _____ (high, low) temperature of metamorphism.

6. If metamorphic temperature gets extremely high, what common volatiles are likely to vaporize away? _____

7. What mineral typically gets left over after metamorphic reactions, goes into hydrothermal solution, and ends up as fill in veins? _____

8. a. What is the chemical composition of the minerals coesite and stishovite? _____
   b. The presence of these minerals in a layer of sediment suggests what type of event has occurred? _____
   c. What type of metamorphism is this? _____

9. What is the minimum temperature range necessary for any metamorphism to occur? _____

10. Of what special use to a geologist is the discovery of the minerals andalusite, kyanite, or sillimanite at a metamorphic site? _____

11. Sillimanite is found in metamorphic rock at site X, andalusite is found at site Y, and kyanite is found at site Z. Arrange the site letters in order of increasing metamorphic temperatures and pressures. _____

12. a. What foliated metamorphic rock, because of its manner of breaking, has served humans well as tombstones, roof shingles, blackboards, flooring, and pool table beds? _____
    b. What is the term for the manner in which this rock breaks? _____

13. The presence of what mineral gives phyllite its typical silky sheen called phyllitic luster? _____

14. Shale can be metamorphosed to produce any of the following rocks, depending on degree (grade) of metamorphism. Arrange the rock names in order, from lowest to highest degree.
    gneiss   migmatite   phyllite   schist   slate
    _____

86 | Chapter 8

15. The diagram below shows the index minerals contained in three adjacent metamorphic zones.

   | Area A | garnet | biotite | chlorite | Area B |

   a. Would the metamorphism have occurred in area A or in area B? _____
   b. Explain how you know this. _____

16. Gneiss is the result of high-grade metamorphism. It may or may not contain mica. Does the absence of mica in gneiss suggest particularly high or relatively low temperature, and why? _____

17. List three ways compositional banding can form.
    _____
    _____
    _____

18. What causes the folds (contorted layers and swirls) in some metamorphic rocks? _____

19. "Hornfels" is the term for the group of metamorphic rocks formed solely by changes in temperature. What two factors determine the mineral content of a specific variety of hornfel? _____

20. If the formula of all quartz is simply SiO_2 (silica), and quartzite is composed solely of quartz, how can quartzite be so many different colors? _____

21. What are the temperature ranges for low-grade, intermediate-grade, and high-grade metamorphism? _____

22. Why do high-grade metamorphic rocks tend to be "drier" than intermediate- or low-grade rocks? _____

23. Why would a water solution have to infiltrate gneiss before it could undergo retrograde metamorphism? _____

24. Why do geologists prefer to interpret a metamorphic facies in order to decide what the metamorphic temperature and pressure were, instead of relying on information suggested by the presence of an index mineral? _____

25. List the seven major metamorphic facies. _____
    _____
    _____

26. Which is more specific in delineating metamorphic conditions, a *grade* determination or a *facies* determination? _____

27. a. What are blueschists? _____
    b. Under what conditions do they form? _____
    c. Where would you find these conditions? _____

28. In what type of areas have geologists found the oldest rocks on Earth (more than 3.9 billion years old)? _____

29. List several locations of large shield areas. _____

Making Use of Diagrams

PROGRADE AND RETROGRADE METAMORPHISM

Use the figure above to answer questions 1 to 4.

1. When shale gets buried and experiences increased temperature and pressure, it transforms to phyllite, then schist, then gneiss. Which path on the figure represents this progression? _____

2. What compound separates out of the minerals and leaves the rock as this happens? _____

3. If a high-grade metamorphic rock is uplifted from great depth to Earth's surface and hot water is not present, which path on the figure represents the reaction? _____ Explain your answer. _____

4. If a high-grade metamorphic rock is uplifted from great depth to Earth's surface, and hot water circulates through the rock during the process, which path on the figure represents the reaction? _____ Explain your answer. _____

ISOTHERMS AND MOUNTAIN RANGES

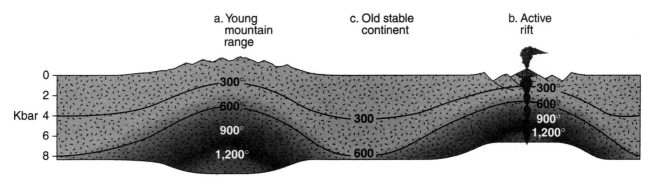

Use the figure above to answer questions 1 and 2.

1. When the surrounding pressure is 4 kbar, what is the approximate Earth temperature
   a. in a young mountain range? _____
   b. in an active rift area? _____
   c. on an old stable continent? _____

2. What pressure is associated with a temperature of 600° on stable continental crust? _____

Short-Essay Questions

1. Why do some rocks (like quartz sandstone and limestone) when metamorphosed produce new rocks with the same mineral composition, and other rocks (like shale) produce new rocks with a variety of new minerals?

2. What is preferred orientation, and how do plastic deformation, pressure solution, rotation of preexisting inequant grains, and development of new inequant grains contribute to it?

3. Slate is often found in association with metasandstone and metaconglomerate. Explain why they're found together and why they all suggest relatively low metamorphic temperatures and pressures.

4. Why does a quartz sandstone break *around* its grains of quartz, but a quartzite breaks *through* its grains of quartz?

Practice Test

1. Choose the listing that shows the rocks in increasing degree of metamorphism (i.e., from lower to higher grade).
   a. migmatite, metaconglomerate, gneiss with mica, gneiss without mica
   b. metaconglomerate, gneiss with mica, gneiss without mica, migmatite
   c. meteaconglomerate, gneiss with no mica, gneiss with mica, migmatite
   d. gneiss with no mica, gneiss with mica, migmatite, metaconglomerate
   e. It makes no difference; they're all examples of high-grade metamorphism.

2. What is the protolith of marble?
   a. quartz sandstone
   b. chert
   c. phyllite
   d. gypsum
   e. limestone

3. Which term has nothing to do with minerals in a vein?
   a. "metasomatism"
   b. "joints"
   c. "quartz"
   d. "mylonite"
   e. "hydrothermal solutions"

4. Which of the following locations could not possibly be part of a shield?
   a. Hawaii
   b. Canada
   c. northern Europe
   d. Siberia
   e. India

5. Which of the following statements is FALSE? A metamorphic rock
   a. may be composed of different minerals than its protolith.
   b. may have different texture than its protolith.
   c. cannot be formed below 500°C.
   d. may have formed from shearing stress only.
   e. may have formed by reaction with hydrothermal solution.

6. Which of the following statements is FALSE? Extremely high heat of metamorphism
   a. may drive off volatiles like carbon dioxide and water.

b. causes high-grade metamorphic rocks to be drier than low grade.
   c. is necessary to form migmatite.
   d. is necessary to elongate the large clasts in conglomerate to make metaconglomerate.
   e. causes mica to react, lose its identity, and become feldspar.

7. Mylonite
   a. forms by recrystallization in a fault zone.
   b. is produced by shear stress on softened rock.
   c. has pronounced foliation parallel to the direction of faulting.
   d. forms due to dynamic metamorphism.
   e. All of the above are true statements.

8. Which of the following statements is FALSE? Shock metamorphism
   a. can be caused by firing high-speed projectiles into quartz sandstone.
   b. can be caused by meteorite impact.
   c. causes quartz to recrystallize into coesite and stishovite.
   d. probably hasn't happened on the Moon because no coesite or stishovite have been found in Moon regolith.
   e. happened to rock worldwide at the time of the extinction of the dinosaurs.

9. Pick out the rock which is nonfoliated.
   a. hornfels
   b. phyllite
   c. slate
   d. schist
   e. gneiss

10. Quartzite
    a. is basically a solid mass of interlocking quartz crystals.
    b. breaks around the separate grains of quartz that make it up.
    c. is always either white or gray.
    d. always shows strong compositional banding.
    e. All of the above are true statements.

11. Which of the following statements is FALSE? Retrograde metamorphism
    a. may happen as erosion and uplift bring rock closer to Earth's surface.
    b. requires the addition of hot water to replace water driven off during prograde metamorphism.
    c. happens when hot water rises through basalt on the sea floor near the mid-ocean ridge.
    d. happens under conditions of decreasing temperature and pressure.
    e. requires very high pressures but very low temperatures to occur.

12. Which of the following statements is FALSE? Metamorphic facies
    a. are identical to metamorphic mineral assemblages.
    b. are groups of metamorphic rocks indicative of specific ranges of pressure and temperature.
    c. each contain several metamorphic rock types.
    d. are subdivided into seven major categories.
    e. are more reliable indicators of metamorphic conditions than index minerals are.

13. Which of the following statements about metamorphic rocks and their characteristic environments is TRUE?
    a. Gneiss is found in metamorphic aureoles surrounding plutons.
    b. Mylonites are found along convergent plate boundaries where magma is rising.
    c. Slate, phyllite, schist, and gneiss are found in areas of continental collision.
    d. Blueschists are found in the mid-ocean ridge.
    e. All of the above.

14. Which of the following statements is FALSE? Shields
    a. are the oldest areas on continents, composed of rock more than 540 million years old.
    b. are composed of extensive areas of sedimentary layers laid down on lava flows.
    c. are large areas where the ground surface is metamorphic rock.
    d. make up large areas of Canada, northern Europe, and South America.
    e. All of the above are true statements.

15. If there are three concentric rings of metamorphosed rock surrounding a pluton, and the rocks contain the index minerals andalusite, kyanite, and sillimanite, the innermost ring will be the one with andalusite, the middle ring will contain kyanite, and the outer ring will contain sillimanite.
    a. True
    b. False

16. Normal stress is a push (compression) or pull (tension) perpendicular to a surface.
    a. True
    b. False

17. Recrystallization occurs because thermal energy causes atoms to vibrate rapidly, break existing chemical bonds, and migrate to new positions on the crystal lattice where they are more stable under the hotter conditions.
    a. True
    b. False

18. Roofers use the rock schist to shingle roofs because its foliation, called schistosity, causes it to break in convenient size.
    a. True
    b. False

19. The random motion of atoms causes the eventual migration of atoms throughout a surrounding material in a process called preferred orientation.
    a. True
    b. False

20. Schist may form from a variety of protoliths that contain atoms necessary to make mica.
    a. True
    b. False

21. Amphibolite is a dark-colored metamorphic mafic rock that is rich in amphibole and quartz.
    a. True
    b. False

22. Folds in metamorphic rock like gneiss happen slowly, over hundreds of thousands of years, as huge masses of softened rock flow slowly and at differential speeds within the mass.
    a. True
    b. False

23. Quartzite is a favorite material of sculptors because of its uniform texture, wide range of colors, and relative softness.
    a. True
    b. False

24. Metamorphic zones are regions between isograds named after the index minerals that occur within them.
    a. True
    b. False

25. Blueschist is a common metamorphic rock of continental shields.
    a. True
    b. False

26. If a pluton melts mafic country rock, the resulting metamorphic rock is called amphibolite.
    a. True
    b. False

27. Metamorphic aureoles typically contain nonfoliated rock like hornfels, because the pluton intrusion provides heat but not the shearing stress necessary for foliation.
    a. True
    b. False

28. In the United States you can see Precambrian metamorphic rocks in places where rivers have sliced deeply into Earth's surface, such as in the Grand Canyon.
    a. True
    b. False

ANSWERS

Completion

1. metamorphic rock
2. metamorphism
3. protolith
4. metamorphic mineral assemblage
5. agents of metamorphism
6. recrystallization
7. compositional banding
8. metasomatism
9. veins
10. metamorphic conditions
11. stable
12. unstable
13. differential stress
14. normal stress
15. compression
16. tension
17. shear stress
18. equant grains
19. platy grains
20. elongate grains
21. plastic deformation
22. pressure solution
23. preferred orientation
24. foliation
25. slate
26. slaty cleavage
27. metasandstone
28. metaconglomerate
29. phyllite
30. phyllitic luster
31. schist
32. schistosity
33. amphibolite
34. gneiss
35. migmatite
36. hornfels
37. quartzite
38. foliated quartzite
39. marble
40. dolomitic marble
41. foliated marble
42. low-grade metamorphic rocks
43. intermediate-grade metamorphic rocks
44. high-grade metamorphic rocks
45. index minerals
46. isograds
47. metamorphic zone
48. prograde metamorphism
49. retrograde metamorphism
50. environment of metamorphism
51. metamorphic aureole
52. contact metamorphism and thermal metamorphism
53. dynamic metamorphism
54. mylonite
55. dynamothermal metamorphism and regional metamorphism
56. shields
57. basement rock

Matching Questions

ENVIRONMENTS OF METAMORPHISM

1. a
2. f
3. b
4. d
5. c
6. e

Short-Answer Questions

1. a. sedimentary   b. igneous   c. metamorphic

2. It's composed of a metamorphic mineral assemblage, and it displays metamorphic foliation.

3. heat, pressure, hot groundwater, and differential stress

4. High temperatures cause more intense molecular motion, so chemical bonds break more frequently and atoms are freed to migrate and establish new crystal structure.

5. high

6. carbon dioxide and water vapor

7. quartz

8. a. SiO_2 (silica), which is the same as quartz
   b. an impact event   c. shock metamorphism

9. 250–300°C

10. Although they are all forms of Al_2SiO_5, each forms under its own unique temperature and pressure conditions. Therefore the existence of one of them tells the geologist exactly what the metamorphic conditions were.

11. Y, Z, X

12. a. slate   b. slaty cleavage

13. mica

14. slate   phyllite   schist   gneiss   migmatite

15. a. area A
    b. because garnet is the index mineral that designates the highest temperature and pressure, so the metamorphic agents would be closest to it, not to the minerals indicating lower temperatures and pressures

16. extremely high, because at very high temperatures mica looses its identity and reacts to form feldspar

17. During recrystallization at very high temperature, light-colored minerals separate from dark-colored minerals; compositional contrasts between beds in the sedimentary protolith are retained during metamorphism; and igneous rock can intrude like small sills between layers of the protolith.

18. The heat of metamorphism makes the rock so soft it flows slowly and differentially, moving faster in some parts than in others, thus producing the bends and swirls.

19. the composition of the protolith and the temperature of the metamorphism

20. Minor amounts of impurities in quartz can color it almost anything (refer back to Chapter 5), so the quartzite it becomes when metamorphosed can be any color.

21. low grade, < 320°C; intermediate grade, > 320°C, < 500°C; high grade, > 500°C

22. High heat drives out volatiles like water.

23. High temperature drove off water during the formation of gneiss; for the metamorphic process to be "correctly" reversed (retrograde metamorphism), water must be put back into the system.

24. One mineral may exist under a variety of metamorphic conditions; existence of a facies (assemblage of minerals) is limited to only one set of conditions.

25. zeolite, hornfels, greenschist, amphibolite, blueschist, eclogite, and granulite

26. "Grade" is a more general term, has only three divisions (low, intermediate, and high), and can usually be determined in the field. A "facies" designation is more precise, since there are seven varieties, and it usually requires microscopic examination.

27. a. rocks that contain a blue-colored type of amphibole called glaucophane
    b. conditions of very high pressure but relatively low temperature
    c. in subduction zones

28. shields

29. Canada, South America, northern Europe, Africa, India, and Siberia

Making Use of Diagrams

PROGRADE ANE RETROGRADE METAMORPHISM

1. prograde path

2. water

3. None, since there's no water present to replace that lost during prograde metamorphism, no metamorphism takes place and there's obviously no representative pathway on the figure.

4. Retrograde path: Uplift results in reduced pressure and temperature, and with hot water present to replace that lost during prograde metamorphism, retrograde metamorphism can occur.

ISOTHERMS AND MOUNTAIN RANGES

1. a. 600°   b. 600°   c. 300°

2. 8 kbar

Short-Essay Questions

1. Rocks that are composed of only one chemically simple mineral (like sandstone which is all quartz, or lime-

stone which is all calcite) have no other option but to produce a metamorphosed version of the same mineral (quartzite and marble respectively). Rocks of more complex mineral and chemical composition have their elements freed and separated during metamorphism, and there may be several potential recombinations of the ingredients for a resultant metamorphic rock.

2. Preferred orientation in a metamorphic rock is the condition where all the inequant (not the same length in all directions) grains of the rock are aligned in the same direction, parallel to each other. Plastic deformation, which is the slow flow of mineral grains, takes compressed, flattened disks and aligns them all perpendicular to the direction of compression. Pressure solution, which is the dissolving of soluble mineral grains first where there is the most pressure, allows the migration of these dissolved mineral grains to areas of less pressure and their subsequent recrystallization and preferred alignment perpendicular to the applied pressure. Preexisting inequant mineral grains and newly formed inequant mineral grains rotate so as to align themselves perpendicularly to the direction of metamorphic stress. Thus all four processes contribute to the orientation of all mineral grains perpendicularly to the direction of metamorphic stress.

3. Shale, sandstone, and conglomerate are often found in sequence in sedimentary beds. Slate is shale that has been only slightly metamorphosed. (Continued metamorphism would turn the slate into phyllite, then schist, then gneiss, and finally migmatite.) Metasandstone and metaconglomerate are still recognizable as the original rocks, sandstone and conglomerate. The change has been simply the flattening of their clasts. From all of the above it's logical to conclude that the entire area, originally sedimentary, has been touched lightly by metamorphic conditions.

4. The metamorphic process enlarges existing quartz crystals of the sandstone. This basically eliminates any open pore spaces that had existed and any cement that was there. The rock no longer consists of grains of rock in cement but is a solid mass of quartz with no structural reason for preferred breakage within weaker cement.

Practice Test

1. b Metaconglomerate is barely metamorphosed; higher heat changes mica to feldspar; and migmatite has been so strongly heated, some rock has melted and become new igneous rock.

2. e A protolith is the parent rock that is metamorphosed into a different rock; metamorphosed limestone is marble.

3. d "Mylonite" is the term for metamorphic rock formed in fault zones.

4. a Shields are the oldest areas of Earth's surface; Hawaii is a recently formed hot-spot island.

5. c The minimum temperature for high-grade metamorphism is 500°C; low-grade metamorphism can begin between 250 and 300°C.

6. d Metaconglomerate is an example of very mild metamorphism.

7. e Statements a to d are true.

8. d Coesite and stishovite were found in Moon regolith that was brought back to Earth by astronauts; this suggested that shock metamorphism had occurred on the Moon.

9. a Hornfels is nonfoliated rock.

10. a Quartzite is a nonfoliated metamorphic rock that may be white, gray, purple, or green; it breaks across, not around, its component grains.

11. e Statement e applies to the creation of rare blueschist metamorphic rock, not to retrograde metamorphic rocks.

12. a The different rocks in a metamorphic facies do not all have the same mineral assemblages.

13. c Gneiss requires shearing stress to form, and this is not produced by plutons; mylonites form due to shearing stress only, not the heat of convergent boundaries; and blueschists form where there's extreme pressure and low heat, which does not describe mid-ocean ridge conditions

14. b Any sedimentary layers or lava flows that might have been present have eroded away, exposing the deep, ancient metamorphic rock of the shield.

15. False The sillimanite, which represents the highest grade, will be closest to the pluton, which is the heat source, and the andalusite, which represents the lowest grade, will be farthest from it.

16. True

17. True

18. False They use slate because its slaty cleavage causes it to break in good shingle size.

19. False This is a definition of diffusion.

20. True   Many different types of rocks are composed of atoms that can be metamorphosed into the mineral mica, which is a chief component of the metamorphic rock schist.

21. False   Amphibole has very little quartz.

22. True   The process of metamorphism occurs only in the solid state, and rock flows slowly in the solid state.

23. False   All of the above is true of marble, not quartzite.

24. True   Index minerals are those that indicate the varying conditions within the different zones of the metamorphosed region.

25. False   Blueschist is a rare rock containing glaucophane, a blue-colored type of amphibole; it is formed under high-pressure, low-temperature conditions such as are found in accretionary prisms adjacent to subduction zones.

26. False   What a sneaky question! If a pluton melts the country rock, you're no longer dealing with a metamorphic situation.

27. True

28. True   The Grand Canyon cuts deep into Earth's surface and thus exposes very old rock.

CHAPTER 9 | The Wrath of Vulcan: Volcanic Eruptions

GUIDE TO READING

In this chapter the author explains that there are many possible manifestations of volcanic activity, and numerous variables to consider when trying to understand what has occurred or predict what will occur at some specific location. The nature of volcanic activity depends on the chemistry of the situation (chemical compositions, temperatures, and pressures) and the environment in which it's occurring (submarine or subaerial/not underwater).

There's lots of vocabulary to learn and many classification schemes to use. Keep reminding yourself that scientists establish classification schemes to organize data, discern patterns, and thus reach an understanding of why things happen as they do. Look for these patterns; make logical associations of such things as lava types, cone types, and styles of eruption. But be aware that volcanoes are the result of the interplay among so many factors they're not always perfectly predictable. They've been known to fool even the smartest volcanologists.

The chapter begins with discussion of the three types of volcanic products: lavas, pyroclastic debris, and gases. It continues with the vocabulary of volcanic structures: "magma chambers," "chimneys," "fissures," "craters," "calderas," and "resurgent domes." With these basic terms to build on, the author discusses the three basic types of cones (shield, cinder, and composite or stratovolcanoes), the differences between summit eruptions and flank eruptions, and eruptive styles (effusive, explosive, and phreatic.)

Any history buff will feel right at home in this chapter because woven throughout it are accounts of historic volcanic eruptions that have affected human society, often in devastating fashion. Pay particular attention to the geographic locations of these eruptions; they're determined by plate tectonics. The phenomena of hot spots, mid-ocean ridges, convergent margins, and continental rifts reappear, this time with emphasis on their association with volcanic activity.

The chapter continues with societal issues:

- What are the positive effects of volcanic activity?
- What specifically are the dangers? (lava; ash, pumice, and lapilli falls; pyroclastic flows; blast; landslides; lahars; earthquakes; tsunamis; gas)
- Can humans protect themselves and their property? (prediction and control)
- How is climate affected?
- How long have humans been affected? (The record goes way back in both the popular and religious literature.)

The author ends with some really far out material, literally. He addresses the question "Is Earth the only planet blessed (or cursed) with volcanism?" Read on to learn the answer.

LEARNING ACTIVITIES

Completion

Test your recall of new vocabulary terms.

_____ 1. a vent or fissure that erupts molten rock, or the mountain this activity creates

_____ 2. molten rock that flows like a liquid on Earth's surface

_____ 3. "fire pieces," erupted from a volcano, that solidify by the time they fall

_____ 4. vapors that escape from magma and emerge from a volcano

_____
_____ 5. two different adjectives that are applied to lava with high silica content

Chapter 9

_____ 6. two different adjectives that are applied to lava with medium silica content
_____

_____ 7. two different adjectives that are applied to lava with low silica content
_____

_____ 8. natural glass, resulting from either very fast cooled lavas or very silicic lavas

_____ 9. the description of how smoothly and easily a material flows, that is, whether it resists flowing

_____ 10. an area of volcanic rock (can be shaped like broad sheets or narrow arms) that formed when extruded lava cooled

_____ 11. a type of basaltic lava flow that has a smooth but wrinkled (rope-like) surface

_____ 12. a type of basaltic lava flow consisting of sharp, angular pieces, like a pile of rubble

_____ 13. a tunnel-like structure created when an arm of flowing lava cools and solidifies on the outside, continues to flow through the inside, and eventually drains out, leaving empty space

_____ 14. a phenomenon that occurs when a thick lava flow cools, contracts, and solidifies as polygon-shaped columns (often hexagonal)

_____ 15. blobs of lava that have cooled quickly in seawater

_____ 16. the bulbous structure formed when rhyolitic lava cools above the vent that released it

_____ 17. a cone of glassy volcanic fragments that collects around a submarine vent

_____ 18. marble-sized pieces of pyroclastic debris

_____ 19. plum- or walnut-sized pieces of pyroclastic debris

_____ 20. basketball-sized (or larger) pieces of pyroclastic debris

_____ 21. teardrop-shaped glass beads resulting from falling basaltic lapilli

_____ 22. long strings of glass resulting from falling, stretched basaltic lapilli

_____ 23. pyroclastic debris pieces that were the "froth" (gas-filled bubbles) of lava

_____ 24. tiny flakes of glass (< 2-mm diameter) that cooled instantly when lava was blasted out of the volcano

_____ 25. accumulations of pyroclastic grains (ash or lapilli) that remain unconsolidated after falling

_____ 26. accumulations of pyroclastic grains that are cemented together

_____ 27. the volcanic rock formed of solidified pyroclastic flows

_____ 28. a glowing cloud or avalanche of volcanic ash and very hot air

_____ 29. a slurry of volcanic ash, debris, and water that races down valleys or river channels on a volcanic mountain (or, very simply, a volcanic mud and debris flow)

_____ 30. small holes in lava that are "frozen" gas bubbles

_____ 31. an open space underground, or an area in which the rock is highly fractured, in which magma has accumulated

_____ 32. a pipe-shaped conduit through which lava leaves a volcano

_____ 33. a long, linear crack through which lava leaves a volcano

_____ 34. the circular depression at the top of a volcanic cone

_____ 35. volcanic eruptions in which material leaves through the summit crater

_____ 36. volcanic eruptions in which material leaves through chimneys or fissures on the sides of the volcano

_____ 37. a circular depression, much larger than a crater, formed after an eruption, when much of the mountain collapses into the drained magma chamber

_____ 38. a volcanic mound that grows within a caldera

_____ 39. volcanic mountains that are broad cones with gently sloping sides

_____ 40. volcanic mountains with fairly steep sides, composed of piles of tephra

_____ 41. the steepest angle at which unconsolidated material can sit without sliding

_____ 42. the classic volcano, large, steep sided, long lived, composed of alternating layers of lava and tephra

_____ 43. the character of a volcanic eruption—what kind of material is produced, how violent the action is, and so on

_____ 44. volcanic eruptions in which the main product is lava and there is minimal explosive activity

_____ 45. volcanic eruptions in which the main products are pyroclastic materials and there is dramatic, explosive activity

_____ 46. a special case of pyroclastic eruption in which water enters the magma chamber and the resulting steam pressure blasts magma outward

_____ 47. a local accumulation of lava which "puddles up" because it has been erupted so quickly and which eventually drains as streams of fast-moving, very fluid lava

_____ 48. lava flows that are broad sheets covering large areas, resulting from high-temperature, low-viscosity basalt welling up through rifts

_____ 49. a column of superhot water shooting out of a vent along a mid-ocean ridge, which is dark with mineral grains

_____ 50. volcanic eruptions, issuing from long, linear cracks, that result in either curtains of lava or linear chains of small cinder cones

_____ 51. the chain of volcanoes, thousands of miles long, encircling the Pacific Ocean because of the subduction zones there

_____ 52. the average time period between successive eruptions of a volcano

_____ 53. volcanoes that have erupted within the last few centuries

_____ 54. volcanoes which haven't erupted for hundreds or even thousands of years but for which the reason for volcanism still exists, so there may be future eruptions

_____ 55. volcanoes which erupted in the past but for which the geologic cause for volcanism no longer exists, so no future eruptions can occur

_____ 56. maps for which volcanic hazards and the topography of the region have all been considered and which show volcanic danger zones specifically where lava or debris flows or lahars, and so on, may be expected

_____ 57. the dark areas on the Moon's surface, which are regions covered by flood basalts

_____ 58. giant sea waves caused by underwater explosions and landslides (or, as you'll learn in the next chapter, by ocean-bottom earthquakes)

Matching Questions

PLATE TECTONICS AND VOLCANISM

Match the specific volcanic location with the plate tectonics description.

a. Aleutian Islands
b. Basin and Range Province
c. Cascade Range
d. Deccan Plateau, India
e. Columbia River Plateau
f. Hawaii (the big island)
g. Iceland
h. Karroo Plateau, South Africa
i. Kilauea
j. Mauna Loa
k. mid-ocean ridges
l. Parana Basin, Brazil
m. Snake River Plain
n. Surtsey
o. Yellowstone

_____ 1. Thousands of basalt flows piled up to allow this oceanic hot spot to emerge from the sea and form an island composed of huge overlapping shield cones, including the tallest mountain on Earth. Name the island and the mountain.

_____ 2. This continental hot spot is a caldera. During millions of years it has explosively erupted everything from basaltic lava to rhyolitic pyroclastic debris. Currently it's noted for its geysers and other geothermal features.

_____ 3. This track of a hot spot is an elongate area of volcanic rock created over millions of years as the North American continent moved across the Yellowstone hot spot.

_____ 4. Enormous flood basalts erupted along fissures 15 million years ago and covered this area in the U.S. Northwest. The fissures existed because of continental rifting.

_____ 5. Gas bubbles in these flood basalts, erupting from continental rift fissures, froze into vesicles which, over geologic time, became lined with amethyst crystals from percolating groundwater.

_____ 6. This divergent plate boundary can claim the greatest number of submarine volcanoes on Earth. It's associated with black smokers and many strange forms of life.

_____ 7. This hot spot–mid-ocean ridge combination boasts frequent fissure eruptions, basalt flows that get ripped apart by faulting, curtains of fountaining lava, chains of cinder cones, and even flooding caused by melted glacial ice.

_____ 8. Phreatic eruptions at this hot spot–mid-ocean ridge combination produced an island that emerged from the sea and, due to its tough exterior of hardened lava, has survived wave action.

_____ 9. This volcanic island arc owes its existence to subduction of the Pacific Plate under the North American Plate. It's part of the "Ring of Fire."

_____ 10. This range of mountains, a continental volcanic arc, makes a beautiful (but dangerous) backdrop for several cities in the U.S. Northwest.

_____ 11. continental rifting of this area is slowing tearing apart Nevada, Utah, and Arizona, producing layers of basalt and layers of ignimbrite as it does so.

ERUPTIVE STYLE AND RESULTING PHYSICAL FORM

Match the physical form of the volcano to its eruption style.

a. caldera
b. cinder cone
c. composite cone (or stratovolcano)
d. shield cone
e. submarine volcano

_____ 1. An effusive eruption of basalt produces this volcanic mountain.

_____ 2. A relatively small eruption of pyroclastic grains produces this type mountain.

_____ 3. Volcanoes that alternate between periods of effusive eruptions and large-scale pyroclastic eruptions have this structure.

_____ 4. Extremely violent, explosive eruptions produce this volcanic feature.

_____ 5. Basaltic flow, which ordinarily spreads out like a pancake, piles up around the vent with this kind of volcano.

DESCRIPTIONS OF HISTORIC ERUPTIONS

Match the volcanic eruption with its description. (Treat this exercise not just as a history-geography review [which of course it is] but as an invitation to review and make logical associations between circumstances, volcanic activity, and volcanic products.)

a. Crater Lake
b. Krakatau
c. Lake Nyos
d. Mt. Pelée
e. Mt. St. Helens
f. Montserrat
g. Nevado del Ruiz
h. Parícutin
i. Tambora, Indonesia
j. Vesuvius
k. Vulcano

_____ 1. By the Mediterranean Sea, Bay of Naples, Italy, 79 A.D.: Volcanic ash buried the Roman town of Pompeii; victims' bodies decayed inside the consolidated ash (tuff) and were discovered centuries later.

_____ 2. On a Mediterranean island, west of Italy: The island's name came from the ancient Roman god of fire.

_____ 3. On an island in the Caribbean, 1902: A mountain gave fair warning, then released one of the most famous pyroclastic flows in history; 28,000 people in the city died, 2 survived.

_____ 4. On another island in the Caribbean, 1990s: A volcano threatened, then erupted many pyroclastic flows; luckily this time the people have been evacuated.

_____ 5. In the U.S. Northwest, 6,850 years ago: An explosive eruption produced a caldera that today is a well-known tourist attraction of the Cascade Range.

_____ 6. In the U.S. Northwest, 1980: A volcano that had last erupted in 1857 gave a few months warning, then erupted again; an earthquake triggered a landslide which allowed a horizontal explosive eruption and accompanying ash falls and lahars.

_____ 7. On an island between Indonesia and Sumatra, 1883: Several phreatic explosions climaxed with the largest explosion in recorded history; a killer tsunami was generated, a submarine caldera formed, and the ash released produced spectacular sunsets worldwide for years.

_____ 8. In Mexico, February 20, 1943: This classic cinder cone began in a cornfield; its birth was witnessed by many, and it's 9-year lifetime was carefully monitored.

_____ 9. In Africa, 1986: This lake, which sits on an active hot spot, burped carbon dioxide gas that suffocated more than 1,700 people as they slept.

_____ 10. In Columbia, South America, 1985: The eruption of this mountain created a lahar that buried the population of Armero in hot volcanic mud.

_____ 11. In Indonesia, 1815: The eruption of this volcano put so much dust in the sky stars were dimmed, 1816 became known as the year without a summer, sunsets were fabulous worldwide, and authors were inspired to write gloomy poems and novels (like Frankenstein).

Short-Answer Questions

Answer the following briefly, with a few words, numbers, or phrases:

1. In terms of plate tectonics, where are volcanoes located? _____
2. What are the three different types of volcanic materials? _____
3. How are the silica content and the temperature of lava related to its viscosity? _____
4. Which major type of lava is involved in each of the following scenarios?
   a. Initially this lava mounds up around the vent, then advances a few meters per day as a lumpy mass with a bulbous snout, until it extends about 10 km from the vent. _____
   b. A relatively thin flow moves freely and quickly and reaches a few hundred kilometers distant from its vent. _____
   c. This flow, about 1 km long, has a very broken, blocky surface. _____
5. Why does lava cooled under water look so different from subaerial flows? _____
6. Ash, pumice, cinders, lapilli, bombs, block, Pele's tears, and Pele's hair are all examples of which type of volcanic product? _____
7. Pop looses its fizz after the bottle is opened, and volcanic gases come out of magma as it reaches Earth's surface. Both are examples that gases come out of liquid solution when pressure gets _____ (lower, higher).
8. Name the volcanic gas described.
   a. It was responsible for the tragedy at Lake Nyos, Africa. _____
   b. It smells like rotten eggs. _____
   c. It forms a strong acid in solution. _____
   d. It is an important component of weather phenomena. _____
9. What is Loihi? _____
10. What organism is at the bottom of the food chain for giant clams, giant worms, and shrimp that live by black smokers, and what is unusual about this organism's food? _____

11. Where are most of Earth's submarine volcanoes? _____
12. Where are most of Earth's subaerial volcanoes? _____
13. Relatively quiet effusive eruptions do sometimes spray lava upward in impressive fountains. What produces the force that causes the spraying? _____
14. Name two geographic locations in the United States where the softer exteriors of volcanic mountains have eroded away, leaving the harder solidified lava of the vents and radiating dikes dramatically exposed. _____
15. It's true volcanoes can be hazardous to life. It's also true volcanism has contributed positively to Earth throughout geologic time. List six good things volcanoes have done for Earth. _____
16. List several volcanic products or events that are hazardous to human life and property. _____
17. Name the U.S. scientist (better known as a founding father) who correctly linked volcanic activity to climate change, and when did he do this? _____
18. What do the following three eruptions have in common weatherwise? Karkatau, Indonesia (1883); El Chichon, Mexico (1982); Pinatubo, Philippines (1991) _____
19. What do geologists find in ancient ice from cores drilled in Greenland and Antarctica that suggests extensive volcanic activity? _____
20. What volcanic product preserved the footprints of prehistoric humans, and where did this happen? _____
21. What Mediterranean civilization is suspected of having been the victim of a cataclysmic eruption of the island volcano Thera more than 3,000 years ago? _____
22. What lunar feature is evidence of past (but not present) lunar volcanism? _____
23. What is the name and planetary location of the largest mountain found in our solar system? _____
24. Where is the only known currently active extraterrestrial volcano? _____

Short-Essay Questions

1. Why are highly viscous lavas associated with volcanoes that erupt explosively?
2. Does the existence of a caldera necessarily mean a cataclysmic volcanic eruption occurred?
3. Why aren't submarine volcanoes considered in the typical cinder-shield/composite classification scheme?
4. Which type of volcanic cone erodes the fastest and why: composite, cinder, or shield?
5. Kauai, the most northwesterly of the Hawaiian Island chain, is a beautiful rugged place, with steep cliffs and numerous river valleys carved into its hillsides. From this information give two reasons you should know the volcanoes there stopped erupting a long time ago.
6. Why shouldn't airplanes fly through clouds of fine volcanic ash?
7. Why are major volcanic eruptions accompanied by earthquakes?
8. The amount of erosion is a key to classifying volcanoes as active, dormant, or extinct. Write a brief explanation of why this is so.
9. List and briefly discuss four factors that a monitoring team would check to predict if a volcanic eruption is imminent.
10. Although it's not possible to prevent volcanic eruptions or turn them off once they begin, there are some things people can do to help prevent loss of life and property. List three.

Practice Test

1. The presence of pillow lava is evidence of what volcanic circumstance?
   a. a submarine volcano
   b. high-silica-content magma
   c. an explosive eruption
   d. a phreatic eruption
   e. caldera formation

2. Black smokers
   a. are hydrothermal vents that parallel the mid-ocean ridge.
   b. are dark because they contain specks of sulfide minerals.
   c. provide an environment for giant clams, worms, and shrimp.
   d. provide an environment for bacteria that eat sulfide minerals.
   e. All of the above are true.

3. A glowing cloud of hot gases and volcanic pieces
   a. was named in Hawaii because it's so common there.
   b. is called a pyroclastic flow or nuée ardente.
   c. is so rare none occurred all last century.
   d. is associated with shield cones.
   e. All of the above are true.

4. A lahar is
   a. a glowing cloud of pyroclastic grains and hot gas.
   b. a cinder cone.
   c. an enlarged volcanic crater.
   d. a volcanic mudflow and debris flow.
   e. A lava lake in a summit crater.

5. The enlarged volcanic craters at Yellowstone, Crater Lake, and Krakatau are
   a. evidence of safe, nonexplosive activity.
   b. called lahars.
   c. filled with columnar joints.
   d. formed by both explosion outward and collapse inward of the mountain.
   e. evidence of low silica content.

6. Volcanic mudflows and debris flows
   a. occur only in arctic regions because snow supplies the water for mud.
   b. leave messes but rarely kill people because they flow slowly.
   c. are called nuée ardentes.
   d. have often resulted in damage and death to humans and their property.
   e. are extremely rare.

7. Which of the following is NOT a form of lava?
   a. aa
   b. andesite
   c. obsidian
   d. lapilli
   e. pahoehoe

8. Lava of low viscosity
   a. could logically build a shield cone.
   b. indicates an area has little potential for explosive eruptions.
   c. is logically basalt.
   d. has low silica content.
   e. All of the above are true.

9. Vesuvius erupted—a very violent explosion—in 79 A.D. and buried the residents of Pompeii in ash. On the basis of this information, which of the following would you LEAST expect to find at Vesuvius?
   a. basalt
   b. lapilli
   c. rhyolite
   d. cinders
   e. pyroclastic flow deposits

10. Lava is
    a. always produced during any volcanic activity.
    b. molten rock that cools and hardens on Earth's surface.
    c. all the same chemical composition, whether it's rhyolite or basalt.
    d. the most life-threatening volcanic product of any eruption.
    e. All of the above are true.

11. Rhyolite lava
    a. has more silica than basalt lava does.
    b. indicates the tendency for explosive activity.
    c. may form hyaloclastite around submarine vents.
    d. may form a lava dome above the vent.
    e. All of the above are true.

12. Pick out the FALSE statement. Carbon dioxide
    a. is exhaled by humans as part of normal respiration.
    b. is a common volcanic product that can suffocate life.
    c. is a poisonous gas; even a little can kill you.
    d. was released at Lake Nyos, Cameroon, Africa, and killed many people.
    e. is released by magma as it rises and the pressure gets lower.

13. Mt. St. Helens erupted in 1980, sending out gas and pyroclastics. Geologists find evidence of many eruptions of St. Helens, dating back 37,000 years. The mountain is huge and is composed of layers of lava and pyroclastics. Which of the following is most likely true?
    a. When lava flows, it will be basalt.
    b. It will not erupt again.
    c. It's a stratovolcano (or composite cone).
    d. It's on a transform plate boundary.
    e. It's a cinder cone.

14. This stratovolcano erupted in 1980 in the United States. Although it warned of its impending eruption, it did still kill people, partly because it first blew sideways. Its previous eruption was in the 1850s. The volcano is
    a. Surtsey.
    b. Mt. St. Helens.
    c. Mt. Ranier.
    d. Yellowstone.
    e. Mt. Pelée.

15. This volcano warned of its impending eruption for several months, but early in 1902 it killed thousands with a pyroclastic flow. It's a popular vacation spot in the Caribbean.
    a. Surtsey
    b. Paricutin
    c. Vesuvius
    d. Mt. Pelée
    e. Pinatubo

16. In the 1990s a California camper in a volcanic area died after he ignored posted warnings and slept in a campground among standing but dead trees. He was not scalded, he was not poisoned, and the campground did not smell bad. What volcanic gas killed the camper and the trees?
    a. water vapor
    b. sulfur dioxide
    c. carbon dioxide
    d. hydrogen sulfide
    e. carbon monoxide

17. Individual volcanoes erupt fairly regularly, and if the recurrence interval is known, their eruptions can be accurately predicted.
    a. True
    b. False

18. Renewed eruption in a caldera can produce a resurgent dome.
    a. True
    b. False

19. Most subaerial volcanoes on Earth lie along divergent plate boundaries.
    a. True
    b. False

20. Devil's Tower, Wyoming, and Shiprock, New Mexico, are landforms that exist because softer volcano exteriors erode faster than solidified lava in fissures and vents.
    a. True
    b. False

21. Lava is the greatest volcanic hazard to human life because it travels so fast.
    a. True
    b. False

22. Many planes have crashed because volcanic ash sucked into their engines, blocked the engine's air flow, and shut it down.
    a. True
    b. False

23. Pyroclastic flows bring instant death to any life caught in them.
    a. True
    b. False

24. A tsunami is a tidal wave caused by the proper lineup of the Moon and the Sun.
    a. True
    b. False

25. Short-term prediction of earthquakes is usually possible; short-term prediction of imminent volcanic eruption has had no proven success.
    a. True
    b. False

100 | Chapter 9

26. Volcanic danger assessment maps are valuable tools in protecting human life when volcanic eruption threatens.
    a. True
    b. False

27. It is not possible to divert lava flows because they're too hot and powerful.
    a. True
    b. False

28. Volcanic ash and sulfur-dioxide gas can absorb solar radiation and result in cooler average temperatures worldwide.
    a. True
    b. False

29. Active volcanoes have been confirmed on Venus, Mars, and Jupiter's moon Io.
    a. True
    b. False

30. Label the following:
    B if it's associated with basaltic chemistry
    R if it's associated with rhyolitic chemistry
    _____ a. aa
    _____ b. pahoehoe
    _____ c. highest silica content
    _____ d. fastest flowing
    _____ e. most viscous
    _____ f. pillow lava
    _____ g. shield cones
    _____ h. lava domes
    _____ i. columnar jointing
    _____ j. lava tubes

28. pyroclastic flow (or nuée ardente)
29. lahar
30. vesicles
31. magma chamber
32. chimney
33. fissure
34. crater
35. summit eruption
36. flank eruption
37. caldera
38. resurgent dome
39. shield volcanoes
40. cinder cones
41. angle of repose
42. stratovolcanoes (or composite cones)
43. eruptive style
44. effusive eruptions
45. explosive eruptions
46. phreatomagmatic eruption
47. lava lake
48. flood basalts
49. black smoker
50. fissure eruptions
51. ring of fire
52. recurrence interval
53. active volcanoes
54. dormant volcanoes
55. extinct volcanoes
56. volcanic danger assessment maps
57. mare
58. tsunamis

Matching Questions

PLATE TECTONICS AND VOLCANISM

1. f; j   5. l   9. a
2. o   6. k   10. c
3. m   7. g   11. b
4. e   8. n

ERUPTIVE STYLE AND RESULTING PHYSICAL FORM

1. d   3. c   5. e
2. b   4. a

DESCRIPTIONS OF HISTORIC ERUPTIONS

1. j   5. a   9. c
2. k   6. e   10. g
3. d   7. b   11. i
4. f   8. h

ANSWERS

Completion

1. volcano
2. lava
3. pyroclastic debris
4. volcanic gas
5. silicic and rhyolitic
6. intermediate and andesitic
7. mafic and basaltic
8. obsidian
9. viscosity
10. lava flow
11. pahoehoe
12. aa
13. lava tube
14. columnar jointing
15. pillow lava
16. lava dome
17. hyaloclastite
18. lapilli
19. cinders
20. bombs
21. Pele's tears
22. Pele's hair
23. pumice
24. volcanic ash
25. tephra
26. tuff
27. ignimbrite

Short-Answer Questions

1. at plate boundaries, rifts, and hot spots
2. lava, pyroclastic debris, and volcanic gas
3. The lower the silica content and hotter the lava, the lower its viscosity is (i.e., the more smoothly it flows).
4. a. andesitic flow   b. basaltic flow   c. rhyolitic flow
5. because it cools so quickly (Water transfers heat away from anything much more quickly than air does.)
6. pyroclastic debris
7. lower
8. a. carbon dioxide   b. a sulfur gas   c. a sulfur gas   d. water

9. the youngest of the Hawaiian Islands, forming southeast of Hawaii and still below sea level
10. Bacteria; they eat sulfide minerals.
11. along the mid-ocean ridge system
12. along subduction zones
13. bursting gas bubbles in the lava
14. Devil's Tower, Wyoming, and Shiprock, New Mexico
15. Produced Earth's crustal rock; contributed atmospheric gases; possibly served as the birthplace for life (black smokers); hosted isolated populations on volcanic islands which evolved and thus increased the diversity of life; contributed to fertile soil; and provided mineral and energy resources.
16. lava; ash, pumice, and lapilli falls; pyroclastic flows; blast; landslides; lahars; earthquakes; tsunamis; volcanic gases
17. Benjamin Franklin, in 1789
18. Average global temperatures dropped noticeably for months after each of them.
19. high concentration of sulfuric acid from volcanic sulfur gases
20. fossilized volcanic ash layer in the East African Rift Valley
21. Minoan civilization
22. the dark areas called mares, which are flood basalts
23. the volcano Olympus Mons on Mars (not currently active)
24. on Io, one of Jupiter's moons

Short-Essay Questions

1. Viscous lavas have high silica content. Silica molecules form microscopic chains that literally trap gas. Trapped gas builds up pressure until the material holding it eventually explodes.

2. No, calderas result from the collapse of mountain material into an emptied magma chamber. It's true this might happen after an explosive eruption, in which much mountain material is also blasted away (as at Krakatau, Indonesia, and Crater Lake, Oregon), but it might also be a quiet sinking down (as at Kilauea, Hawaii) simply because support has been removed from below.

3. Mounds of underwater volcanic products grow in irregular shapes and are changed frequently by massive landslides. They just don't fit the subaerial classification scheme.

4. Cinder cones erode more quickly than either composite or shield cones. Their unconsolidated cinders, ash, and other pyroclastic pieces wash away more readily than the lava-encrusted flanks of shield or composite cones do.

5. First, since it is the farthest northwest of the island chain, it moved off the Hawaiian hot spot long ago, so its source of volcanism is gone. Second, the rugged landscape is evidence erosion has been taking place unopposed by the addition of recently erupted material.

6. Although fine ash may not even be visible to the pilot, its effects can be damaging and even disastrous. Angular particles abrade turbine blades, lowering engine efficiency. Ash and sulfuric acid (from sulfur gases present) etch windows and damage the fuselage. Ash sucked into engines melts, then freezes as a glassy coating on turbines where it blocks air flow and can shut down the engine.

7. Magma moving underground opens new cracks along which there is movement, and gaseous magma in magma chambers explodes and sends shock waves through the surrounding earth. These underground disturbances are transferred to the surface as earthquake (seismic) waves, with all the usual associated hazards.

8. Active volcanoes typically have symmetrical shapes, possibly some coating of newly erupted lava or pyroclastics, sparse vegetation, and little evidence of erosion. Dormant volcanoes are obviously dissected by erosion, but they still have enough symmetry to look like what they are—volcanic cones. Extinct volcanoes are so thoroughly dissected by weathering they've lost the "cone look" of a volcano.

9. A monitoring team would look for:
   Changes in heat flow: Rising magma heats surrounding rock, and the heat makes its way to the surface. It may even melt surface snow and ice.
   Changes in shape: Rising magma fills and inflates the magma chamber; this causes the land surface above to bulge.
   Earthquake activity: Moving magma and bursting gas bubbles create cracks and explosions, all of which register as earthquakes.
   Increases in gas and steam emission: Magma releases gases as it rises into areas of lower pressure, and it heats groundwater. Increased gas pressure, steam emissions, or the appearance of a hot spring all signal the approach of magma to the surface.

10. a. Prepare danger assessment maps and act on the information supplied when appropriate.

b. Evacuate the population when eruption is imminent.

c. Divert or slow or stop the lava flows. Diversion can be done by the breaking up of existing hardened lava channels by use of explosives or by the construction of dams and channels. Slowing or stopping the flow can be done by cooling it with seawater.

Practice Test

1. a Pillow lava is a low-silica-content, basaltic lava associated with underwater effusive eruptions.

2. e Statements a to d are all true.

3. b A glowing cloud of gases and pieces is not uncommon with explosive eruptions; shield cones, including those of Hawaii, are associated with effusive, not explosive eruptions.

4. d Choice a defines a pyroclastic flow (nuée ardente); c defines a caldera; and b and e are known simply by the names given in the question.

5. d Columnar jointing and low silica content are associated with relatively safe effusive eruptions; the enlarged craters mentioned are called calderas and were formed by explosive eruptions.

6. d Volcanic mudflows and debris flows, called lahars, are fairly common, fast-moving, deadly events associated with volcanic eruptions worldwide; volcanic outgassing alone provides enough moisture to form mud, and melted snow just adds to the danger.

7. d Lapilli are marble-sized pyroclastic pieces, explosively ejected; the others are all forms of molten rock that flow from a volcanic opening, that is, lavas.

8. e Statements a to d are all true.

9. a Basalt lava is associated with nonviolent, effusive eruptions; rhyolite lava and the pyroclastic products all indicate a rock chemistry that results in explosive eruptions.

10. b Many types of lava exist, with different chemical compositions; often volcanic eruptions don't produce any lava, and when it is present, it usually moves slowly enough that people can get out of its way.

11. e Statements a to d are all true.

12. c Carbon dioxide is not a poison; it simply keeps oxygen out, so life suffocates.

13. c The longevity, large size, and alternating eruptions of pyroclastics and lava are classic attributes of stratovolcanoes (composite cones) that form along convergent plate boundaries where subduction is occurring.

14. b The descriptive information all fits Mt. St. Helens, but none of the others.

15. d The descriptive information all fits Mt. Pélee, but none of the others.

16. c Carbon dioxide is the only volcanic gas that fits the description. The sulfur gases have bad odors; water vapor could kill only if it were scalding hot; and carbon monoxide is a poisonous gas, but it would change to carbon dioxide in open spaces.

17. False The recurrence interval for an individual volcano can vary from a few years to a few centuries or even longer.

18. True A small mound that grows in a newly formed caldera is called a resurgent dome.

19. False They lie along convergent plate boundaries and form both continental and island volcanic arcs.

20. True Erosion of soft exterior rocks has exposed the frozen lava that filled the vents at Devil's Tower and Shiprock.

21. False Lava often travels no faster than several feet per hour, and even the fastest moves no faster than about 20 mph; people can usually get out of the way.

22. False Volcanic ash is a hazard and it can shut down engine function, but so far no planes have crashed.

23. True

24. False A tsunami is a giant wave caused by an explosive volcanic eruption in the sea (or by ocean-floor earthquakes, as you'll learn in the next chapter); it has nothing to do with tides.

25. False Just the reverse is true.

26. True Since we can't prevent volcanic eruptions, the most effective precaution is to compile larger assessment maps that show the probable paths of lava flows, lahars, debris flows, and pyroclastic flows.

27. False They've been successfully diverted on several documented occasions.

28. True Ash and gas may circle the planet and stay suspended in the stratosphere for months, lowering the average global temperature by as much as 6°C.

29. False Active volcanoes have been confirmed only on Io; Venus and Mars have volcanic features but no current activity has been observed.

30.
a. B
b. B
c. R
d. B
e. R
f. B
g. B
h. R
i. B
j. B

CHAPTER 10 | A Violent Pulse: Earthquakes

GUIDE TO READING

The topic of this chapter, earthquakes, is often a favorite of readers. After all, quakes are dramatic events, potentially dangerous, and many people live in or visit areas where a quake might happen at any moment. Those who've experienced major quakes aren't always so fond of them. Many testify it's profoundly disturbing when the good old solid dependable Earth shakes and crashes around you.

Nevertheless, with such widespread interest, everybody kind of knows what an earthquake is. But there's more to defining an earthquake than most people realize, and the chapter begins with more to say about the essentials (the what, why, how, where, and when of quakes) than you'd expect. There's seismicity, focus, epicenter, foreshocks, aftershocks, and stick-slip behavior. Most significant quakes are caused by movement along faults, so you'll read about hanging walls, foot walls, normal faults, reverse faults, thrust faults, strike-slip faults, displacement, active faults, inactive faults, blind faults, and fault scarps and traces. Faults happen because rock is stressed, so stresses (compressive, tensile, and shear) and the strains and deformation they produce (elastic, brittle, and ductile) are examined.

Keeping track of seismic activity is a worldwide concern. You read about the instrument that detects and records quakes (the seismograph) and how to interpret the seismograms it produces (using arrival times and travel-time curves). Earthquakes come in all sizes, from too small to be detected by humans to real monsters. This chapter deals with three commonly used scales that rate quake size more precisely than this, the Mercalli, Richter, and seismic-moment magnitude scales).

When the question arises "Why do quakes occur where they do," guess what the answer is? By now you shouldn't be surprised: plate tectonics. The information in this section isn't new; it's just presented in a way to point out the seismic connections.

The last part of the chapter deals with how quakes affect society. Types of damage from earthquakes vary. Naturally there's ground shaking and displacement, and depending on the location, there can also be seiches, landslides, avalanches, liquefaction, fire, tsunamis, and even widespread disease.

What can people do about earthquakes? We can't stop them. We've had limited success in predicting them (short-term and long-term prediction, recurrence intervals, and seismic gaps). Earthquake zoning and engineering seem to be the best ways to protect human life and property. The chapter ends with a discussion of what society as a whole can do and what you as an individual can do to protect against quake dangers. It also reminds us that no matter what we do, plate tectonics will continue to shift the world and earthquakes will continue to shake it.

Completion

Test your recall of new vocabulary terms.

_____ 1. shaking of the Earth produced by the rapid release of energy, usually because of a sudden slip along a fault

_____ 2. geologists who study earthquakes

_____ 3. a fracture within the Earth along which sliding has occurred

_____ 4. the place within the Earth where an earthquake originates, usually because rock has ruptured and slipped, and from which earthquake energy radiates outward in all directions

_____ 5. the point on the surface of the Earth that is directly above an earthquake's focus

_____ 6. the mass of rock which sits above a sloping fault plane, from which, if you could stand on the fault plane, you'd hang a light

_____ 7. the mass of rock which is below a sloping fault plane, on which, if you could stand on the fault plane, you'd place your feet

_____ 8. faults in which the hanging wall has apparently slipped down along the slope of the fault plane (Even if the footwall did the actual moving, classification is based on what the hanging wall appears to have done in relation to the footwall.)

_____ 9. faults in which the hanging wall has apparently moved upward along a steeply sloping fault plane (Even if the footwall did the actual moving, classification is based on what the hanging wall appears to have done in relation to the footwall.)

_____ 10. faults in which the hanging wall has apparently moved upward along a gently sloping fault plane (Even if the footwall did the actual moving, classification is based on what the hanging wall appears to have done in relation to the footwall.)

_____ 11. faults in which one side has moved relative to the other along an imaginary line that shows the direction the fault trends across Earth's surface (the strike line) and for which there's no vertical motion, just horizontal motion along the strike line

_____ 12. the distance between the two ends of a rock layer or feature that has been cut across and offset due to movement along a fault, which is a measure of the amount of slippage that has occurred

_____ 13. faults that have moved recently or are likely to move in the near future

_____ 14. faults that can generate earthquakes

_____ 15. faults that have no history of recent movement and are unlikely to move again in the near future

_____ 16. the intersection of a fault plane and the ground surface

_____ 17. the little cliff or step formed when one side of a normal or reverse fault moves in respect to the other side

_____ 18. faults that don't intersect the ground surface so are usually unknown until there's movement along them

_____ 19. the push, pull, or shear on a material because of force applied to it

_____ 20. the change in shape of a material in response to an applied stress

_____ 21. a change in the shape of an object that disappears when the stress that caused it is removed

_____ 22. the resistance to sliding on a surface

_____ 23. the typical type of movement along a fault; not constant but sporadic stop-and-go, as stress builds up enough to cause sliding and friction quickly stops the sliding

_____ 24. smaller earthquakes that precede a major earthquake

_____ 25. smaller earthquakes that occur for two or three days after a major earthquake

_____ 26. cracking and fracturing of a material in response to applied stress

_____ 27. bending and slow flowing of a material in response to applied stress

_____ 28. very slow, steady movement along a fault that does not result in earthquakes

_____ 29. the form in which earthquake energy travels away from the quake focus

_____ 30. seismic waves of two varieties, P-waves and S-waves, that travel through the interior of the earth

_____ 31. seismic waves of two varieties, R-waves and L-waves, that travel along the ground surface

_____ 32. waves in which particles of material move back and forth parallel to the direction in which the wave itself moves

_____ 33. waves in which particles of material move back and forth perpendicularly to the direction in which the wave itself moves

_____ 34. an instrument that detects and records ground motion caused by seismic waves (Two versions of this instrument exist, which detect motion in different planes.)

_____ 35. a basic law of motion: The tendency of an object at rest to remain at rest, and an object in motion to remain in motion.

_____ 36. the paper record made of an earthquake by a seismograph

_____ 37. the instant at which the first earthquake wave appears on a seismogram

_____ 38. a network of over 2,000 seismograph stations that supply real-time data to seismologists worldwide

_____ 39. a type of graph that uses the time delay between the arrival times of P- and S-waves to determine the epicenter location

_____ 40. a procedure that plots distances from three different seismograph stations to an earthquake epicenter to graphically determine the epicenter location

_____ 41. a scale to assess earthquake intensity in terms of the amount of damage it causes

_____ 42. a scale to express the strength of an earthquake in terms of the largest ground motion it causes, as represented on a seismogram

_____ 43. the movement of Earth's surface due to the passage of seismic waves through it

_____ 44. a scale to quantify the strength of an earthquake by considering four factors: the amount of slip, length of rupture, depth of rupture, and rock strength

_____ 45. fairly long, narrow regions on Earth's surface where almost all earthquakes occur

_____ 46. earthquakes that occur along plate boundaries because of plate interactions there

_____ 47. earthquakes that have no physical or causal association with plate boundaries

_____ 48. all earthquakes whose foci lie within 20 km of Earth's surface

_____ 49. all earthquakes whose foci are between 20 and 300 km deep in Earth

_____ 50. all earthquakes whose foci are between 300 and 670 km deep in Earth

_____ 51. the plane which marks the interface of a subducting plate with the overlying plate and which appears as a downward-sloping band of seismicity

_____ 52. ridges of sediment created by glacier movement and deposited along glacial edges when ice melts

_____ 53. seismic events caused by human activity

_____ 54. water waves sloshing back and forth in a confined area such as a lake or pool, the motion of which started because of ground shaking

_____ 55. tumbling, flowing, slipping, and sliding of unconsolidated earth materials down slopes

_____ 56. the process in which ground shaking turns areas of wet, clay-rich, reasonably solid sediment into a jello-like mass or slurry of clay and water

_____ 57. small hills of sand created when subsurface sand layers liquefy during earthquakes and erupt through surface holes or cracks

_____ 58. giant ocean waves generated by sea-floor surface displacement due to earthquakes or by submarine landslides or volcanic explosions

_____ 59. predictions that an earthquake will happen at a locality within a few decades or a few centuries

_____ 60. predictions that an earthquake will happen at a locality within a few hours or weeks

_____ 61. studies of geographic localities that lead to earthquake predictions for the region

_____ 62. the average time between successive earthquake events at some specific location

_____ 63. areas along an active fault where obvious movement and/or earthquakes should be occurring but aren't

_____ 64. the design of buildings to withstand the anticipated shaking produced by earthquakes

_____ 65. the procedure of determining areas that are particularly vulnerable to earthquake-caused collapse and passing sensible building codes based on this knowledge

Matching Questions

STRESSED ROCKS

Fill in the blanks from the choices offered.

a. brittle
b. ductile
c. earthquake
d. elastic
e. fault
f. no

_____ 1. At first there is deformation in the form of a slight lengthening of the rock in one direction. The stress stops, and the rock resumes its original shape. What type of strain occurred?

___ 2. The same stress begins again, and this time continues and causes small cracks to appear in the rock. What type of deformation is this?

___ 3. The small cracks begin to interconnect and eventually become a fracture, sliding occurs along the fracture, and vibrations are generated. What type of structure has been created?

___ 4. What event has occurred?

___ 5. Some types of rock, like gypsum, would not break under the above circumstances but would instead slowly flow and assume a permanent change of shape. What type of deformation would this be?

___ 6. Is an earthquake likely to happen in connection with the above deformation?

SEISMIC WAVES

Answer questions 1 to 9 with the correct seismic wave or waves.

L—Love   P—primary   R—Rayleigh   S—secondary

___ 1. Which can be used to investigate Earth's interior?
___ 2. Which cause most surface damage?
___ 3. Which travels fastest?
___ 4. Which travels slowest?
___ 5. Which typically shows the largest amplitude on a siesmogram?
___ 6. Which is a compressional body wave?
___ 7. Which is a shear body wave?
___ 8. Which causes the surface to ripple vertically, up and down?
___ 9. Which causes the surface to ripple horizontally, back and forth?

MEASURING EARTHQUAKE MAGNITUDES

Label each descriptive phrase with the appropriate scale.

M—Mercalli intensity scale   R—Richter magnitude scale
S—seismic-moment magnitude scale

___ 1. is based on the amplitude of the largest ground motion as recorded on a seismograph.

___ 2. is giving the "big picture"—how much energy was released by the entire event, not just how large it was at its point of maximum strength.

___ 3. is based on the subjective assessment of human reactions and on the amount of damage to man-made structures.

___ 4. does not involve the use of a scientific instrument to apply.

___ 5. is a logarithmic scale.
___ 6. would have an m = 5 reading.
___ 7. has ratings that are always written as Roman numerals.
___ 8. has never had an earthquake rating greater than 8.9.
___ 9. requires all seismogram recordings to be adjusted (by chart reference) to equate to what the numbers would have been if the seismograph-epicenter distance were 100 km.
___ 10. equals the amount of slip × length of rupture × depth of rupture × rock strength.
___ 11. would have a reading of 9.3.
___ 12. has numerical values that vary inversely with distance from the quake.

Short-Answer Questions

Answer with a few words, numbers, or short sentences, or choose from the words in parentheses.

1. a. How many detectable earthquakes happen annually? _____
   b. How many of these cause damage and human casualties? _____

2. List eight different possible causes of earthquakes.
   _____
   _____
   _____

3. The majority of earthquake foci lie on _____ planes because the majority of quakes are caused by movement along _____.

4. a. For a single earthquake event, displacement may range from a minimum of _____ centimeters to a maximum of about _____ m.
   b. Cumulative movement along large faults may be as great as _____ km.

5. How can earthquake faulting be responsible for creating huge mountains if the maximum vertical displacement for each quake ranges from only centimeters to meters? _____
   _____

6. a. If P-waves in an area travel at 8 km/s, what would be a logical speed for S-waves in that area? _____
   b. What would be a logical speed for the surface waves of that area? _____

7. Any seismograph operates on which basic physics principle? _____

8. Which part of a seismograph remains fixed in space, unmoving, during a quake, the frame of the instrument with a recording sheet attached, or the recording device (pen or laser beam)? _____

9. a. The New Madrid Fault, in the Missouri-Tennessee-Kentucky area, caused huge earthquakes in 1811–12. Today seismographs there record dozens of quakes annually, but usually they're too small to be felt by residents. Should the fault be classified as active or inactive? _____
   b. The New Madrid Fault isn't associated with any current plate boundary, and so quakes there should be called _____ earthquakes.

10. How great a magnitude must an earthquake have before it can be recorded globally? _____

11. Give the range of numbers on the Richter scale that would fit each of the following earthquake categories:
    a. great   _____
    b. major   _____
    c. strong   _____
    d. moderate   _____
    e. light   _____
    f. minor   _____

12. An earthquake of 7.4 Richter releases how much more energy than a quake of 4.4 Richter? _____

13. Why do shallow-focus earthquakes cause so much more damage than intermediate- and deep-focus earthquakes do? _____

14. Theoretically earthquakes should not occur deeper than 300 km, even on cold subducting ocean plate, yet earthquake foci are recorded to depths of 670 km on such plates. What is a possible explanation? _____

15. Fault traces appear as zones of broken-up rock on Earth's surface; this indicates they are a weaker surface than the rest of the surroundings. Because of this, what is the topography (ups and downs of the land surface) likely to be along the fault? _____

16. What is the effect of earthquake shaking on a healthy human body? (I.e., how badly can just the shaking hurt a person?) _____

17. List some quake-related phenomena that can put your life in danger. _____

18. Why bother to predict earthquakes, since the best we can do with any reliability is to say there's a certain probability that an area will have a quake within a designated period of time (usually a few decades)? _____

19. Scientists agree that earthquakes are not going to cause California to fall into the ocean, but what is a valid earthquake-related danger to residents of the California coast? _____

20. Are the terms "tsunami" and "tidal wave" synonyms? Why or why not? _____

21. Parkfield, California, had earthquakes of magnitude 5–6 range (Richter reading) in 1857, 1881, 1901, 1922, 1934, and 1966. On the basis of this information the U.S. Geological Survey has stated that the recurrence interval is 20–22 years.
    a. Which year is an anomaly in this cycle? _____
    b. Which year would have fit the sequence better? (By the way, so far the latest predicted 5–6 magnitude quake has not occurred.) _____

PLATE TECTONICS

Earthquakes and Plate Boundaries

Transform Faults

1. Most transform faults are found in oceanic lithosphere where they link _____.

2. Motion along these faults is usually _____ (normal, reverse, thrust, strike-slip).

3. Two transform faults that cut across continental lithosphere, one in California and one in New Zealand, are _____.

4. All earthquakes along transform faults are _____-focus quakes.

Mid-ocean Ridges

1. Mid-ocean ridges are what type of plate boundary? _____

2. Of the four possible types of fault motion (normal, reverse, thrust, or strike-slip), name the type that may develop because of the following circumstances:
   a. horizontal motion along the transform faults _____
   b. stretching across the ridge _____

3. All earthquakes associated with mid-ocean ridges are _____-focus quakes.

Convergent Plate Boundaries

1. _____-focus quakes are located in both the subducting and overriding plates.
2. The subducting plate may also generate _____-focus quakes.
3. The _____ stress between the overriding and subducting plate and the associated _____ type (normal, reverse, thrust, or strike-slip) of fault motion are responsible for the truly great (greater than 8 Richter) quakes.
4. Four geographic locations where major convergent boundary earthquakes have occurred are _____ _____.

Continental Rifting

1. Earthquakes that happen in this type of region are always _____-focus quakes.
2. Three geographic locations where rifting is occurring today are _____ _____.

Collision Zones

1. Crustal compression in these places produces _____ (normal, thrust, reverse, or strike-slip) faults.
2. The Lisbon, Portugal, quake of 1755 happened because which two continents were colliding? _____

Intraplate Quakes

1. What percentage of annual earthquake energy release occurs here? _____
2. These are all _____-focus quakes.
3. What is significant about the age of the faults associated with intraplate quakes? _____
4. Name five areas of North America where significant intraplate quakes have occurred. _____ _____

SECONDARY EFFECTS OF EARTHQUAKES

The secondary effects of a large earthquake are often more devastating than the quake itself. The following accounts of some aspects of significant earthquakes emphasize this point. The questions asked don't require profound scientific thought, just a little bit of textbook science combined with common sense and some imagination.

1. Downtown San Francisco is built on bay fill (basically wet clay). The downtown area of San Francisco was more heavily damaged than surrounding areas by shaking in 1906, and the Marina District (near downtown San Francisco) had numerous buildings collapse in the Loma Prieta earthquake of 1989. In neither quake was the epicenter in downtown San Francisco. Why was there such heavy building damage in the downtown area? (This is not referring to the fires; that's another issue.)

2. In 1906 the main water supply for San Francisco was the San Andreas Lake, which is located in a valley. The fires burned for three days. Why weren't they put out sooner?

3. The epicenter of the great Alaska quake of 1964 was on the bottom of Prince William Sound, a large body of water not too far from Anchorage. The gas and electric systems of Anchorage, Alaska, are constructed so as to shut down when a certain level of shaking is reached. Anchorage did not burn after the quake. On the other hand, much of Valdez, Alaska, which sits at the southern end of the trans-Alaska oil pipeline at the tip of a narrow harbor on Prince William Sound, was destroyed by water and fire. Why was there such a difference in destruction between Anchorage and Valdez?

4. The great quake in Tokyo in 1923 hit just before noon. Most residents cooked over open fires. The city was constructed chiefly of wood and paper. The Imperial Hotel, designed by the U.S. architect Frank Lloyd Wright to be earthquake proof, had a large water-lily pond in front of it. What caused the most destruction after the quake? Do you think the hotel survived?

5. A 6.8 quake in an Armenian city in 1988 killed more than 20,000 people, yet the 7.1 Loma Prieta (California) quake of 1989 killed fewer than 70. There were no fires or floods or other special circumstances involved in either; there were in both just cities with buildings hit by quakes. How could the smaller quake have killed so many more people?

6. Much of the construction in poorer areas of China is done using loess, which is wind-blown silt material that is made into a mud-brick building material. It is not reinforced in any way. Death tolls in China for even moderate quakes are usually very high. Explain.

7. Turn-of-the-(19th)century buildings in southern California often had brick awnings over doorways. Was this a good idea?

8. You probably haven't heard of the Landers Quake of 1992 in the Mojave Desert of California, yet it was a major one (7.6), larger than either the Loma Prieta or the Northridge quakes. Why isn't it well known?

Making Use of Diagrams

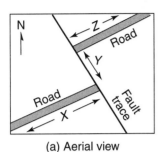
(a) Aerial view

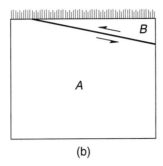
(b)

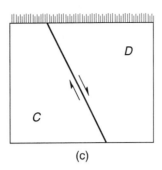
(c)

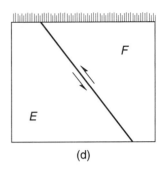
(d)

TYPES OF FAULTS

Referring to the diagram above, answer by number or letter.

_____ 1. Which diagram illustrates a normal fault?

_____ 2. Which diagram illustrates a reverse fault?

_____ 3. Which diagram illustrates a thrust fault?

_____ 4. Which diagram illustrates a strike-slip fault?

_____ 5. Which diagram(s) illustrate(s) squeezing and shortening of the crust?

_____ 6. Which diagram(s) illustrate(s) pulling apart and stretching of the crust?

_____ 7. Which letters show the hanging walls?

_____ 8. Which letters show the foot walls?

_____ 9. What letter labels displacement along a fault?

_____ 10. What is the strike of the strike-slip fault shown? (N, S, E, W, NW, NE)

2. On the diagram below, basement rock of Precambrian age sits at the surface, right next to layers of Mesozoic age sedimentary rocks. Which type of fault is it? _____ (normal, reverse, thrust, or strike-slip)

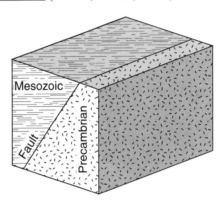

Briefly explain how you decided this. _____
_____
_____
_____
_____

MORE FAULTS—BLOCK DIAGRAMS

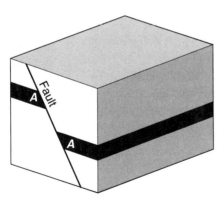

1. On the diagram above, bed A began as a continuous horizontal layer. Faulting occurred to produce the results shown. Which type of fault is it? _____ (normal, reverse, thrust, or strike-slip)

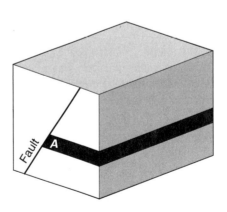

3. Draw bed A in an appropriate position on the hanging wall above to show this is a reverse fault.

A (Very) Simple Self-Demonstration about Time

The average significant earthquake lasts about 1 min. How long is a minute? That depends. A minute of ecstasy is way too short; a minute of terror lasts for an infinity. Experiment with circumstances in between these two extremes, and you decide how long a minute of serious earthquake shaking really is. Do something that will make you mildly uncomfortable; don't breathe for 1 min.

Could you do it? How uncomfortable did you get? Imagine the world shaking and collapsing around you during that minute and think how uncomfortable *that* would make you. Also think of this: sometimes people are lucky, and major quakes are shorter than the average 1 min. The Loma Prieta quake (near San Francisco) in 1989 lasted only 15 s. But sometimes people are not so lucky; the 1964 quake in Alaska lasted 3 to 4 mins, and the Tokyo quake of 1923 lasted for 5 mins. So geologically speaking earthquakes are extremely short events, but people caught in them sometimes feel they last forever.

Short-Essay Questions

1. The Northridge earthquake of January 17, 1994, was caused by movement along a blind fault. Briefly discuss how this might have influenced the degree of preparedness of the city.

2. Discuss briefly the two fundamental factors that must be overcome before movement along a fault can cause an earthquake.

3. a. Does a seismic gap area pose a very high or a very low seismic risk? Briefly explain.
   b. Does a fault creep area pose a very high or a very low seismic risk? Briefly explain.

4. Why do aftershocks happen? How large are they? Can they do much damage? How long do they persist after the main shock?

5. Earthquake faulting can happen only in the upper 15–20 km of continental crust, but it can happen all the way down into the mantle part of oceanic lithosphere. Why is there such a difference?

6. What do seismographs have to do with keeping the peace in the atomic age?

7. Review the information in the text about the Denver earthquakes (pages 292–93). Use this information to present a logical explanation of why there was an unprecedentedly strong earthquake (6.5 Richter) in west-central India in 1967 after water had been impounded in a reservoir there in 1964.

8. Explain why earthquakes in South America occur in the following pattern: a strip of shallow-focus quakes parallels and is close to the west coast; a strip of intermediate-focus quakes lies next to this first strip, but farther inland; and a third strip of deep-focus quakes parallels the first two strips and is farthest inland. Sketch a cross section of the Pacific Ocean and west coast of South America to illustrate your answer. On your sketch show depths in kilometers, and label the Wadati-Benioff zone.

9. Explain how a killer tsunami generated by a quake in Alaska can travel unnoticed across the Pacific Ocean and yet kill people hours later in Hawaii.

10. a. Discuss some lines of evidence geologists look for in the field to determine if there has been recent faulting in an area.
    b. Discuss some geologic field evidence geologists use, in addition to the relatively short historic record, to figure the recurrence interval of large earthquakes in an area.

11. What surface changes do geologists look for in an area to supply clues to whether an earthquake is imminent there? (Unfortunately these have *not* made short-term prediction reliable.)

12 Since reliable short-term earthquake prediction is currently not possible, earthquake preparedness seems a more reasonable approach to protecting human life and property. Discuss factors that influence how much damage an earthquake does and what can be done, by society and by individuals, to minimize such damage.

Practice Test

1. An earthquake with a 7.2 Richter scale reading
   a. would be so large that a quake this size would occur only once a century.
   b. would have surface waves 10,000 times the amplitude of a 4.2 quake.
   c. would have almost 36,000 times the energy release of a 4.2 quake.
   d. is one of the largest possible and is classed as a great quake.
   e. would logically have a reading of about XII on the Mercalli scale.

2. Which statement is TRUE?
   a. P, S, L and R are all body waves so they're all used to study Earth's interior.
   b. The focus is the point on Earth's surface directly above the epicenter.
   c. Underground nuclear explosions create seismic waves that can be differentiated from natural seismic waves.
   d. Water leakage from reservoirs cannot trigger quakes.
   e. S-waves travel twice as fast as P-waves.

3. A tsunami
   a. is a special kind of tidal wave caused by the gravitational attraction of the Sun, not the Moon.
   b. can get big, but never bigger than 30 ft high.
   c. is dangerous near its source but dies out within about 200 mi.
   d. may be just a broad, gentle swelling out at sea but grows as it approaches shore.
   e. All of the above are true.

4. This diagram

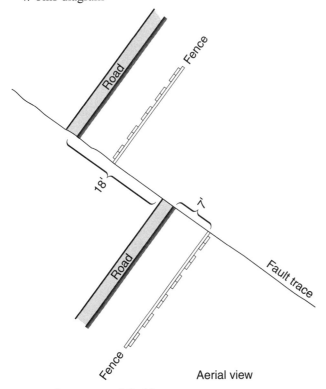

Aerial view

   a. shows normal faulting.
   b. shows a displacement of 18 ft.
   c. shows a recurrence interval of 7 ft.
   d. illustrates how fault scarps are formed.
   e. All of the above are true.

5. If you live along a fault where creep is occurring,
   a. you are very likely to experience a severe quake in the near future.
   b. you could not be in California, because there's no creep along the San Andreas Fault.
   c. you must be living along the San Andreas Fault; it's the only place creep occurs.
   d. fault scarps will build up slowly, and you'll have to trim them to prevent slippage.
   e. you'll likely be bothered by broken foundations, but you won't experience a large quake.

6. Which statement is TRUE?
   a. The seismic-moment magnitude scale has made the Richter scale obsolete; Richter readings are no longer used.
   b. The seismic-moment magnitude scale takes into account the area size and characteristics of rock affected.
   c. All types of building construction are equally vulnerable to earthquake damage.
   d. Earthquakes never have foci deeper than about 100 km (60 mi).
   e. Earthquakes can be prevented by injecting sticky fluid along existing fault planes.

7. Earthquake prediction is not highly reliable, but geologists do know
   a. quakes will never happen in seismic gaps.
   b. short-term predictions usually do turn out to be correct.
   c. northern California, not southern California, is the area most likely to have the "Big One."
   d. more earthquakes happen along plate boundaries than happen at intraplate locations.
   e. All of the above are true.

8. Which of the following statements is FALSE? Earthquakes in Denver that occurred in a series
   a. were triggered by the injection of waste fluid down a deep well.
   b. happened because frictional resistance along existing faults was lowered by fluid injected in a well.
   c. were termed seiche earthquakes, because their magnitudes oscillated regularly.
   d. demonstrated that people can sometimes influence seismicity.
   e. basically resulted from changes in groundwater pressure.

9. Which statement is TRUE?
   a. Earthquakes in California are the result of reverse faulting along the San Andreas Fault.
   b. Earthquakes in California are the result of widening along the San Andreas Fault which will eventually cause western California to sink into the ocean.
   c. Earthquakes in California are always above 6.5 Richter because the San Andreas is such a large fault.
   d. Blind faults that don't intersect the surface can cause earthquakes in unexpected places.
   e. P-waves are the most destructive of seismic waves because they travel fastest.

10. If you read that an earthquake was rated 9.4, you know
    a. this is a Richter scale reading, because that's the only scale that goes this high.
    b. this is a Mercalli reading because Richter and seismic-moment magnitude readings always use Roman numerals.
    c. this is a seismic-moment magnitude reading, and it was an extremely strong quake.
    d. the quake must have happened in northern California because that area has stronger quakes than anywhere else.
    e. it must be a typographical error; there is no scale that rates quakes that high.

11. Which statement is TRUE? Liquefaction
    a. can cause clay-rich sediment to turn into an unstable slurry of clay and water.
    b. is the sudden loss of strength of some soils that happens because of earthquake shaking.
    c. caused great damage in the Alaska quake of 1964 and the Loma Prieta quake of 1989.
    d. can affect sand layers below ground surface and cause them to erupt as sand volcanoes.
    e. All of the above are true.

12. On the diagram shown,

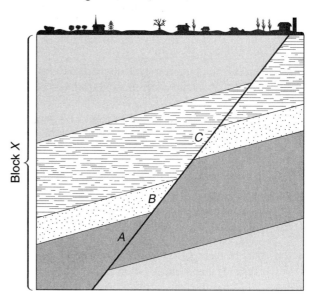

   a. Block *X* is the footwall.
   b. Block *X* is the hanging wall.
   c. there's no way to judge which is the hanging wall.
   d. the displacement of layer *B* shows this is a thrust fault.
   e. the displacement of layer *B* shows this is a strike-slip fault.

13. Which statement is TRUE?
    a. R- and L-waves are surface seismic waves that do the most damage.
    b. S-waves are compressional body waves.
    c. Surface waves are the first to show up on a seismogram recording of a quake.
    d. P-waves are shear body waves.
    e. All of the above are true.

14. The locations of major earthquakes and volcanic eruptions
    a. are not similar to each other.
    b. are usually along the boundaries of crustal plates.
    c. are usually toward the centers of crustal plates.
    d. always mark places where plates are diverging, never where they're converging.
    e. always mark places where plates are converging, never where they're diverging.

15. The reading XIX is logical for a medium-sized quake on the Mercalli scale, which runs from X to XXX.
    a. True
    b. False

16. Plotting the foci of earthquakes, showing their progression from shallow to intermediate to deep as you move eastward across South America, is really drawing the profile of a subducting ocean plate.
    a. True
    b. False

17. Induced seismicity can happen in an area when a reservoir is built there, leaks, and introduces water into existing fault planes.
    a. True
    b. False

18. It's true tsunamis are large waves, but none bigger than 10 m has ever been recorded.
    a. True
    b. False

19. Parkfield, California, has been monitored closely because it seems to have a recurrence interval of 20 to 22 years; in 1993 geologists were delighted to see the latest quake occur right on schedule.
    a. True
    b. False

20. Roughly 80% of the earthquake energy released on Earth comes in the continental collision zone where the Himalayas are still growing; the remaining 20% is scattered at random locations worldwide.
    a. True
    b. False

21. Although the risk is small, disastrous earthquakes can happen in regions that are not seismic zones.
    a. True
    b. False

22. It takes less energy to activate an old fault than to create a comparably sized new one, so old faults must still be treated as areas of weakness vulnerable to earthquakes.
    a. True
    b. False

23. If a rock undergoes enough stress to produce elastic strain, an earthquake always happens.
    a. True
    b. False

24. Friction occurs between sliding surfaces because no surface can be perfectly smooth.
    a. True
    b. False

25. Earthquake faulting occurs to a greater depth in ocean crust than in continental crust because cold ocean crust needs to subduct to warm up enough to flow rather than break and quake.
    a. True
    b. False

26. A seismograph is an earthquake recording instrument that operates on the principle of friction.
    a. True
    b. False

27. The New Madrid, Missouri, quakes of 1811–12 and the Charleston, South Carolina, quake of 1886 were both large intraplate quakes.
    a. True
    b. False

28. Deep-focus quakes occur in the Wadati-Benioff zones of divergent plate boundaries.
    a. True
    b. False

29. Interpret the travel-time curve shown.

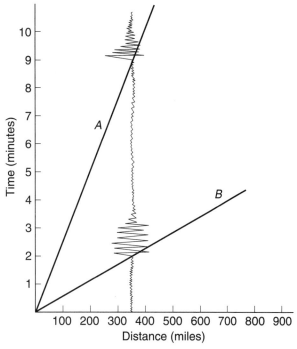

P-wave arrived at 4:18 P.M.
S-wave arrived at 4:25 P.M.

a. How many minutes between the arrival of the first P-wave and the arrival of the first S-wave? _____

b. Which line on the graph represents the S-wave, upper line *A* or lower line *B*? _____

c. How far away is the earthquake epicenter? _____ mi

d. What time did the earthquake occur? _____

e. Mark each statement as true or false.

In order to determine the epicenter of the quake, after getting the answers above a seismologist would:

i. Plot a circle on a map, using his location as the center and the epicenter distance as the radius. _____

ii. Need similar data from a minimum of four other stations. _____

iii. Need data about the L-waves before he could determine the epicenter. _____

ANSWERS

Completion

1. earthquake
2. seismologists
3. fault
4. focus (or hypocenter)
5. epicenter
6. hanging wall
7. footwall
8. normal faults
9. reverse faults
10. thrust faults
11. strike-slip faults
12. displacement
13. active faults
14. earthquake faults
15. inactive faults
16. fault trace (or fault line)
17. fault scarp
18. blind faults
19. stress
20. strain
21. elastic strain
22. friction
23. stick-slip behavior
24. foreshocks
25. aftershocks
26. brittle deformation
27. ductile deformation
28. fault creep
29. seismic waves (or earthquake waves)
30. body waves
31. surface waves
32. compressional waves
33. shear waves
34. seismograph (or seismometer), two varieties: vertical and horizontal motion
35. inertia
36. seismogram
37. arrival time
38. worldwide seismic network
39. travel-time curve
40. triangulation
41. Mercalli intensity scale
42. Richter magnitude scale
43. ground motion
44. seismic-moment magnitude scale
45. seismic belts (or seismic zones)
46. plate-boundary earthquakes
47. intraplate earthquakes
48. shallow-focus earthquakes
49. intermediate-focus earthquakes
50. deep-focus earthquakes
51. Wadati-Benioff zone
52. moraines
53. induced seismicity
54. seiche
55. landslides
56. liquefaction
57. sand volcanoes (or sand blows)
58. tsunamis
59. long-term predictions
60. short-term predictions
61. seismic-risk (or seismic-hazard) assessment
62. recurrence interval
63. seismic gaps
64. earthquake engineering
65. earthquake zoning

Matching Questions

STRESSED ROCKS

1. d
2. a
3. e
4. c
5. b
6. f

SEISMIC WAVES

1. P and S
2. R and L
3. P
4. R and L
5. first surface waves to arrive, either R or L
6. P
7. S
8. R
9. L

MEASURING EARTHQUAKE MAGNITUDES

1. R
2. S
3. M
4. M
5. R
6. R
7. M
8. R
9. R
10. S
11. S
12. M

Short-Answer Questions

1. a. about 1 million   b. a few hundred
2. formation of a new fault and movement along it
   movement along an existing fault
   sudden change in the arrangement of minerals in an atom
   cracking of a volcano as it fills with magma
   volcanic eruption
   large landslide
   meteor impact
   underground nuclear explosion
3. fault; faults
4. a. a few; 10   b. hundreds or even greater than 1,000
5. The cumulative effect can result in a displacement of thousands of kilometers, and Earth has had a lot of time to get things done.
6. a. 8 km/s $\times$ 60% = 4.8 km/s
   b. 4.8 km/s $\times$ 55% = 2.6 km/s
7. inertia
8. the recording device (pen or laser beam)
9. a. active   b. intraplate
10. greater than 3 or 4 on the Richter scale
11. a. 8 and larger   b. 7.0–7.9   c. 6.0–6.9
    d. 5.0–5.9   e. 4.0–4.9   f. less than 3.9
12. From 4.4 to 7.4 is three steps on the logarithmic Richter scale, so 33 times the difference between each full number difference or 33 $\times$ 33 $\times$ 33 = 35,937 times more energy.
13. Because they're so close to the surface, they don't lose much of their energy getting to it as would be true of deeper-focus earthquakes.
14. the sudden collapse of minerals as a result of pressure, subsequent rearrangement into new minerals that take up less space, and the resulting shock waves generated during shape change
15. a deep valley
16. The shaking may knock you down and increase the force of gravity on your body, but it's highly unlikely to kill you or even break any bones. The mental effect may be worse; it can be terrifying and disorienting to have the dependable solid Earth shaking and collapsing around you.
17. flying debris, collapsing walls and roofs and bridges, and disappearing roadways
18. Urban planners and civil engineers can create sensible building codes for the region and decide whether to build especially vulnerable structures like nuclear power plants, hospitals, or dams in areas prone to earthquakes.
19. landslides in the steep cliff areas of the California coast
20. No, true tidal waves are much smaller and are the result of changing tides, caused mainly by the attraction of the Moon's gravity; tsunamis are giant sea waves caused by submarine earthquakes, landslides, or volcanic eruptions.
21. a. 1934   b. 1944

PLATE TECTONICS

Earthquakes and Plate Boundaries

Transform Faults

1. segments of ocean ridges
2. strike-slip

3. the San Andreas Fault and the Alpine Fault

4. shallow

Mid-ocean Ridges

1. divergent

2. a. strike-slip   b. normal

3. shallow

Convergent Plate Boundaries

1. shallow

2. both intermediate- and deep

3. shear; thrust

4. southern Alaska, Japan, the west coast of South America, and Mexico

Continental Rifting

1. shallow

2. the East African Rift, Basin and Range Province (Nevada, Utah, and Arizona), and the Rio Grande Rift (New Mexico)

Collision Zones

1. thrust

2. Africa and Europe

Intraplate Quakes

1. about 5%

2. shallow

3. They're ancient fault zones, some as old as Precambrian.

4. New Madrid, Missouri; Charleston, South Carolina; eastern Tennessee; Montreal, Quebec; Adirondack Mountains, New York

SECONDARY EFFECTS OF EARTHQUAKES

1. The bay fill downtown is extremely susceptible to liquefaction. When shaking starts, buildings there act as if they're sitting on jello.

2. The San Andreas Lake is located in a valley sitting right on the San Andreas Fault. Displacement along the fault in 1906 was more than 20 ft. Water pipes coming from the lake were broken as a result of this displacement, so there was no water available to put out the fires.

3. With gas and electricity shut off automatically by the shaking, the odds of fire were greatly reduced. Anchorage had other problems, but it did not burn. Valdez fell victim to both tsunami and fire. The fires started when oil spewed all over from pipes that were delivering oil to ships in the harbor.

4. Much of the city was destroyed by firestorm. Yes, the hotel did survive, in part because it had its own adequate supply of water to put out the fires.

5. Construction in Armenia was shoddy. California enforces building codes written specifically for earthquake zones.

6. The unreinforced loess mudbrick, like unreinforced concrete, experiences total collapse when shaken. People inside such buildings don't stand a chance.

7. Definitely not. Quakes shake them apart, and being right over escape routes (doorways) was certainly not a good idea. As quakes took down the brick awnings, people were wise enough not to rebuild them.

8. The answer lies in the location, the Mojave Desert. There are few people and few structures there to get damaged, so the quake received relatively little publicity.

Making Use of Diagrams

TYPES OF FAULTS

1. c
2. d
3. b
4. a
5. b and d

6. c
7. *B*, *D*, and *F*
8. *A*, *C*, and *E*
9. *Y*
10. NW

MORE FAULTS—BLOCK DIAGRAMS

1. normal

2. normal; The Precambrian basement rock, by age and by definition of the word "basement," belongs under Mesozoic age rock, yet it sits adjacent to it. Therefore the Mesozoic age rock, which composes the hanging wall, moved down in relation to the Precambrian footwall, and this makes it a normal fault.

3.

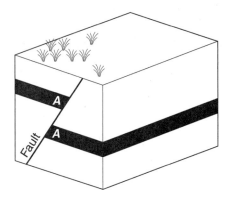

Short-Essay Questions

1. The designation "blind fault" means it doesn't show at the surface, so until the Northridge quake happened the existence of that fault wasn't known. From this aspect, those responsible for public safety were not at fault by ignoring a serious potential hazard. It is common knowledge California has numerous strong earthquakes, and seismologists do know there are many faults in California. Some are famous (like the San Andreas), but many are unknown blind faults. It just makes sense for all parties in California, from individuals through city governments, to build with earthquakes in mind and have emergency procedures in place. If you compare data that show quake severity and death tolls in California with data that show quake severity and death tolls in other parts of the world, it's obvious California deals with its earthquakes better than most.

2. A fault is not in constant motion. Even under stress it usually remains motionless, held in place by *friction*. More specifically, fault plane surfaces are not perfectly smooth, and even tiny bumps on each surface lock into each other and resist sliding. This resistance to sliding, called friction, must be overcome. The other factor that must be overcome is *inertia*, which is the tendency of an object at rest to remain at rest, an object in motion to remain in motion. Even when stress grows strong enough to temporarily overcome both frictional resistance and inertia, and movement occurs, the sliding slows and stops within a few seconds or minutes, defeated again by the frictional resistance.

3. a. Seismic gaps are areas of high seismic risk; earthquakes should be happening there, but they're not. Stresses keep building up and eventually will be strong enough to cause a quake, and the longer the wait, the greater the chance of a strong quake.

   b. An area that experiences fault creep has very low seismic risk. Stresses are not building up but are being released by constant slow movement along the fault. If you've built on the fault trace, you'll have regular repair bills, but you won't have damaging earthquakes.

4. Aftershocks happen because the main shock rearranges the rock and creates new instabilities; the aftershocks are adjustments to them. They are generally much smaller than the original shock and may continue for 2 or 3 days. Although even the largest are usually no more than one-tenth the size of the main shock, they often do damage, because they serve as the "last straw" that brings down already damaged structures.

5. Earth temperatures increase with depth. At a depth of 15–20 km, continental crustal rock is too warm to break and quake (brittle deformation) as it did up higher and instead undergoes ductile deformation (slow flow to produce change of shape.) Ocean plate is colder than continental plate and therefore must reach greater depth before it warms up enough to change from brittle to ductile mode of deformation.

6. Seismographs record underground nuclear explosions as well as natural earthquakes, so countries can monitor each other to see whether everybody is complying with the nuclear test ban. Scientists can differentiate between natural earthquakes and subsurface nuclear explosions because nuclear explosions show a quick pulse of energy release, while natural earthquakes release energy over longer periods of time and have accompanying foreshocks and aftershocks.

7. All reservoirs leak into the surrounding rock; there's no practical way to avoid this. If an area has existing faults, escaped water can infiltrate the fault planes, exert pressure in all directions (a basic physics principle), and thus counteract the resistance of friction holding the surfaces in place. Slippage along the fault may result, producing an earthquake that might not otherwise have happened. All of the above is not just theoretical speculation; the Denver incidents offered conclusive proof that water introduced along fault planes could trigger earthquakes. The circumstances in India certainly seem to fit the picture of induced earthquakes.

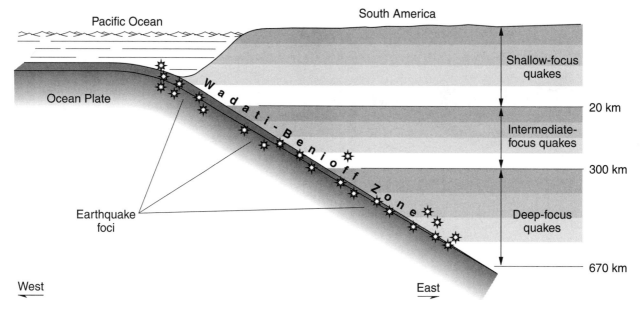

8. The subducting ocean plate gets deeper as it progresses to the east under South America. Earthquake foci are in the interface area between subducting and overriding plates known as the Wadati-Benioff zone. When the ocean plate gets deeper than about 670 km (and "deeper" in this case correlates with "farther east"), the rock no longer undergoes brittle deformation; that is, there are no more earthquake foci.

9. Above the spot on the ocean bottom where a quake and subsequent water displacement has generated a tsunami, the wave may be only a few tens of centimeters high. Ships in the area might not even notice it. This wave can cross the entire ocean basin at speeds comparable to a jet plane, slow (due to friction) as it reaches shallow water, and build up in narrow harbors to heights as great as 70 m.

10. a. To determine if there has been recent faulting in an area, geologists look for young fault scarps, recent sediment layers cut by faults, the offset of stream channels, the creation of small ponds, and the creation of ridges.
    b. To figure recurrence intervals of large quakes in an area, geologists look for layers of sediment produced by sand volcanoes, sequences in the stratigraphic record that show disrupted bedding, and buried soil horizons that have been offset. In order to figure the recurrence interval, they must date these events, and to do this they look for organic matter (such as plant fragments) in the relevant layers and date it by radiocarbon dating.

11. Surface changes geologists look for as clues to imminent earthquake action include:
    a. the bulging or sinking of ground surface, or the bending of straight line features, both of which can be detected by laser survey
    b. changes in water levels in wells, due to changes in the volume of rock pore space
    c. the detection of gas in wells, released from pore spaces as rock cracks open
    d. changes in electric conductivity of the area rock
    e. changes in animal behavior

    Unfortunately short-term earthquake prediction is currently very unreliable, and many geologists question whether it ever will be reliable.

12. Factors that influence how deadly an earthquake will be include the proximity of the epicenter to a population center, the depth of the quake focus, the style of construction in the region, whether the quake has occurred in a region of steep slopes or along the coast, whether buildings were built on solid bedrock or on weak substrate, whether the quake happened when people were inside or outside, and whether public officials provide prompt emergency services.

    In order to minimize death and destruction, wise earthquake zoning and engineering should be practiced. Ways to make buildings less likely to collapse due to the shaking of earthquakes include making them fairly flexible and using supports strong enough to maintain loads far greater than normal circumstances would call for. Bridges can be made stronger by wrapping their columns with steel cables and by bolting bridge spans to the tops of bridge columns. Concrete should be reinforced, steel girders should be used for support in concrete and brick buildings, and brittle roof tiles and loose decorative stone should not be used. Huge open-span roofs should be avoided.

Construction should not be done on land underlain by weak, wet mud, which is vulnerable to liquefaction, or on top of or at the base of steep cliffs. Cities should not be downstream of dams, and buildings should not be built directly on active faults.

Cities should have proper emergency plans. The list goes on and on; unfortunately in today's crowded world it's impossible to locate everyone in earthquake-safe places. Wise individuals who live in earthquake-prone areas should educate themselves on earthquake safety and act accordingly. Briefly this means, if at all possible, get outdoors and stay away from buildings. If you're inside, stand near a wall or in a doorway near the center of the building or crouch under a heavy table. If you're on the road, stay away from bridges.

Practice Test

1. c There's a 3-step difference between a 4.2 and 7.2 quake, so the amplitude difference would be only $10 \times 10 \times 10 = 1,000$, not 10,000, but the energy difference is $33 \times 33 \times 33$, which is almost 36,000; quakes of 7.0–7.9 are major but not great quakes.

2. c L and R are surface waves; the epicenter is above the focus; water leakage can trigger quakes; and S-waves travel at 60% the speed of P-waves.

3. d Tsunamis are caused by volcanic or seismic activity, can get higher than 30 m, and can cross entire ocean basins.

4. b Shows strike-slip faulting; the recurrence interval is the average time between quakes.

5. e Creep releases stresses and avoids quakes.

6. b We can't prevent quakes; we can construct quake-resistant structures; Richter readings remain a valuable way to describe quakes; and foci can occur as deep as 670 km.

7. d Seismic gaps are areas of high seismic risk; accurate short-term predictions are rare; and nobody knows where the "Big One" will strike.

8. c A seiche is an oscillating wave in a confined body of water produced by earthquake shaking; it has nothing to do with this question.

9. d Quakes in California can be large or small and occur along hundreds of faults; motion along the San Andreas Fault is strike-slip, and it's not causing the state to sink; and surface waves (L and R) cause the most damage.

10. c Richter readings go no higher than 8.9; Mercalli readings are written as Roman numerals; and great quakes can occur worldwide.

11. e Statements a to d are all true.

12. b This is a normal fault because the hanging wall, block X, slipped down.

13. a P-waves are compressional body wave that arrive first, followed by S-waves (shear body waves), and then by R- and L(surface)-waves.

14. b Plate boundaries, whether they're converging, diverging, or colliding, are places of dramatic Earth action; some actions happen too slowly to be noticed by humans, but volcanoes and earthquakes are fast and obvious.

15. False The Mercalli scale runs from I to XII; medium would be in the V-VII range.

16. True Earthquake foci lie on the plane between the subducting Pacific Plate and the overriding South American Plate, and this plane reaches deeper as it extends eastward.

17. True Water on the fault planes exerts pressure in all directions and allows slippage along the planes.

18. False Some of the largest tsunamis reach up to 30 m high along ordinary coastlines, and if funneled into narrow bays, have even reached 70 m.

19. False The first part of the question is true, but the latest quake is years overdue.

20. False About 80% of energy is released by the plate boundary quakes of the Pacific Rim; most of the remaining 20% is released in the continental collision area north of Africa and the Indo-Australian plate.

21. True Although disastrous quakes are more common in seismic zones, they can happen anywhere.

22. True Faults are zones of weakness that can be reactivated by relatively little new stress.

23. False   The rock needs to get past this stage of elastic strain, and crack and break (brittle deformation) before a quake can happen.

24. True   There are no perfectly smooth surfaces, so friction always occurs when surfaces move against each other.

25. True   Ocean crust is initially cold and must go deeper than continental crust before it is warm enough to flow rather than break and quake.

26. False   It operates on the principle of inertia.

27. True   Although most quakes occur at plate boundaries, major quakes like the New Madrid quake and the Charleston quake have occurred far from plate boundaries.

28. False   They're convergent, not divergent plate boundaries.

29. a. 7 min
    b. *A*
    c. about 350
    d. 4:16 P.M. The P-wave has been traveling for 2 min, and it arrived at 4:18 P.M.; it's been traveling since it came into existence, at the start of the quake, so 4:18 − 2 min = 4:16, or the S-wave has been traveling for 9 min, arrival time 4:25 − 9 min = 4:16.
    e. i. T   ii. F (Only 3 stations are needed.)   iii. F

CHAPTER 11

Crags, Cracks, and Crumples: Crustal Deformation and Mountain Building

GUIDE TO READING

This chapter, concerning mountains and the geologic reasons they exist, offers a change of pace from the drama and danger of earthquakes and volcanoes in the preceding two chapters. Mountains are certainly not insignificant structures, geologically or aesthetically, but their story is majestic rather than wildly exciting.

Before you begin, you'll find it helpful to review the following terms introduced in Chapter 10:

brittle deformation	reverse faults
compression (compressive stress)	shear stress
ductile deformation	strain
fault	stress
footwall	strike-slip faults
hanging wall	tension (tensile stress)
normal faults	thrust faults

The chapter begins by explaining that with a few rare exceptions (some volcanic mountains that appeared almost overnight), a mountain-building event, an orogeny, goes on for tens of millions of years. An orogeny produces not just uplifted areas of land but also highly deformed rock layers and unique mountain structures. Many of the orogenic processes are reviewed for you. Once again you read about stress (compressional, tensional, and shear), strain, brittle and ductile deformation, joints, folds, and faults (normal, reverse, thrust, strike-slip, footwalls, and hanging walls). You are presented with more details about these topics than before:

- orientation of these geologic structures (strike, dip, bearing, and plunge)
- joint sets
- fault classification (dip-slip, right-lateral, and left-lateral strike-slip)
- details of the fault zone (fault breccia, fault gouge, slickensides, slip lineations, mylonites, and shear zones)
- fault systems (detachment faults, grabens, half grabens, and horsts
- types of folds (drag fold, hinge, limb, axial plane, anticline, syncline, monocline, tight fold, open fold, plunging and nonplunging folds, domes, and basins)
- formation of folds (flex, flow, and buckle)

The very rocks making up an area may be changed by an orogeny. Tectonic foliation may occur to existing rocks, or totally new igneous, sedimentary, and metamorphic rocks may appear.

Once the background of processes and rock types has been established, the author looks at the mountain itself. Why does it stick up above the surrounding crustal surface? This brings up a consideration of crustal roots, buoyancy force, Archimedes' principle, isostasy, isostatic equilibrium, and isostatic compensation.

Even mountains don't last forever. The chapter continues with:

- erosion issues; agents of erosion (water and ice), and climate influences
- features created by erosion, like cuestas (or hogbacks)
- lateral spread, called orogenic collapse
- exposure of a mountain's innards (exhumation)

Why are mountains located where they are? Wouldn't you know it, plate tectonics again! There may be a new term or two introduced here, like accretionary orogens and fault-block mountains, but the concepts are all old acquaintances (subduction, convergent plate boundaries, and continental rifting).

The chapter draws to a close with a few new terms for some continental areas (shields, cratons, and cratonic platforms) and the information that dome and basin formation (epeirogeny) are less exuberant processes of land uplift than are orogenic events.

The last chapter topic is the life story of one particular mountain range, the Appalachian Mountains of North America. Read it carefully to make sure you understand the narrative, but don't overwhelm yourself by trying to memorize this particular sequence of events. Do of course check with your instructor, but unless you live right in these mountains, it's unlikely you'll be expected to recite the history of the Appalachians. Instead, what you should understand is that there's no such thing as a really simple explanation of the formation of any particular mountains. All major mountain ranges are the result of multiple orogenies over long geologic time.

And speaking of time, that's what the next chapter is all about, geologic time.

Completion

Test your recall of new vocabulary terms.

_____ 1. the process in which rocks get squashed, stretched, bent, or broken in response to squeezing, stretching, or shearing

_____ 2. natural cracks in rock along which there has *not* been movement

_____ 3. rock layers that started out flat but were bent or wrinkled by compressive stress

_____ 4. linear ranges of mountains (two different terms)

_____ 5. a mountain-building event

_____ 6. a process that makes a layer of rock become longer

_____ 7. a process that makes a layer of rock become shorter

_____ 8. the change of shape that is the result of shear stress

_____ 9. the depth in continental crust above which rocks deform brittlely, below which they deform ductilely

_____ 10. a push or pull whose strength is equal to mass times acceleration

_____ 11. long planar cracks with similar orientation that occur fairly regularly through a rock body

_____ 12. a group of systematic joints

_____ 13. joints that are filled with minerals (like quartz or calcite) that precipitated out of groundwater

_____ 14. faults in which the sliding has occurred vertically, either up- or downslope on the fault plane

_____ 15. faults in which the sliding has occurred diagonally along the fault plane, with movement along both the strike and dip directions

_____ 16. the category of strike-slip fault in which the block of land across the fault line from the observer has appeared to move to the left

_____ 17. the category of strike-slip fault in which the block of land across the fault line from the observer has appeared to move to the right

_____ 18. the amount of displacement that occurs across a fault plane

_____ 19. folds that happen during faulting or just before layers of rock undergo fault slip

_____ 20. pieces of angular rock fragment, large enough to be seen by the naked eye, produced when rock gets crushed and broken along fault surfaces

_____ 21. the fine powder found on fault surfaces, produced when rock is shattered and ground fine during faulting

_____ 22. fault surfaces that have been polished by the movement of a hanging wall past a footwall

_____ 23. linear grooves on fault surfaces produced when rough surfaces of either hanging or footwalls gouge the opposing surface during faulting

_____ 24. the fine-grained foliated metamorphic rock produced by shear stress in fault zones, where the deformation is ductile and causes large grains to recrystallize into smaller ones

_____ 25. faults in which movement occurs ductilely and mylonites are formed

_____ 26. groups of many related faults

_____ 27. a horizontal fault, found at depth, into which numerous related faults have merged

_____ 28. a block of crust that has dropped down between normal faults to form an elongate basin

_____ 29. the block of crust that remains as a high divider between two grabens

_____ 30. a triangular basin that forms when hanging-wall block slips down along a curved fault plane

_____ 31. the portion of a fold where the curvature is greatest

_____ 32. the sides of a fold that show less curvature than the hinge area

_____ 33. the imaginary plane that cuts down through the hinges of all layers of the fold

_____ 34. upward folds of rocks, with an arch shape and limbs that dip away from the hinge

_____ 35. downward folds of rocks, with a trough shape and limbs that dip toward the hinge

_____ 36. a fold which resembles one stair-step, that is, in which the rock layer simply exists at two elevations, with a gentle slope between

_____ 37. a fold in which the angle between limbs is large

_____ 38. a fold in which the angle between limbs is small

_____ 39. a fold in which the hinge is horizontal

_____ 40. a fold in which the hinge is tilted, causing the rock layers to apparently dive into the Earth

_____ 41. an upward fold that encompasses 360°, so it looks like an overturned bowl

_____ 42. a downward fold that encompasses 360°, so it looks like a bowl sitting right side up

_____ 43. a type of fold-forming event in which layers bend, and slip occurs between them

_____ 44. folds that form because the rock is so soft it behaves like weak, warm plastic

_____ 45. a type of fold-forming event in which end-on compression causes layers to wrinkle up

_____ 46. a type of layering in metamorphic rock created by the alignment of deformed and reoriented grains

_____ 47. the boundary between a mountain range and the surrounding plains

_____ 48. the process that moves Earth's surface vertically from a lower to a higher elevation

_____ 49. the mass of thicker-than-normal crust beneath a mountain

_____ 50. the upward push of one material on another due to different densities

_____ 51. a principle of fluid mechanics that states a fluid buoys up a completely immersed solid so that its apparent weight is lowered by the weight of the displaced fluid and so that an object sinks in a fluid only until it has displaced its own weight of fluid

_____ 52. the condition that exists when the buoyancy force pushing lithosphere up equals the gravitational force pulling it down

_____ 53. the process in which Earth's surface slowly rises or falls to reestablish isostasy

_____ 54. the topographic feature that is a ridge of rock, made up of tilted layers, with a steep face on one side and a gentle slope, parallel to the layers, on the other side

_____ 55. the process in which mountains begin to collapse and spread laterally under their own weight

_____ 56. the exposure of metamorphic and plutonic rocks, once deep in the crust, by erosion and orogenic collapse

_____ 57. the process in which buoyant blocks of crust collide and attach at convergent margins

_____ 58. the name for buoyant blocks headed for convergent margins but not yet attached

_____ 59. exotic terranes that have attached themselves to convergent margins

_____ 60. mountain belts that have grown laterally by the attachment of exotic terranes

_____ 61. a region, rich in both faults and folds, on the continent side of a volcanic arc in which a thrust system has developed above a detachment fault

_____ 62. linear mountain ranges that are typical of rift areas, are composed of tilted rocks, and border deep basins that were produced when crustal rock dropped along normal faults

_____ 63. a land area that consists of crust that hasn't been affected by orogeny for at least 1 billion years

_____ 64. a land area that contains outcrops of Precambrian metamorphic and igneous rock

_____ 65. a land area in which a thin layer of younger rock (Phanerozoic, or post-Precambrian) covers the Precambrian rocks

_____ 66. a geologic event in which vertical movement generates huge but gentle domes and basins

Short-Answer Questions

Answer with a few words, numbers, or short sentences, or choose from the words in parentheses.

1. Allowing for some variations in naming, geographers agree there are about how many major orogenic belts in the world? _____

2. The Sierra Nevadas and the Rocky Mountains of the U.S. West are smaller ranges within the larger mountain belt named _____.

3. After active mountain building (an orogeny) ceases, how long might it take erosion to change the mountains back to sea-level plains? _____

4. List four causes of mountain formation (orogeny). All are plate tectonics events. _____

5. The difference in elevation between the top and bottom of the Grand Canyon is about 1 mi. Except for the inner gorge itself, the canyon consists of normal sedimentary rocks stacked horizontal layer upon horizontal layer. Is the Grand Canyon part of an orogenic belt? Briefly explain your answer. _____

6. Geologists talk of the Ancestral Rockies, and even draw maps of these ancient mountains that eroded away millions of years ago. How do they know they existed and where they were located? _____

7. Joints and faults are produced by _____ (ductile, brittle) deformation; folds are the result of _____ (ductile, brittle) deformation.

8. In terms of chemical bonds, how is brittle deformation different from ductile deformation? _____

9. For each statement in this question, choose "ductile" or "brittle."
    a. In terms of temperature, warm rock tends to undergo _____ deformation, cold rock to undergo _____ deformation.
    b. In terms of pressure, rock at depth and therefore under high pressure undergoes _____ deformation; rock under low pressure undergoes _____ deformation.
    c. In terms of speed of application of pressure, abrupt change of pressure is most likely to cause _____ deformation; slow change tends to cause _____ deformation.
    d. Different rock types exhibit characteristic deformation; halite typically shows _____ deformation, while granite shows _____ deformation.

10. Earthquakes in continental crust happen only _____ (above, below) the brittle-ductile transition zone.

11. Fill in the blanks with "stress" or "force." _____ is the total amount of pressure applied to an object; it's independent of the area it's applied to. The _____ any particular object experiences depends on the area the _____ was distributed over.

12. _____ (Stress, Strain) causes the change in shape called _____ (stress, strain). _____ (Compression, Tension) causes shortening; _____ (compression, tension) causes stretching.

13. Fill in the blanks with "bearing," "dip," "plunge," or "strike."
    When describing the physical orientation of geologic structures, by convention planar structures like faults, sedimentary beds, and joints are described as having _____ and _____; linear features like striations (scratches on rocks) or needle-like crystals are described as having _____ and _____.

14. Joints, which are a type of _____ (brittle, ductile) deformation, develop in response to _____ (compressional, tensional) stress.

15. List three possible geologic situations that can produce tensional stress in rock. _____

16. a. What is the typical thickness of continental crust *not* under mountains? _____
    b. What is the typical thickness of continental crust under mountains? _____

17. Low-density crustal roots under mountains exert a force that helps thrust mountains up. The force exists because of density differences. Name the force.
_____

18. Why does isostatic compensation take a long time to occur? _____

19. Mountain uplift can be described as an example of isostatic _____

20. Name two erosional agents that can carve rugged features (like peaks, knife-edge ridges, steep cliffs, and deep valleys) into areas of uplifted land. _____
_____

21. Name two factors that influence whether the dominant agent of erosion in an area will be ice (glaciers) or running water (streams). _____
_____

22. Fill in the blanks with "equals," "exceeds," or "is less than."
A mountain range can continue to grow only so long as the rate of uplift _____ the rate of erosion. It will maintain its elevation as long as the rate of uplift _____ the rate of erosion. It will diminish in elevation when the rate of uplift _____ the rate of erosion.

23. As a rule of thumb, for every kilometer eroded from the top of a mountain range, the base of the range rises, due to isostasy, by about _____ km.

24. Convergent boundary orogenies can last as long as _____ years.

25. The western half of the North American Cordilleran has grown laterally by adding pieces of land called _____.

26. The Himalayas, Alps, and Appalachian Mountains all resulted from what type of plate tectonics event?
_____

27. Name the geographical feature that exists today because of this sequence of events: India collided with Asia and caused the Himalayas to grow. Rock at depth in the orogen (mountain belt) heated, collapsed, and spread out sideways, to the northeast. _____

28. Name two types of geologic structure produced by relatively mild crustal deformation found in the midwest United States. _____
_____

29. What very new instrument or system allows geologists to detect the shift of continents and the vertical movement of Earth's surface, even movement as small as centimeters per year? _____

BIOLOGY SPECIAL

Fill in the blanks.

The top of Mt. Everest, the highest mountain on Earth, is not a comfortable place for humans to be. Air pressure is so low at its summit (elevation _____ ft) the amount of oxygen per unit volume of air is just _____ of what it is at sea level. As a rule of thumb, average temperature on Earth decreases 3.5°F for every 1,000 ft of elevation gain. If the temperature at sea level is 70°F on a day in May (the month climbers usually climb Everest), what is the temperature on the summit? _____

Making Use of Diagrams

STRIKE AND DIP

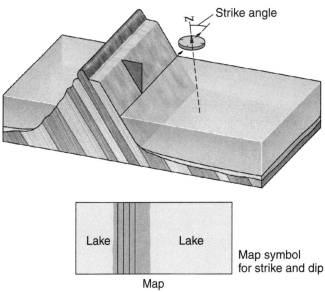

Add the proper notation to show dip of 40°E on the figure shown above.

TYPES OF FAULTS

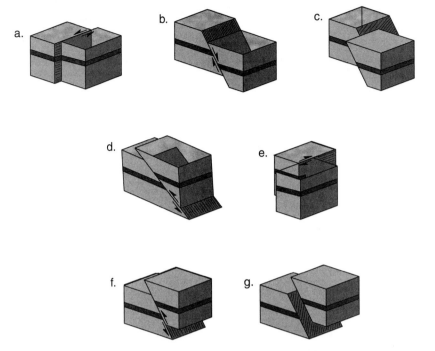

Match the fault types shown in the diagram above to the labels provided below.

_____ 1. dip-slip fault, normal
_____ 2. dip-slip fault, reverse
_____ 3. dip-slip fault, thrust
_____ 4. left-lateral strike-slip fault
_____ 5. oblique-slip fault
_____ 6. right-lateral strike-slip fault

FAULT BLOCKS

Label the fault blocks shown in the figure below:

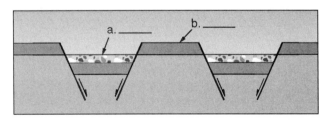

SYNCLINES AND ANTICLINES

Questions 1 to 4 pertain to the figure on the right:

1. For a and b of the figures, add the proper symbolic notation for the nonplunging anticline and the nonplunging syncline.

2. For map view figure c below, add arrow heads to show that this is a plunging anticline.

3. Study of the cross-section view d and map view c shows that an anticline makes a V-shaped pattern pointing in the _____ (same, opposite) direction of its plunge.

4. Label the structures in e and f:
   e. _____ f. _____

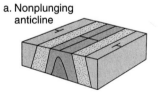
a. Nonplunging anticline

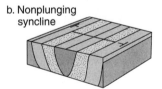
b. Nonplunging syncline

c. Map view

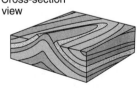
d. Cross-section view

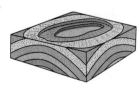
e.

f.

5. Map views of both domes and basins show concentric patterns of layers of rocks. In a dome, the layers grow progressively _____ (older, younger) as you go toward the center. In a basin, the layers grown progressively _____ (older, younger) as you go toward the center.

Short-Essay Questions

1. Dinosaurs became extinct about 65 million years ago. Could they have trod the same mountains we do today? Was the topography of their world the same as it is today?

2. Why can brittle deformation structures (like faults) and ductile deformation structures (like folds) occur in the same rock outcrop?

3. Discuss the wisdom of constructing dams, roads, and buildings on highly jointed rock.

4. Two reasons geologists study faults are to assess seismic risk of an area and to locate subsurface resources. Elaborate on these topics.

5. Explain the nature of igneous activity that accompanies orogenies at:
   a. convergent plate boundaries   b. continental rifts
   c. continental collision zones

6. Why do deep sedimentary basins form along mountain fronts?

7. Briefly discuss the types of metamorphism that can occur in orogenic belts.

8. Discuss reasons geologists believe mountains can never be much higher than Mt. Everest.

9. "Gravity" does not simply mean "things fall down." Instead, it's the attraction of any mass to any other mass. Sir George Everest's experience with the deflection of a plumb bob when surveying in the Himalayas demonstrated this. The deflection wasn't as great as expected, considering the size of the mountains. How did scientist George Airy explain the discrepancy?

Practice Test

1. Which term has nothing to do with true mountains?
   a. "orogeny"
   b. "Cordillera"
   c. "plateau"
   d. "orogen"
   e. "deformation and contortion"

2. What kind of rocks are never found in mountains?
   a. intrusive igneous
   b. clastic sedimentary
   c. chemical sedimentary
   d. contact metamorphic
   e. Any of these could be found in mountains.

3. Which of the following conditions will tend to make rocks change by ductile deformation rather than by brittle deformation?
   a. cool surroundings
   b. slowly applied stress
   c. position fairly close to Earth's surface
   d. granitic composition
   e. All of the above would encourage ductile deformation.

4. Which of the following locations is a good example of the erosion of a joint set?
   a. the Grand Canyon
   b. any cuesta in the Southwest
   c. Ozark Dome, Missouri
   d. Arches National Monument, Utah
   e. the Sierra Nevadas

5. Pick out the TRUE statement. The San Andreas Fault
   a. is a right-lateral strike-slip fault.
   b. has a fault plane than dips 120° to the southeast.
   c. is a planar structure, so its orientation should be expressed in terms of bearing and plunge.
   d. can also properly be called the San Andreas Joint.
   e. All of the above are true statements.

6. If you see small angular fragments of shattered rock along a linear boundary between two masses of rock, you must be looking at
   a. fault gouge
   b. slickensides
   c. fault breccia
   d. mylonites
   e. slip lineations

7. Faulting in the Basin and Range Province of Nevada, Utah, and Arizona
   a. is normal faulting.
   b. is causing the crust to stretch.
   c. has produced valleys that are grabens.
   d. is an example of continental rifting.
   e. All of the above are true statements.

8. Which of the following statements is FALSE?
   a. Major mountain ranges are the result of multiple orogenies over long geologic time.
   b. Mountains continue to uplift as long as the rate of erosion equals the rate of uplift.

c. Domes and basins in the midwestern United States are the result of epeirogeny.
d. Fault-block mountains result when blocks of crust drop downward along normal faults.
e. Small fragments of continental crust that are too buoyant to subduct during a convergence event are called exotic terranes.

9. Which of the following statements is FALSE? Mountains
   a. have roots of continental crust that extend deeper than the level of crust under plains.
   b. are randomly distributed over Earth's surface.
   c. given time can spread laterally under their own weight.
   d. are created during orogenies that may last for millions of years.
   e. are made up of rocks that have undergone ductile deformation and brittle deformation.

10. Deformation
    a. is brittle when many chemical bonds are broken quickly and the rock pieces separate.
    b. produces changes in the shape and orientation of rock grains and rock layers.
    c. is a characteristic of mountain belts.
    d. may produce faults if brittle, and may produce folds if ductile.
    e. All of the above are true statements.

11. Which of the following statements is FALSE? This diagram shows

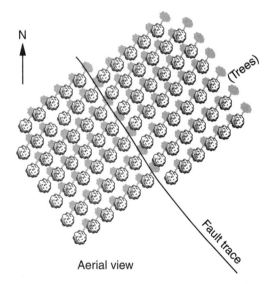

Aerial view

   a. a right-lateral strike-slip fault.
   b. a fault with a northwest strike.
   c. that dip-slip motion has occurred along the fault trace.
   d. offset of a linear feature (the line of trees).
   e. a situation that is not uncommon in California.

12. Which of the following statements is FALSE? The African Rift Valley
    a. is basically a long graben.
    b. is bordered by linear mountain ranges called fault-block mountains.
    c. is produced by compressional stress.
    d. has volcanoes along it because asthenosphere is rising under the rift and supplying heat.
    e. is geologically similar to the Basin and Range Province of Utah, Nevada, and Arizona.

13. Mountains don't get infinitely high or exist forever because
    a. erosion attacks as soon as they begin to rise.
    b. their own weight causes them eventually to collapse.
    c. when they get exceptionally large and old they spread laterally.
    d. they experience orogenic collapse.
    e. All of the above are true statements.

14. A shield
    a. may be part of a craton.
    b. contains Precambrian metamorphic and igneous rocks outcropping at ground surface.
    c. can be associated with a cratonic platform.
    d. has lots of shear zones, flow folds, and tectonic foliation.
    e. All of the above are true statements.

15. You're looking at a fault trace in the field. You observe a polished surface, with linear grooves on it, and fine powder along it. In more technical terms, what are you seeing?
    a. shear zone, fault gouge, and fault breccia
    b. slip lineation, slickensides, and mylonites
    c. slickensides, slip lineations, and fault gouge
    d. slickensides, mylonites, and fault gouge
    e. shear zone, mylonites, and fault breccia

16. Tension produces shear strain.
    a. True
    b. False

17. A horizontal line on Earth's surface has a plunge of 90°.
    a. True
    b. False

18. Joints may get filled with quartz or calcite from groundwater and are then called joint sets.
    a. True
    b. False

19. Even today geologists can only theorize about shifting continents and uplifting mountains; there's no way they can detect such small annual movement.

a. True
b. False

20. Faulting in a thrust-fault system shortens the crust, since slivers of Earth get pushed over each other like overlapping shingles.
    a. True
    b. False

21. An inclinometer is an instrument used to detect gravity anomalies caused by mountain ranges.
    a. True
    b. False

22. Since the Tetons are fault-block mountains, the valley adjacent to their steep faces (Jackson Hole) must be a graben.
    a. True
    b. False

23. Typical continental crust is 35–40 km thick; under mountains it is thinner, only 15–20 km thick.
    a. True
    b. False

24. When the buoyancy force pushing up on lithosphere equals the gravitational force pulling down on it, the situation is said to show isostatic equilibrium.
    a. True
    b. False

25. Isostatic compensation takes a long time because core material must flow to make the adjustment, and it flows slowly because of its great depth.
    a. True
    b. False

26. An asymmetric ridge formed along a tilted rock layer is called a craton.
    a. True
    b. False

27. Exhumation exposes metamorphosed rocks and plutons that once were deep inside mountains.
    a. True
    b. False

28. For each kilometer of mountain top lost to erosion, that same mountain rises 2 km due to isostatic compensation.
    a. True
    b. False

29. Exotic terranes collided with the west edge of the North American Cordillera and became accreted terranes; because the mountains grew laterally this way, they're called accretionary orogens.
    a. True
    b. False

30. The Himalayas, Alps, and Appalachian Mountains are all examples of convergent boundary mountains.
    a. True
    b. False

31. The cratonic platform of the U.S. Midwest contains regional basins and domes that resulted from epeirogeny.
    a. True
    b. False

ANSWERS

Completion

1. deformation
2. joints
3. folds
4. mountain belt (or orogenic belt or orogen)
5. orogeny
6. stretching
7. shortening
8. shear strain
9. brittle-ductile transition
10. force
11. systematic joints
12. joint set
13. veins
14. dip-slip faults
15. oblique-slip faults
16. left-lateral strike-slip fault
17. right-lateral strike-slip fault
18. offset
19. drag folds
20. fault breccia
21. fault gouge
22. slickensides
23. slip lineations
24. mylonite
25. shear zones
26. fault systems
27. detachment fault
28. graben
29. horst
30. half graben
31. hinge
32. limbs
33. axial plane
34. anticlines
35. synclines
36. monocline
37. open fold
38. tight fold
39. nonplunging fold
40. plunging fold
41. dome
42. basin
43. flexing
44. flow folds
45. buckle
46. tectonic foliation
47. mountain front
48. uplift
49. crustal root
50. buoyancy force
51. Archimedes' principle
52. isostasy (or isostatic equilibrium)
53. isostatic compensation
54. cuesta (or hogback)
55. orogenic collapse
56. exhumation
57. accrete
58. exotic terranes
59. accreted terranes
60. accretionary orogens
61. fold-thrust belt
62. fault-block mountains
63. craton
64. shield
65. cratonic platform
66. epeirogeny

Short-Answer Questions

1. a dozen
2. the North American Cordillera
3. about 50 million years
4. subduction at convergent plate boundaries; motion at some special locations along continental transform faults; continental collisions; continental rifting
5. No, there are no deformed rock grains and no contorted rock layers, and these are always present in orogenic belts.
6. Areas of contorted and broken and metamorphosed rocks testify to the existence and location of the ancient mountains.
7. brittle; ductile
8. Brittle deformation: many chemical bonds break at once and the rock falls apart.
   ductile deformation: few chemical bonds break, new ones form quickly, and the rock doesn't separate into pieces.
9. a. ductile; brittle   b. ductile; brittle   c. brittle; ductile   d. ductile; brittle
10. above
11. Force; stress; force
12. Stress; strain; Compression; tension
13. strike; dip; bearing; plunge
14. brittle; tensional
15. cooling and contraction of rock; decrease in compression because erosion removed overlying weight; bending of rock layers
16. a. 35–40 km   b. 50–70 km
17. buoyant force
18. because asthenosphere must flow out of the way when lithosphere sinks and must flow back in under lithosphere when it rises, and asthenosphere flows very slowly
19. compensation
20. ice (glaciers) and water (streams)
21. climate and elevation
22. exceeds; equals; is less than
23. 1/3
24. 200 million
25. accreted terranes
26. continental collision
27. Tibet Plateau
28. regional basins and domes and local zones of folds and faults
29. Global Positioning System (GPS)

BIOLOGY SPECIAL

29, 029;   one-third;   –32°F ($3.5° \times 29 = 101.5°$; $70° - 101.5 \approx -32°$)

Making Use of Diagrams

STRIKE AND DIP

See Figure 11.12 in your text.

TYPES OF FAULTS

1. b   3. d   5. g, c
2. f   4. a   6. e

FAULT BLOCKS

a. graben   b. horst

SYNCLINES AND ANTICLINES

1. See Figure 11.22 in your text.
2. See Figure 11.22 in your text.
3. same
4. e. dome   f. basin
5. older; younger

Short-Essay Questions

1. The topography of Earth back in the time of the dinosaurs was not the same as today's topography. This chapter tells us that 50 million years is enough time to level any mountain range after orogenic processes stop, so dinosaurs couldn't possibly have had the same mountains we do today. Other chapters in the text contribute information too. Plate tectonics tells us that continental masses have shifted position drastically over time, and rocks weather and erode and thus change the topography. There are many examples you could use, but to summarize, Earth is constantly changing and the surface appearance of any particular region changes drastically over geologic time.

2. An orogeny goes on for millions of years, and during this time the rate of deformation can change many times. Periods of slow deformation in an area would yield structures like folds; spurts of deformation would produce faults in the same area.

3. Reservoirs created behind basins always leak water into the surrounding rock. Water flows more easily through jointed rock than through nonjointed rock. Water that has infiltrated joints decreases frictional resistance between adjacent surfaces, making movement more likely. Joint-bounded blocks of rocks separate more easily than nonjointed rock masses, so they don't offer as reliable building support, particularly on steep slopes. Therefore, if at all possible, highly jointed rock should be avoided when constructing anything major that needs strong support.

4. The most common cause of seismicity is movement along faults. Although earthquake prediction is hardly an exact science, the more you know about the fault situation in an area, the better prepared you can be to protect against possible quake damage. Movement along faults changes the shapes and locations of subsurface rock layers and in general complicates the arrangement of rocks underground. Among other complications, faults can displace rocks that contain valuable resources like coal, oil and gas, minerals, and water. The better you understand what happened in the faulting process, the more likely it is you can find the displaced resources.

5. a. Subduction at convergent plate boundaries produces volcanic arcs, both oceanic and continental. Both intrusive and extrusive igneous rocks are created in the arcs.
   b. Thinning lithosphere of rift areas decompresses the upper mantle and generates magma that rises along normal faults and may become dikes, sills, or lava flows.
   c. The last bit of subduction that occurs before continental masses collide generates magmas that are trapped in the collision zone.

6. Mountain belts (orogens) are heavy, and their weight pushes down on the crust and causes it to sink (be depressed). Sediments eroded off the mountains are carried to mountain edges, where they fill the depressed basins along the mountain fronts and weight down the land even more.

7. Igneous plutons generated during the orogeny produce contact metamorphic aureoles along their edges. Thrust faulting buries blocks of crust under overriding hanging walls, and both the heat of burial and the shearing stress produce regional metamorphism.

8. One height-limiting factor is erosion, which attacks mountains as soon as they rise. Although erosion rates do vary, uplift never seems to get so ahead of erosion that these resulting mountains are taller than Everest. Mountain height is also limited by the fact rocks don't have infinite strength. The weight of a mountain pushes down onto itself, forcing its underlying rock to go deeper, where it gets warmer and softer and eventually flows slowly outward. Thus by the time the mountain has reached old age and huge size, it is collapsing on itself and spreading laterally under its own weight.

9. British scientist George Airy correctly suggested that continental crust is thicker than normal under mountains. In other words, mountains have roots; they stick up high and they also extend down low. A mountain range with low-density continental crust roots has less overall mass than a mountain range would have if it had a normal thickness of continental crust underlain by a denser mantle. It therefore exerts less pull than expected and disturbs normal gravity less than expected.

Practice Test

1. c A plateau region is higher than its surroundings, but it doesn't have the deformed and contorted rocks of true mountains.

2. e Any of these could be found in mountains.

3. b All except b would encourage brittle, not ductile deformation.

4. d Arches National Monument is a good example of a joint set.

5. a A strike-slip fault has a nearly vertical fault plane; bearing and plunge are used for linear, not planar features; and a joint doesn't have movement along it.

6. c You are looking at fault breccia.

7. e All of the above are true.

8. b Equal rates of uplift and erosion will maintain mountain height but not increase it.

9. b Except for hot-spot volcanoes, mountains occur in belts associated with plate boundaries.

10. e Statements a to d are all true.

11. c The aerial view doesn't show anything about vertical motion, which is what dip-slip motion is; movement along the San Andreas Fault in California is right-lateral strike-slip.

12. c Rifting and normal faulting produce tensional stress.

13. e Statements a to d are all true.

14. e Statements a to d are all true.

15. c Only choice c lists the correct terms in the proper order.

16. False Tension produces stretching; shear stress (one side moves sideways past the other side) produces shear strain.

17. False If something is horizontal it has 0° plunge; vertical plunge is 90°.

18. False   They're called veins.
19. False   The satellite based Global Positioning System (GPS) detects movement as small as centimeters per year.
20. True   Thrust faulting, which is a low angle variety of reverse faulting, does shorten Earth's crust.
21. False   An inclinometer measures dip and plunge.
22. True   The block of Earth that has moved downward along a fault is called a graben.
23. False   Crust under mountains thickens to 50–70 km.
24. True   A situation in which opposing forces cancel each other is called isostatic equilibrium.
25. False   The asthenosphere makes the adjustment, and it flows slowly because it's highly viscous.
26. False   The definition is of a cuesta (or hogback); a craton is crust that hasn't been affected by orogeny for at least 1 billion years.
27. True   The metamorphosed rocks and plutons in mountain cores are eventually exposed by erosion; this process is called exhumation.
28. False   The general idea is correct, but the figures are wrong; there's a rise of one-third of a kilometer for every kilometer of height lost to erosion.
29. True   Mountains along the west coast of North America are typical accretionary orogens.
30. False   They're continental collision mountains.
31. True   The U.S. Midwest contains structures that are the result of epeirogeny, basins and domes, instead of mountains that are the result of orogeny.

CHAPTER 12 | Deep Time: How Old Is Old?

GUIDE TO READING

This chapter deals with geologic time, from mere seconds to the billions of years in an eon, and examines the means by which geologists puzzle out Earth's history. You learn to decipher the clues Earth offers, to date Earth materials and events, and to match this to an appropriate time scale.

As human culture did, the author starts with small units and builds to bigger time divisions. You read about seconds (which may not be as simple as you think), days, time zones, and Greenwich mean time. When human society advanced enough to "have time on its hands," it used some of it to speculate about our planet. Many cultures asked "How old is Earth?" and "What's been happening to Earth throughout all of its existence?" Some persons earned their places in history by trying to answer these questions. You read about Pope Gregory XIII, Archbishop James Ussher, Leonardo da Vinci, Nicholaus Steno, James Hutton, Charles Lyell, William Smith, John Wesley Powell, Lord William Kelvin, Henri Becquerel, and Ernest Rutherford and their contributions to establishing the time frame of Earth's history.

Telling when something happened is an important part of any historical narrative. Scientists had to figure out not only what came first, last, and in between in Earth's history (relative dating), but they had to apply real numbers (numerical or absolute dating) to Earth's materials and events. Relative dating is based on the application of several commonsense principles; numerical dating requires more science. Therefore Earth happenings were put in proper order before they were dated. You'll read about the commonsense principles of relative dating and work with them in this study guide—principles of uniformitarianism, superposition, original horizontality, continuity, baked zones, cross-cutting relations, inclusions, and fossil succession.

Geologists were quite confident they were getting the events of Earth's history in proper sequence long before they felt much confidence in the numbers they assigned to the events. There were several creative lines of logic applied to the problem; they involved the salinity of oceans, depths of sediments, and temperature of Earth. Unfortunately, new data and newer and better interpretation of old data always showed fatal flaws in these schemes. Finally, during the early 1900s, observations of the statistical regularity of radioactive decay allowed geologists to assign dates to ancient geologic materials and events that are firmly believed in to this day. The dating method is termed radiometric dating. Your author discusses it thoroughly: the actual procedures used, what the special case of carbon 14 dating is all about, the accuracy of the method and the uncertainty of measurement, and the mechanics of radioactive decay (alpha and beta particles, electron capture, half-lives, parent and daughter isotopes, fission, fission track dating, and isotope ratios).

Several other nonradioactive procedures have played their parts in dating Earth events. Your author talks briefly about dendrochronology (tree ring dating) and about seasonal influences that result in rhythmic layering of sediments, glacial ice, mineral precipitation, and organic productivity. Rock layers, some with fossils in them, read like pages in a book to reveal Earth's history. Sometimes there are breaks in the rock record—pages missing—called unconformities, which often can be accounted for by finding the missing pages (rock layers) elsewhere in the world (a procedure called correlation).

As time passed and communications got better, correlations worldwide became complete enough to compile a geologic column showing all (or almost all) of Earth's history as written in the rocks. Improved communications also resulted in the development of a dated geologic time scale.

Its organization is a bit lacking because it grew by bits and pieces over more than a century, but its terminology is essential to any discussion of Earth's history. This chapter presents the largest, most basic divisions of the scale (Precambrian, Hadean, Archean, Proterozoic, Phanerozoic, Paleozoic, Mesozoic, and Cenozoic). Chapter 13 will go into greater detail.

In a study of geology you get used to hearing about millions and billions of years. You may be very comfortable with the words and know how many zeros go with each, but as humans we all lead lives that revolve around smaller figures and much less time. Therefore your author concludes the chapter with an analogy that tries to fit these immense numbers onto a time frame we can feel. He equates all of Earth history to one calendar year. It is a humbling paragraph.

Completion

Test your recall of new vocabulary terms.

_____ 1. the span of time since Earth's formation

_____ 2. 60 of these make 1 h

_____ 3. the time it takes for Earth to rotate once on its axis

_____ 4. 15°-wide strips of longitude in which all clocks keep the same standard time

_____ 5. the time at the astronomical observatory in Greenwich, England

_____ 6. any indication of life in the past (Latin for "dug up")

_____ 7. the present is a key to the past

_____ 8. one of the numerous principles that are applied to determine the relative ages of rock layers

_____ 9. a comparison of ages, older versus younger

_____ 10. age of something in years

_____ 11. fragments of an older rock included in an igneous intrusion

_____ 12. pieces of one rock that are incorporated into another (not as specific as "xenoliths")

_____ 13. the sequence of geologic events that took place in an area over time

_____ 14. the group of fossil species contained in a layer of sedimentary rock

_____ 15. the idea that a species exists during a limited time interval, becomes extinct, and is followed by a new species, so that fossils are found in limited strata

_____ 16. all the levels of strata (and all the time this represents) that contain fossils of a particular species

_____ 17. a break in the rock record that indicates missing time

_____ 18. the missing time that's indicated by an unconformity

_____ 19. a drawing that shows the sequence of strata at a location and their relative thicknesses

_____ 20. a recognizable unit of rock that represents the environment of an area during some past time interval

_____ 21. the boundary surface between two formations

_____ 22. the matching of similar strata across distances in order to establish age relationships

_____ 23. the correlation based on matching similar rock types at the locations involved

_____ 24. the correlation of widely dispersed areas based on their similar fossil assemblages

_____ 25. any rock layer that is unique enough to be useful for correlation purposes

_____ 26. a map that shows the distribution of stratigraphic formations at Earth's surface

_____ 27. a compilation of stratigraphic columns gathered worldwide to produce the most comprehensive rock record of Earth history possible

_____ 28. the geologic time unit of highest rank (example: the Phanerozoic, which includes the Paleozoic, Mesozoic, and Cenozoic)

_____ 29. geologic time divisions that are smaller than an eon (examples: Paleozoic, Mesozoic, and Cenozoic)

_____ 30. geologic time divisions that are smaller than an era (examples: Triassic, Jurassic, and Cretaceous subdivisions of the Mesozoic)

_____ 31. geologic time divisions that are subdivisions of a period

_____ 32. the development of many species from one species

_____ 33. the science of dating geologic events by analysis of radioactive decay events

_____ 34. the number of protons in the nucleus of an atom

_____ 35. the number of protons plus neutrons in the nucleus of an atom

_____ 36. varieties of atoms of an element that differ only in their numbers of neutrons

_____ 37. varieties of atoms that are unstable and destroy themselves by radioactive decay

_____ 38. a radioactive isotope of an element that undergoes decay

_____ 39. the isotope that is the product of radioactive decay of a parent isotope

_____ 40. the time needed for half of any mass of radioactive material to decay into its daughter product

_____ 41. a measure of the limits of accuracy of a radiometric date, expressed as a percentage of the reported date

_____ 42. the temperature below which isotopes are locked into a crystal lattice and so are not free to move around

_____ 43. the visible rings of growth in tree trunks due to seasonal variations

_____ 44. the visible differences in layers of sediment and layers of glacier ice due to seasonal variations

_____ 45. a scientist who studies tree rings and thus determines dates and past climates

_____ 46. the study of magnetic field reversals over time

_____ 47. a line through a crystal lattice that marks the path of an atomic particle released during radioactive decay

_____ 48. a dated column that shows the names and ages for all geologic time intervals

Matching Questions

FAMOUS PERSONS WHO STUDIED GEOLOGIC TIME

Match the person with the contribution.

a. Henri Becquerel
b. Leonardo da Vinci
c. James Hutton
d. Lord William Kelvin
e. Charles Lyell
f. John Wesley Powell
g. William Smith
h. Nicholaus Steno (Niels Stenson)
i. Archbishop James Ussher

_____ 1. By adding up generations of patriarchs in the bible, he figured the birth date of Earth was October 23, 40004 B.C.E.

_____ 2. More famous for his art than his science, in 1500 this great Italian Renaissance man proposed the organic origin of fossils.

_____ 3. In 1669 he published a book, *Forerunner to a Dissertation on a Solid Naturally Occurring Within a Solid,* which correctly explained how fossils form; he also formulated some of the basic principles of relative dating.

_____ 4. Acclaimed as the "father of geology," his work in the late 1700s introduced the principle of uniformitarianism.

_____ 5. He compiled the ideas of early geologists like Hutton, Smith, and Steno into the book *Principles of Geology*, which popularized geology in the middle 1800s.

_____ 6. While surveying the English countryside, he deduced the significance of finding fossils in certain stratigraphic layers and not in others; in 1815 he created the first geologic map.

_____ 7. In 1869 he and his companions made the first geologic exploration of the Grand Canyon of the Colorado River.

_____ 8. Basing his calculations on how long it would take Earth to cool down from Sun temperature, this famous physicist of the late 1800s concluded Earth was 20 million years old.

_____ 9. A physicist, in about 1900 he discovered the phenomenon of radioactivity.

PRINCIPLES USED TO DETERMINE RELATIVE AGES

These commonsense ideas have rather imposing titles because when they were proposed (from the mid-1600s through the mid-1800s), they were grand ideas, ahead of their time. Match the principle with its description.

a. Baked Contacts
b. Cross-Cutting Relations
c. Inclusions
d. Original Continuity
e. Original Horizontality
f. Superposition
g. Uniformitarianism

_____ 1. Sedimentary layers begin as continuous expanses of sediment.

_____ 2. Sedimentary layers start out flat.

_____ 3. The oldest sedimentary bed is the one on the bottom.

_____ 4. The present is a key to both the past and the future.

_____ 5. An igneous intrusion is younger than the rock it has intruded and metamorphosed.

_____ 6. If one geologic feature cuts across another, the feature doing the cutting is younger than the one it cuts across.

_____ 7. Rocks containing inclusions (whether the inclusions are xenoliths in igneous intrusions or igneous pebbles in sedimentary layers) are younger than the inclusions.

MAJOR DIVISIONS OF GEOLOGIC TIME

Label this simplified chart that shows the major subdivisions of geologic time with the eon and era names given. Details will be added in Chapter 13.

a. Archean
b. Cenozoic
c. Hadean
d. Mesozoic
e. Paleozoic
f. Phanerozoic
g. Precambrian
h. Proterozoic

Time Chart

Note: The oldest is always written at the bottom (just like sedimentary rock layers).

_____ 1. recent life
_____ 2. middle life
_____ 3. ancient life
_____ 4. visible life (named this because finally organisms had hard parts that could get preserved as fossils)
_____ 5. beginning life (a misnomer)
_____ 6. first life forms actually found here
_____ 7. Earth's beginnings (refers to hell)
_____ 8. all time from Earth's beginning up to the Phanerozoic

CARBON 14 DATING

Fill in the blanks, using the choices offered.

carbon 12;   decreases;   eating;   increases;   is; is not;   nitrogen 14;   photosynthesis;   plants;   remains; starts;   stays the same   stops

Organisms get carbon 14 into their systems because they eat, or, more scientifically, because green _____ extract carbon 14 from the atmosphere through the process of _____, and they are the basis of the food chain. All during an organism's lifetime, the carbon 14 in its system _____ decaying, but the amount of carbon 14 _____ because it is constantly replaced by eating. On dying, replacement of carbon 14 _____ and the amount of carbon 14 in the organism _____. Scientists determine when the creature died by interpreting the ratio of carbon 14 to _____.

Short-Answer Questions

Answer with a few words, numbers, or short sentences, or choose from the words in parentheses.

1. Define a second
   a. as a fraction of a year. _____
   b. in terms of behavior of the cesium atom. _____
   _____
   _____

   Your instructor will probably not expect you to memorize these numbers but will probably expect you to realize a second, properly defined, is a precise measurement when based on an "astronomic clock" (the year) and an extremely precise measurement when based on an "atomic clock" (the cesium atom).

2. Earth is essentially a sphere that completes one rotation on its axis every day. In doing this, through how many degrees does it turn (i.e., how many degrees in a circle)? _____
   How many hours are there in 1 day? _____
   Divide your first answer by your second answer. _____
   This shows that Earth turns through _____ of longitude every _____ h of time, so this longitude range equates to one time zone.

3. Earth rotates from west to east; this means locations to the east of your location are _____ (ahead of, behind) you in time.

4. When the time in Greenwich, England (0° longitude), is 6 A.M. Thursday, what is the time in Denver, Colorado (105° west longitude)? _____
   _____

5. What significant time-keeping event happened in 1883? _____
   _____

6. There is a tongue-in-cheek megalith in Nebraska called Carhenge, where old cars are set upright in a circle. What famous megalith is this copying, and what is the significance of the stone (or car) placements? _____
   _____

7. What principle gives scientists the right to interpret the past and to predict the future? _____
   _____

8. When discussing the dating of geologic events, the terms "numerical age" and "absolute age" have the same meaning. Why do some geologists favor "numerical" over "absolute"? _____
   _____
   _____

9. a. Name the location along the Scottish coast where Hutton puzzled over the unusual positions of some rock layers and formulated some now-famous ideas to explain the situation. _____

b. What is the geologic term for the rock configuration that puzzled Hutton? _____

10. Sedimentary beds deposited at one location during a given interval of geologic time may be totally different from the beds deposited at another location during the same time interval. Why? (Keep it simple; it's just common sense.) _____

11. Why is there no place on Earth that provides a complete rock record of Earth's history? _____

12. Name three things that period names (the subdivisions of eras) may refer to. _____

13. What famous biologist was influenced by Charles Lyell's book, *Principles of Geology*? _____

14. a. The first simple life forms (bacteria) appeared in what eon? _____
    b. The first complex invertebrates appeared in what eon? _____
    c. The first abundance of shelled invertebrates appeared at the time boundary between what two geologic time intervals? _____
    d. The first vertebrates (fish) appeared in what period? _____
    e. The first land plants appeared in what period? _____
    f. The first reptiles appeared in what period? _____
    g. The first dinosaurs appeared in what period? _____
    h. The first mammals appeared in what period? _____
    i. The first birds appeared in what period? _____

15. Sometimes a form of life was so dominant it characterized an era. For example, the _____ era is known as the Age of Dinosaurs, and the _____ era as the Age of Mammals.

16. Choose the proper answer for the following statements concerning radiometric decay:

    SS—Stays the Same   D—Decreases   I—Increases

    As time passes and a radioactive isotope continues to decay,
    a. the half-life period of the isotope _____.
    b. the actual number of atoms decaying in a half-life period _____.
    c. the percentage of radioactive isotope destroyed each half-life period _____.
    d. the amount of parent isotope _____.
    e. the amount of daughter isotope _____.

17. List five radioactive decay pairs that have been particularly useful to geologists in dating rocks. _____

18. List four steps in the procedure of radiometric dating. _____

19. Radiometric dating of a material gave a date of 300 million years old, with an uncertainty of measurement of 0.5%. What is the proper way to state the age of this material? _____

20. The real meaning of a radiometric date is different for different rock types. Radiometric dating of
    a. an igneous rock tells you _____.
    b. a metamorphic rock tells you _____.
    c. a sedimentary rock tells you _____.

21. What type of material is the carbon 14 dating process used for? _____

22. a. The half-life of carbon 14 is _____ years.
    b. It can be used to date back to only about _____ years ago.

23. List four phenomena, all of which exhibit seasonal changes, that are used to date geologic (and anthropologic) events. _____

24. How do the blocking temperatures of minerals compare to the melting temperatures of those same minerals? _____

25. What are the oldest living trees, and how old do they get? _____

26. a. Name the element that has proven useful in determining past global temperatures because the ratio of two of its isotopes varies with temperature changes. _____
    b. Where do geologists find the ancient samples of this element that they use to decipher past climates? _____

27. How far back in time have geologists gone with their analyses of Greenland ice cores? _____

28. a. What is the name of the field of study that produced a reference column of Earth's magnetic-field reversals? _____
    b. How far back in time does the reference column go? _____

29. State the flaws in each of the following early dating techniques used to figure Earth's age:
    a. Divide thickness of the thickest available column of sediment by rate of deposition of sediments. _____

138 | Chapter 12

b. Divide current salinity of the ocean water by the rate of salt flowing into oceans yearly. _____

c. Figure how long it would take Earth to cool down from its starting temperature (Sun temperature) to its current temperature. _____

30. Give the locations where the following rocks were found, and the dates assigned to each:
    a. oldest rocks on Earth? _____
    b. oldest mineral grains in a rock _____
    c. oldest rocks in our solar system _____

31. If Earth really is 4.6 billion years old, why haven't scientists found any rock samples this old? _____

32. Many different analogies have been offered to try to make humans understand the enormity of geologic time. One very popular one compares Earth's entire time of existence to 1 calendar year. Give dates or times for the following significant events, based on this 1-year analogy.
    a. The first life form (bacteria) appears. _____
    b. Pangaea forms. _____
    c. The first mammals and the first dinosaurs appear. _____
    d. Dinosaurs become extinct. _____
    e. The first human-like ancestors appear. _____
    f. Our species, *Homo sapiens*, appears. _____
    g. All recorded human history occurs. _____

Making Use of Diagrams

TELLING TIME, GEOLOGICALLY SPEAKING

There are two general ways to tell time geologically; you'll work with both methods in this exercise. Relative dating is a comparison; something is younger than or older than something else. Numerical dating (absolute dating) gives a numerical age to a geologic material or event. Relative dating methods are based on a few simple geologic principles and a lot of common sense. Some of these methods were devised more than three centuries ago. Numerical dating is based on the process of radioactive decay, and therefore wasn't developed until after the discovery of radioactivity in about 1900.

RELATIVE DATING

Examing the Rocks

The following are bare bones exercises in putting geologic happenings in chronological order. You don't have to interpret such things as baked zones around igneous intrusions or explain when the tilting of layers took place. All you are asked to do is *decide the sequence of events and record them by letters given*. You'll need to think about what you're looking at; is it a sedimentary layer that was laid down or an intrusion that had to intrude into something already there or an event that affected the rock that was already there? Apply the principles of relative dating (superposition, cross-cutting relations, etc.) and think real world geology. Start at the bottom and work up. List the layers on lines provided, putting *the oldest at the bottom*.

1.

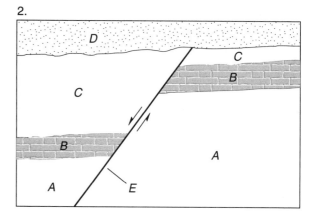

1. _____ youngest

   _____

   _____ oldest

   A and *B* are both igneous intrusions. Which letter (*A*, *B*, or *C*) represents a dike? _____

2.

2. _____ youngest

   _____

   _____

   _____

   _____ oldest

   The fault (letter _____) is younger than _____ but older than _____.

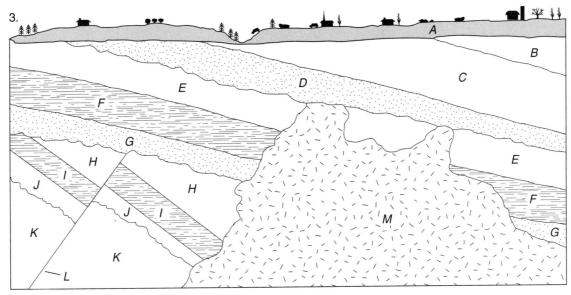

3. (13) _____ youngest (9) _____ (6) _____ (3) _____
   (12) _____ (8) _____ (5) _____ (2) _____
   (11) _____ (7) _____ (4) _____ (1) _____ oldest
   (10) _____

4. (13) _____ youngest (6) _____
   (12) _____ (5) _____
   (11) _____ (4) _____
   (10) _____ (3) _____
   (9) _____ (2) _____
   (8) _____ (1) _____ oldest
   (7) _____

Note the inclusion of a piece of rock K within rock J. This tells you rock J is _____ (older than, younger than) rock K. Also note the pieces of rock J within sedimentary layer H. This tells you rock J is _____ (older than, younger than) rock H. Note how you cannot determine the age of intrusion I very precisely. The best you can say is intrusion I could have happened any time after _____.

140 | Chapter 12

THE ULTIMATE RELATIVE DATING EXERCISE

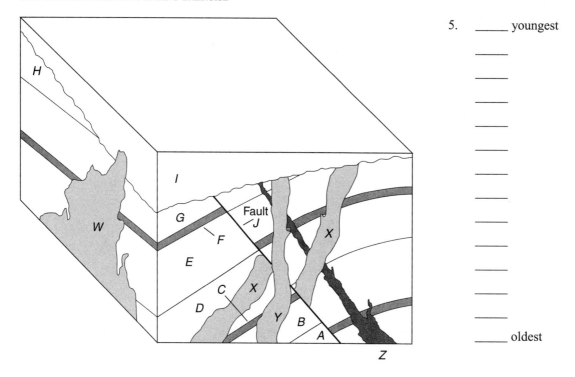

5. _____ youngest

_____
_____
_____
_____
_____
_____
_____
_____
_____
_____
_____
_____
_____ oldest

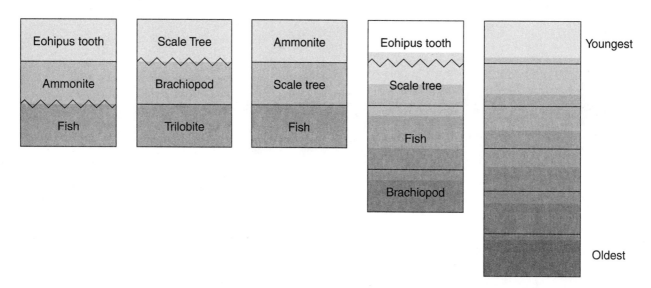

FOSSIL CORRELATION

The four columns above represent fossil-bearing sedimentary rock layers at four different sites. The sedimentary layers are continuous except at the zigzag lines, which represent unconformities. Compare the four different sedimentary sequences, determine the relative ages of the fossils, and list them in the proper global order in the time scale to the right. Remember that an unconformity is a break in the record, so you can't trust the sequence of fossil succession across it.

NUMERICAL (ABSOLUTE) DATING

Radioactive Decay

Radioactive decay happens because radioactive isotopes give off some form of radiation and transform themselves into different elements. The following exercise guides you through the decay process of uranium (U^{238}), as it gives off a sequence of alpha and beta particles and eventually becomes lead (Pb^{206}). It looks long and complicated, and the process itself takes a long time (half-life of 4.5 billion years), but working through the process involves only very simple addition or subtraction. To do the exercise, keep in mind:

An *alpha particle* is a unit of two protons and two neutrons that's emitted from the nucleus. This causes the atomic number (count the protons) to go _____ (up, down) by _____ and the atomic weight (count the protons plus neutrons) to go _____ (up, down) by _____.

A *beta particle* is a nuclear electron, which means a neutron has split and released its negative (electron) component. This causes the atomic number to go _____ (up, down) by _____ but the atomic weight is unchanged (because the weight of an electron is negligible).

The first line is done for you. For each succeeding line, fill in the one piece of missing information—the atomic number, the atomic weight, or the alpha or beta emission.

Notation used:

$_{\text{atomic number}}\textbf{Symbol}^{\text{atomic weight}}$ Example: $_{92}U^{238}$

Start here

$_{92}U^{238} \xrightarrow{\alpha} {}_{90}Th^{234}$ $\_Po^{\_\_} \rightarrow {}_{82}Pb^{\_\_}$

$_{90}Th^{234} \xrightarrow{\beta} \_Pa^{234}$ $_{82}Pb^{\_\_} \xrightarrow{\beta} \_Bi^{\_\_}$

$\_Pa^{234} \xrightarrow{\beta} {}_{92}U^{\_\_}$ $\_Bi^{\_\_} \rightarrow {}_{84}Po^{214}$

$_{92}U^{\_\_} \rightarrow {}_{90}Th^{230}$ $_{84}Po^{214} \rightarrow \_Pb^{210}$

$_{90}Th^{230} \xrightarrow{\alpha} \_Ra^{\_\_}$ $\_Pb^{210} \rightarrow \_Bi^{210}$

$\_Ra^{\_\_} \xrightarrow{\alpha} \_Rn^{\_\_}$ $\_Bi^{210} \xrightarrow{\beta} \_Po^{\_\_}$

$_{86}Rn^{222} \rightarrow {}_{84}Po^{218}$ $\_Po^{\_\_} \xrightarrow{\alpha} {}_{82}Pb^{206}$

(go to top of right hand column)

Half-Lives

The Process: Ratios, Parent to Daughter

The rate of radioactive decay is measured in terms of half-lives, which require a study of the relative amounts of the beginning radioactive element (parent isotope) and the newly formed element (daughter isotope).

For this exercise assume you are following the progress of 64 atoms of radioactive parent isotope as they go through a series of half-life time periods.

1. When decay starts:

    Number of atoms of parent _____

    Number of atoms of daughter _____

    Ratio parent : daughter _____

2. After one half-life:

    Number of atoms of parent _____

    Number of atoms of daughter _____

    Ratio parent : daughter _____

3. After two half-lives:

    Number of atoms of parent _____

    Number of atoms of daughter _____

    Ratio parent : daughter _____

4. After three half-lives:

    Number of atoms of parent _____

    Number of atoms of daughter _____

    Ratio parent : daughter _____

5. After four half-lives:

    Number of atoms of parent _____

    Number of atoms of daughter _____

    Ratio parent : daughter _____

6. After five half-lives:

    Number of atoms of parent _____

    Number of atoms of daughter _____

    Ratio parent : daughter _____

7. After six half-lives:

    Number of atoms of parent _____

    Number of atoms of daughter _____

    Ratio parent : daughter _____

Half-Life Problems

1. A scientist dating material by the radiometric method finds there is a ratio of 1:15, parent material to daughter material, present. How many half-lives have occurred? _____

2. In September 1991, hikers in the Alps found a body that obviously was very old. Dating it by the carbon 14 method (half-life carbon 14 = 5,730 years) showed a ratio of 1:1, parent to daughter product. (This is a slight simplification of the true data concerning this real-life find known as the Ice Man.) How old was the corpse? _____

UNCONFORMITIES

The figures below illustrate different situations that are unconformities. Use a colored pen to draw a line along the unconformable surface in each figure, and label what type of unconformity is shown:

angular unconformity   disconformity   nonconformity

Short-Essay Questions

1. a. Explain the principle of uniformitarianism.
   b. This principle can be worded "the present is a key to the past." Is it also proper to say "the present is a key to the future"?
   c. Why did acceptance of uniformitarianism cause scientists of the early 1800s to reevaluate their beliefs about the age of Earth?

2. A geologist finds pillow lava high in the Himalayas and states that, despite its current elevation, that rock was once sea floor. How could this be so?

3. The geologic time chart is not a good example of a carefully thought out, orderly plan. For instance, only some eras show subdivisions called epochs; Proterozoic, which means "beginning life," was not the time when life began; and Tertiary and Quaternary are mentioned, but not Primary or Secondary. Why is something as important as the time scale so flawed? (To answer, start with information in the chapter and then exercise your own common sense and knowledge of human nature.)

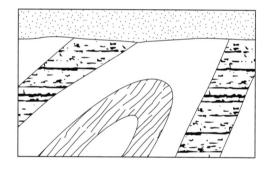

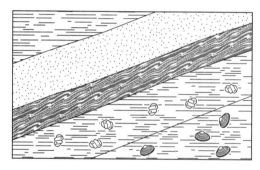
 Cenozoic corals
 Paleozoic clams

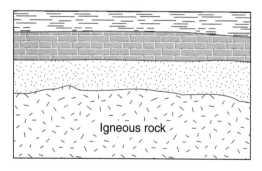

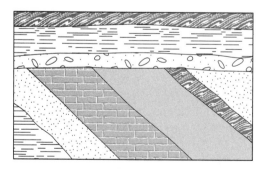

Practice Test

1. Uniformitarianism
   a. means Earth's surface—its oceans, continents, and atmosphere—have always been the same and always will be the same.
   b. is an outdated theory that states catastrophes like major volcanic eruptions no longer happen.
   c. is illustrated by scientists' seeing pillow lava form only under water, then theorizing that pillow lava found high in the mountains today did nevertheless form under water.
   d. means neither the scientific past nor future should be predicted because physical laws change over time.
   e. is an idea that was formulated by the great physicist Becquerel in about 1900.

2. What specifically must have happened in the reaction $_{88}Ra^{226} \rightarrow {}_{86}Rn^{222}$?
   a. alpha emission
   b. beta emission
   c. gamma emission
   d. electron capture
   e. fission

3. Potassium, atomic number 19, decays to produce argon, atomic number 18. What must happen in the decay process?
   a. alpha emission
   b. beta emission
   c. gamma emission
   d. electron capture
   e. fission

4. The uranium decay series you did as a learning activity was a long sequence of alpha and beta emissions. For this question, figure what happens for only a part of the series. Start with uranium, atomic number 92, atomic weight 234. Follow it through 5 alpha emissions and 2 beta emissions. The result is
   a. lead (atomic number 82, atomic weight 220).
   b. bismuth (atomic number 83, atomic weight 214).
   c. polonium (atomic number 84, atomic weight 214).
   d. polonium (atomic number 84, atomic weight 209).
   e. thorium (atomic number 90, atomic weight 218).

5. Which of the following shows four time divisions listed from oldest to youngest?
   a. Precambrian, Mesozoic, Archean, Cenozoic
   b. Precambrian, Paleozoic, Mesozoic, Cenozoic
   c. Phanerozoic, Precambrian, Mesozoic, Cenozoic
   d. Hadean, Cenozoic, Mesozoic, Proterozoic
   e. Archean, Paleozoic, Hadean, Proterozoic

6. Choose the proper listing of names to fit the following three descriptions: Age of Mammals, Age of Dinosaurs, and longest geologic time period.
   a. Mesozoic, Cenozoic, Paleozoic
   b. Cenozoic, Hadean, Paleozoic
   c. Mesozoic, Cenozoic, Proterozoic
   d. Cenozoic, Mesozoic, Precambrian
   e. Phanerozoic, Mesozoic, Precambrian

7. Back in the Middle Ages, the alchemist's dream was to change mercury into gold. Mercury has atomic number 80 and atomic weight 201. Gold has atomic number 79 and atomic weight 197. Nowadays we realize this could happen if there were
   a. one alpha emission only.
   b. one electron capture only.
   c. one beta emission only.
   d. one alpha emission and one beta emission.
   e. one electron capture and one alpha emission.

8. Which of the following is NOT a method to determine numerical age?
   a. carbon 14 dating
   b. radiometric decay of uranium to lead
   c. fission-track dating
   d. dendrochronology
   e. cross-cutting relations

9. Which of the following statements is FALSE?
   a. Scientists can date any radioactive material back only three half-lives; after this there's not enough parent material left to measure.
   b. Carbon 14 dating is used to date anything that was once alive and has not been petrified.
   c. Radiometric dating can be used only in a closed system, where neither parent nor daughter material has escaped.
   d. Radiometric dating of rocks is limited almost exclusively to igneous rocks.
   e. The process of metamorphism resets the clock for radiometric dating.

10. Which of the following statements is FALSE?
    a. A formation is the name of a rock layer identified by such factors as rock type and approximate geologic age.
    b. An unconformity is a break in the rock record that indicates the area was under water for millions of years.
    c. The generally accepted age of Earth is 4.6 billion years.
    d. A fossil is an indication of life in the past.
    e. If there are several layers of sedimentary rock, the oldest layer of rock will be on the bottom.

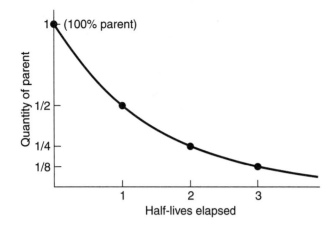

11. Which of the following statements is FALSE? On the diagram shown above,
    a. the area under the curve represents the amount of parent material present.
    b. if the half-life of the parent material is 4,000 years, 3 on the horizontal axis of the graph represents 12,000 years.
    c. if 2 g of parent material are present after three half-lives, there will also be 14 g of daughter material present.
    d. there are equal amounts of parent and daughter material present after the passage of two half-lives.
    e. the ratio of parent to daughter isotopes changes with the passage of each successive half-life.

12. If the time in Greenwich, England (longitude 0°), is 11 P.M. Monday, the time in Baghdad, Iraq (longitude 45° east), must be
    a. 2 A.M. Tuesday.
    b. 8 P.M. Monday.
    c. 3 A.M. Monday.
    d. 3 P.M. Monday.
    e. The correct data are not given; latitude numbers are needed.

13. Which of the following statements is FALSE? The principle of
    a. superposition says in a sequence of sedimentary beds, the youngest is on top.
    b. original continuity says sedimentary layers began as continuous expanses of sediment.
    c. inclusions says rock containing inclusions is older than the inclusions.
    d. cross-cutting relations says the feature doing the cutting is younger than the feature it cuts.
    e. original horizontality says sedimentary layers started out flat.

14. A rock is given a radiometric date of 300,000,000 years, with an uncertainty of measurement of 1%. This means the rock's age
    a. has a 1% chance of being 300,000,000 years old.
    b. is anything between 3,000,000 and 300,000,000 years old.
    c. is anything between 3,000,000 and 303,000,000 years old.
    d. is anything between 297,000,000 and 303,000,000 years old.
    e. will remain unknown; there's only a 1% chance any date given will be correct.

15. Carbon 14 becomes part of organisms because green plants extract it from the atmosphere for their photosynthesis, and green plants get consumed by animals.
    a. True
    b. False

16. A disconformity is a type of unconformity in which sedimentary rocks overlie either igneous or metamorphic rocks.
    a. True
    b. False

17. The oldest rock material found anywhere in our solar system is zircon mineral grains from sandstones in Australia.
    a. True
    b. False

18. If you equate all Earth history to 1 calendar year, all human history occupies the week from Christmas to New Year's Eve.
    a. True
    b. False

19. A time zone is a 15°-wide band of latitude in which all clocks keep the same standard time.
    a. True
    b. False

20. It's not possible to say just when an individual radioactive atom will decay, but it is possible to say when half of an existing quantity of radioactive material will be gone.
    a. True
    b. False

21. Correlation matches up rock layers across distances on the basis of similar sequences of rock layers and similar fossils in the layers.
    a. True
    b. False

22. In the 1600s James Hutton figured out how fossils can occur in rocks.
    a. True
    b. False

23. The boundary surface between two stratigraphic formations is called a key bed.
    a. True
    b. False

24. Numerical (or absolute) dating is just a comparison of age; relative dating assigns numbers.
    a. True
    b. False

25. The first half-life period of carbon 14 is 5,730 years; the second half-life period of carbon 14 is half of this, and the number of years is cut in half for each succeeding half-life period.
    a. True
    b. False

26. A radioactive isotope of the element potassium decays to produce argon. If the ratio of argon to potassium is found to be 7 : 1, how many half-lives have occurred? _____

27. Charcoal (burned wood) that was used to make prehistoric drawings on cave walls in France was scraped off and analyzed. The results were 4 mg carbon 14 (parent isotope) and 60 mg nitrogen (daughter isotope). The half-life of carbon 14 is 5,730 years. How old are the cave drawings? _____

28. On the diagram shown, several surfaces between rock layers are identified by letters. Two of these surfaces are unconformities. Identify them by letter: surface _____ and surface _____.

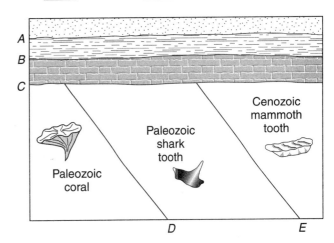

29. Number the events in the figure at the bottom of the page in the proper order of occurrence. Make the first (oldest) event number 1.
    _____ a. faulting (fault F)
    _____ b. deposition of layer D
    _____ c. deposition of layer E
    _____ d. deposition of layer A
    _____ e. deposition of layer B
    _____ f. intrusion of batholith G
    _____ g. Deposition of layer C
    _____ h. intrusion of dike H

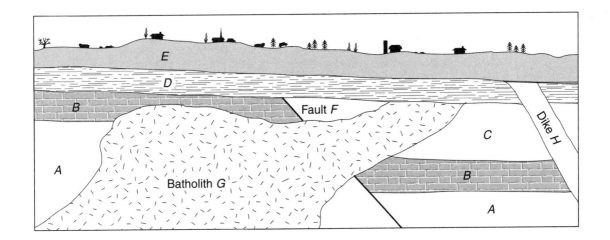

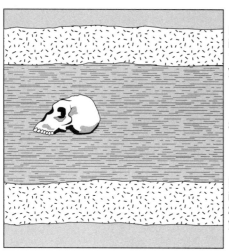

30. A hominid (human family) skull was found in a shale layer between two fine-grained igneous rock layers. There were no baked zones on the rocks above the igneous layers. The skull had been 100% fossilized and changed to stone; there was no original skull material left. On the basis of this information and the diagram, decide which of the following statements are true.

    Letters of all true statements: _____

a. The skull should be dated by the carbon 14 method, since this is the method used for organic material older than 1 million years.
b. The igneous layers can be dated by radiometric methods; the shale layers cannot be dated radiometrically.
c. The skull could logically be 2.5 million years old.
d. The skull could logically be 1.75 million years old.
e. The skull could logically be 0.9 million years old.
f. The igneous layers must be lava flows, because they are fine-grained rock and there are no baked zones on the rock above them.
g. The skull itself cannot be dated, because it's no longer organic material and it is not igneous material. The age of the skull is determined by its position between the igneous layers.

31. The four columns below represent fossil-bearing sedimentary rock layers at four different sites. The sedimentary layers are continuous except at the zigzag lines, which represent unconformities. Compare the sites, determine the relative ages of the fossils, and list them in the far right column.

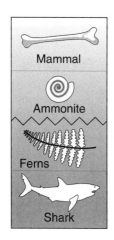
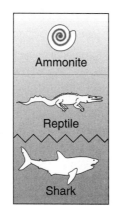

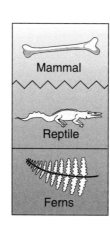

ANSWERS

Completion

1. geologic time
2. second
3. day
4. time zones
5. Greenwich mean time
6. fossil
7. principle of uniformitarianism
8. principle of superposition
9. relative age
10. numerical age (or absolute age)
11. xenoliths
12. inclusions
13. geologic history
14. fossil assemblage
15. principle of fossil succession
16. range of a fossil
17. unconformity
18. hiatus
19. stratigraphic column
20. formation (or stratigraphic formation)
21. contact
22. correlation
23. lithologic correlation
24. fossil correlation
25. key bed
26. geologic map
27. geologic column
28. eon
29. eras
30. periods
31. epochs
32. diversification
33. radiometric dating (or geochronology)
34. atomic number
35. atomic weight
36. isotopes
37. radioactive isotopes
38. parent isotope
39. daughter isotope
40. half-life
41. uncertainty (of a radiometric date)
42. blocking temperature
43. growth rings
44. rhythmic layering
45. dendrochronologist
46. magnetostratigraphy
47. fission track
48. geologic time scale

Matching Questions

FAMOUS PERSONS WHO STUDIED GEOLOGIC TIME

1. i
2. b
3. h
4. c
5. e
6. g
7. f
8. d
9. a

PRINCIPLES USED TO DETERMINE RELATIVE AGES

1. d
2. e
3. f
4. g
5. a
6. b
7. c

MAJOR DIVISIONS OF GEOLOGIC TIME

1. b
2. d
3. e
4. f
5. h
6. a
7. c
8. g

CARBON 14 DATING

plants; photosynthesis; is; stays the same; stops; decreases; carbon 12

Short-Answer Questions

1. a. 1/31,556,925.947 of the year 1900
   b. the time needed for the magnetic field of a cesium atom to flip polarity 9,192,631,770 times

2. 360°; 24 h; 15; 15°; 1

3. ahead of

4. 105°/15 = 7; Greenwich is 7 time zones (7 h) into the future from Denver, so the time in Denver is 11 P.M. Wednesday.

5. Countries worldwide agreed to keep standard time, set in relation to Greenwich, England, and based on a 15°-longitude time zone.

6. Copying Stonehenge, England; the alignment of various stone columns (or cars) with the rising Sun or a star defines specific days of the year.

7. principle of uniformitarianism

8. "Absolute" implies constant, forever, never changing. As geologic dating techniques have grown more sophisticated, some of the dates assigned to events have changed, so calling a date that is subject to change "absolute" is not quite truthful.

9. a. Siccar Point   b. "angular unconformity"

10. Sedimentary beds are the result of depositional environments, and environments are not the same all over.

11. because there's no place that hasn't, over time, experienced nondeposition or erosion (i.e., every stratigraphic column contains unconformities)

12. geographic locations where there are good outcrops of these rocks, characteristics of the rock, and Latin roots of the word that describe the rock

13. Charles Darwin

14. a. Archean   b. Proterozoic   c. Precambrian and Cambrian   d. Ordivician   e. Silurian   f. Pennsylvanian   g. Triassic   h. Triassic   i. Cretaceous

15. Mesozoic; Cenozoic

16. a. SS   b. D   c. SS   d. D   e. I

17. $U^{238} \to Pb^{206}$; $U^{235} \to Pb^{207}$; $K^{40} \to Ar^{40}$; $Rb^{87} \to Sr^{87}$; $Sm^{147} \to Nd^{143}$

18. Collect unweathered rock; separate minerals; extract parent and daughter isotopes by acid solution or laser evaporation; and analyze by spectrograph to obtain parent/daughter ratio.

19. (300,000,000 × 0.005 = 1,500,000) date of material is 300 + or − 1.5 million years.

20. a. when the magma or lava cooled to form a solid, cool igneous rock
    b. when the rock cooled from the high temperature of metamorphism to a value below the blocking temperature
    c. when the minerals making up the grains of the rock came into being, not when the sample became a sedimentary rock

21. anything composed of organic material

22. a. 5,730 years   b. 70,000 years

23. tree ring dating (dendrochronology); sediment layers in rivers; organic productivity of lakes and oceans; and growth rate of chemically precipitated sedimentary rocks, like travertine

24. Typically blocking temperatures are hundreds of degrees cooler than melting temperatures.

25. bristle cone pines; almost 4,000 years old

26. a. oxygen   b. water in the ice caps of Greenland

27. 160,000 years

28. a. magnetostratigraphy
    b. to about 4.5 million years ago

29. a. There's no place on Earth with an uninterrupted column of sediment (i.e., no place without unconformities); sedimentary rocks can change to metamorphic rocks; the assumption of constant rate of sediment deposit is invalid.
    b. Ocean salinity remains generally constant because excess salts precipitate out.
    c. This method neglects the increase of Earth's temperature due to radioactive decay.

30. a. Wyoming, Canada, Greenland, and China; 3.96 billion years
    b. zircon crystals in Australian sandstone; 4.1–4.2 billion years
    c. some meteors and Moon rocks; 4.6 billion years

31. Earth's crust started out and remained too hot for the radiometric clock to start for roughly 1/2 billion years (i.e., it didn't get below the blocking temperature).

32. a. February 21   b. December 7   c. December 15
    d. Christmas day   e. December 31, 3 P.M.
    f. 11 P.M., December 31   g. last 30 s of the year

Making Use of Diagrams

RELATIVE DATING

1.

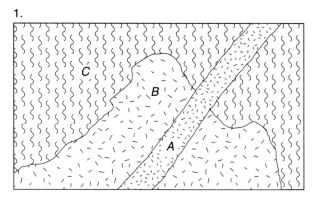

1. _A_ youngest

   _B_

   _C_ oldest

   A and B are both igneous intrusions.

   Which letter (A, B, or C) represents a dike? _A_

2.

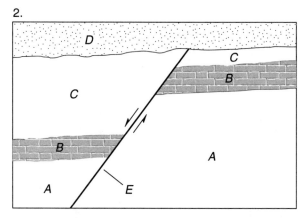

2. _D_ youngest

   _E_

   _C_

   _B_

   _A_ oldest

   The fault (letter _E_) is younger than _C_ but older than _D_.

Deep Time: How Old Is Old? | 149

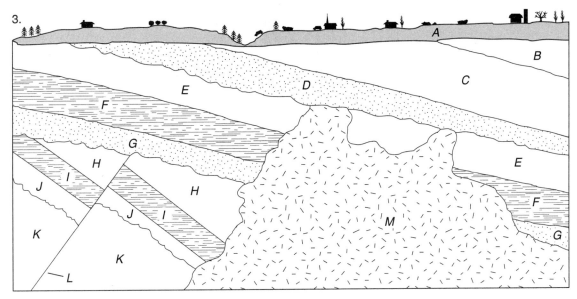

3. (13) _A_ youngest (10) _D_   (7) _F_ (4) _H_ (1) _K_ oldest
   (12) _B_   (9) _M_   (6) _G_ (3) _I_
   (11) _C_   (8) _E_   (5) _L_ (2) _J_

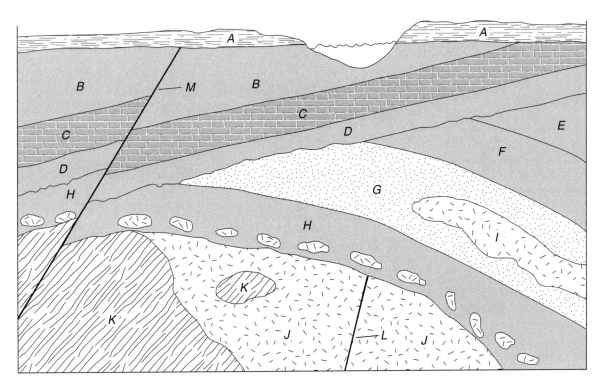

4. (13) _A_ youngest (10) _C_ (7) _F_ (4) _H_ (1) _K_ oldest
   (12) _M_   (9) _D_   (6) _I_ (3) _L_
   (11) _B_   (8) _E_   (5) _G_ (2) _J_

younger than; older than; G

THE ULTIMATE RELATIVE DATING EXERCISE

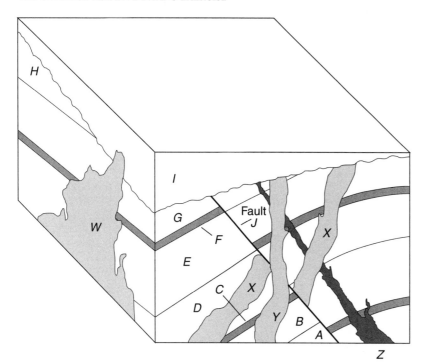

5. _W_ youngest
_I_
_Y_
_J_
_X_
_Z_
_H_
_G_
_F_
_E_
_D_
_C_
_B_
_A_ oldest

FOSSIL CORRELATION

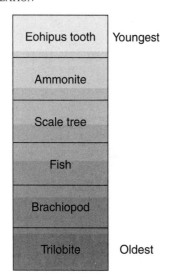

NUMERICAL (ABSOLUTE) DATING

Radioactive Decay

down; 2; down; 4; up; 1
See the figure in the column on the right for the radioactive decay of U^{238}.

$$_{92}U^{238} \xrightarrow{\alpha} {_{90}Th^{234}}$$
$$_{90}Th^{234} \xrightarrow{\beta} {_{91}Pa^{234}}$$
$$_{91}Pa^{234} \xrightarrow{\beta} {_{92}U^{234}}$$
$$_{92}U^{234} \xrightarrow{\alpha} {_{90}Th^{230}}$$
$$_{90}Th^{230} \xrightarrow{\alpha} {_{88}Ra^{226}}$$
$$_{88}Ra^{226} \xrightarrow{\alpha} {_{86}Rn^{222}}$$
$$_{86}Rn^{222} \xrightarrow{\alpha} {_{84}Po^{218}}$$
$$_{84}Po^{218} \xrightarrow{\alpha} {_{82}Pb^{214}}$$
$$_{82}Pb^{214} \xrightarrow{\beta} {_{83}Bi^{214}}$$
$$_{83}Bi^{214} \xrightarrow{\beta} {_{84}Po^{214}}$$
$$_{84}Po^{214} \xrightarrow{\alpha} {_{82}Pb^{210}}$$
$$_{82}Pb^{210} \xrightarrow{\beta} {_{83}Bi^{210}}$$
$$_{83}Bi^{210} \xrightarrow{\beta} {_{84}Po^{210}}$$
$$_{84}Po^{210} \xrightarrow{\alpha} {_{82}Pb^{206}}$$

Half-Lives

The Process: Ratios, Parent to Daughter

1. 64; 0; 1 : 0
2. 32; 32; 1 : 1
3. 16; 48; 1 : 3
4. 8; 56; 1 : 7
5. 4; 60; 1 : 15
6. 2; 62; 1 : 31
7. 1; 63; 1 : 63

Half-Life Problems

1. 4
2. ~5,730 years (because one half-life has passed)

UNCONFORMITIES

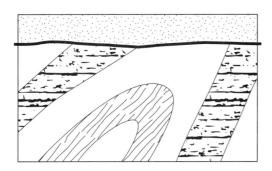

angular unconformity

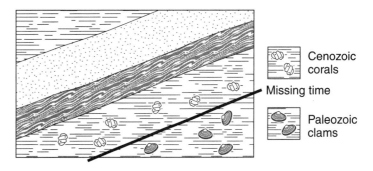

disconformity

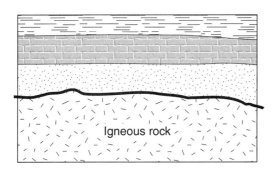

nonconformity

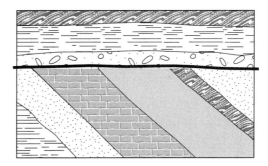

angular unconformity

Short-Essay Questions

1. a. The same physical and chemical processes of today operated in the past and will operate in the future. Gravity will always act like the gravity we know, oxidation will always cause the combustion of flammable material (provided there's enough heat and oxygen in the surroundings), radioactive decay will always happen, and so on.

   b. Since the processes are constant and expected to continue to be constant, the expression "The present is a key to the future" is a truthful statement. In fact, without belief in uniformitarianism, scientists have no right to interpret the scientific past or to predict the scientific future. Since scientists routinely do both, they are demonstrating a belief in the principle, even if they avoid saying (and spelling) the word.

   c. We observe the physical Earth today and see the very slow rate at which many Earth-affecting processes operate. Since the processes have been constant over time, it is obvious Earth needed a very long time to achieve its current condition, much longer (and so to be of much greater age) than was first suggested.

2. Pillow lava today forms only on the ocean bottom. The same processes that produce it today produced it in the past, so despite the fact it has been greatly uplifted, it still must have formed on the ocean bottom. This logic demonstrates the use of the principle of uniformitarianism.

3. The geologic time chart developed haphazardly between 1760 and 1845. During this time geologists gathered new data, changed their interpretations of

existing information, and often renamed time intervals. In those days communications weren't very good, so geologists worldwide could neither share knowledge nor make joint decisions. As you might suspect, human nature and egos did sometimes play a role, and if name duplications or diverse interpretations occurred, the people involved didn't always want to abandon their own ideas.

Practice Test

1. c Features of the Earth are constantly changing; catastrophic events still happen; the present is a key to the past and to the future; Becquerel discovered radioactivity whereas Hutton formulated uniformitarianism.

2. a The atomic number went down 2 and the atomic weight went down 4; this signifies alpha emission.

3. d The atomic number went up 1; this means a proton captured an electron and became a neutron.

4. c The alpha emissions cause the atomic number to go down a total of 10 and the atomic weight to go down a total of 20; the beta emissions cause the atomic number to go up 2 and the atomic weight is unchanged; net result, atomic number down 8, to 84, and atomic weight down 20, to 214. Polonium^{214} wins.

5. b The correct order is Precambrian, Paleozoic, Mesozoic, Cenozoic.

6. d Mammals were the dominant life form during the Cenozoic, and dinosaurs were dominant during the Mesozoic; the Precambrian represents more than seven-eighths of all Earth history.

7. d One alpha emission takes the atomic number down 2, to 78, and the atomic weight down 4 to 197. One beta emission doesn't affect the atomic weight but raises the atomic number by 1, up to the required 79.

8. e This is a method used for relative, not numerical, dating.

9. a After three half-lives there's still 1/8 of the radioactive parent material left, which is easily detected. Carbon 14 can be detected for about 12 half-life periods.

10. b An unconformity is a break in the rock record indicating erosion or nondeposition; being under water would be an excellent environment for deposition.

11. d Equal amounts of parent and daughter material exist only after one half-life.

12. a There is 45/15 = 3 h difference; Baghdad is to the east of England, so time there is into the future.

13. c The principle of inclusions states that the inclusions are older than the surrounding rock.

14. d Since 1% of 300 million is 3 million, the rock's age is 300 million, plus or minus 3 million.

15. True Carbon 14 forms when nitrogen 14 in the atmosphere captures an electron; green plants extract carbon 14 and through photosynthesis incorporate it into the food chain.

16. False This is the definition of a nonconformity.

17. False The zircon grains are the oldest Earth rock material; some meteorites and Moon rocks are older (4.6 billion years old).

18. False Far less; only the last 30 seconds of the year.

19. False A time zone is a 15°-wide band of longitude; latitude has nothing to do with time keeping.

20. True The half-life concept doesn't apply to individual atoms, just to quantities of atoms; it is a statistical truth, not an observation of individual atom behavior.

21. True Correlation of rocks or of any other materials is a matching to achieve a more complete picture over a wide area.

22. False Nicholaus Steno figured this out.

23. False It's called a contact.

24. False The reverse is true.

25. False The half-life period of any element stays constant.

26. 3 half-lives (at end of one half-life, 1:1; two half-lives, 1:3, three half-lives, 1:7)

27. 22,920 years   (4 mg : 60 mg = 1:15; this ratio means 4 half-lives; 4 × 5,730 years = 22,920 years)

28. C (It's an angular unconformity.); E (It's a disconformity; the entire Mesozoic is missing.)

29. a. 4   e. 2
    b. 6   f. 5
    c. 8   g. 3
    d. 1   h. 7

30. b, d, f, g

31. Mammal
    Ammonite
    Reptile
    Ferns
    Shark
    Trilobite

CHAPTER 13 | A Biography of Earth

GUIDE TO READING

The last chapter dealt with how Earth's history has been worked out; this one deals with the story itself. It begins by pointing out that the history of Earth should include information about orogenies, changing shapes of continents, changing positions of continents, past environments, past climates, changes in sea level, and the evolution of life. The chapter organizes all of this information in terms of the major divisions of geologic time: the Precambrian (from Earth's beginning 4.6 billion years ago up to 545 million years ago) and its subdivisions (Hadean eon, Archean eon, and Proterozoic eon) and the rest of all time, the Phanerozoic, including its subdivisions (the Paleozoic era, including the Cambrian, Ordovician, Silurian, Devonian, Carboniferous, and Permian periods; the Mesozoic era, including the Triassic, Jurassic, and Cretaceous periods; and the Cenozoic era, with special attention paid to the Pleistocene Ice Age of the Quaternary period). That's a lot of names, but do learn them. They're basic to any discussion of Earth history.

The rest of Chapter 13 is the story itself; highlights of 4.6 billion years of change. This presents a problem. Even the most concise summary of 4.6 billion of anything is going to be long. What is essential to learn? The author has chosen major milestones in Earth's development that any well-informed general geologist should know. This guide narrows the focus even further. Still, basic literacy means different things to different students, to different teachers, and in different situations. For example, the particular facts required of you could depend on the location of your school. If you're in the Denver area, no doubt you'll have to know about the Ancestral Rockies, the Laramide Revolution, and the Western Interior Seaway. If you're located in Boston or Seattle, other particulars may seem more important to you or to your teacher. This chapter presents a wonderful opportunity to bring basic concepts to life by indulging your curiosity about the history of your present (or hometown) surroundings.

Whatever your goal, it's easy to get mired down in all the details and proper nouns. To avoid this, as you study, try to picture our planet as it was in the past—a truly alien world. It's not dull stuff; it's been the subject matter for popular comics and blockbuster movies for decades. Who hasn't heard of the Flintstones or Jurassic Park? Asteroid encounters have been exciting from Superman days up to *Deep Impact*. People everywhere wonder why the dinosaurs became extinct; maybe after reading this chapter you'll also wonder why the trilobites became extinct. As you read, remind yourself that this is not a fanciful movie plot; it's all based on scientific data and reasoning. Truth really is stranger than fiction.

LEARNING ACTIVITIES

Completion

Test your recall of new vocabulary terms.

_____ 1. depressed areas on the plains adjacent to a mountain front

_____ 2. the expanse of hot, molten ultramafic rock that rose from the mantle and flooded Earth's surface during Hadean times

_____ 3. large blocks of low-density materials that coalesced to become continents

_____ 4. long-lived blocks of durable continental crust

_____ 5. simple cells that lack complex internal structure

_____ 6. blue-green algae

7. carbon that has been incorporated into living tissue

8. layered mounds formed when sediment sticks to a sticky, mucous-like substance on the surface of colonies of cyanobacteria

9. a broad, low-lying region of exposed Precambrian rocks

10. a relatively thin layer of sedimentary rock that covers basement rock

11. mountain-building events that cause continental areas to grow by accretion of terranes

12. the first supercontinent of Proterozoic times

13. the second supercontinent of Proterozoic times, which resulted from the rearrangement of Rodinia

14. the huge collision event that was the last step in the formation of Rodinia in the Proterozoic

15. a thick accumulation of sediment along a tectonically inactive coast

16. past geography of Earth

17. organisms that make their own food, and in the process remove carbon dioxide, water, methane, and ammonia from the atmosphere and return oxygen to it

18. cells with complex internal structure, capable of building multicellular organisms

19. rare and unusual fossil assemblages of 670 million-year-old shell-less marine invertebrates, found in outcrops in southern Australia and England

20. the four continental pieces resulting from the splitting of Pannotia in the early Paleozoic

21. shallow seas that transgressed onto continents many times in Earth's past

22. the collision event at the end of the early Paleozoic that marked the beginning of the Appalachian orogen

23. a period of extensive diversification of life at the beginning of the Paleozoic

24. small teeth-like fossils of early Paleozoic times

25. the first large animals, which were sea-floor grazers of the Paleozoic

26. a 2 m long predator of Paleozoic seas

27. a reef-forming sponge

28. floating animals containing a serrated mineral band that appeared in Ordovician seas

29. a collision orogeny in the Devonian period that involved the Avalon continental block and provided material for the Catskill delta of the northeast and the Old Red Sandstone of the United Kingdom

30. the orogeny of the late Devonian in which a large island arc collided with Nevada

31. plants that have veins for transporting water and food

32. the late Paleozoic collision event that was the final step in forming the supercontinent Pangaea and the Appalachian and Ouachita Mountains

33. a wide band of deformation on the west side of the Appalachian Mountains

34. a distinctive style of deformation in which thrust faults push thin slabs of sedimentary strata above Precambrian basement

35. late Paleozoic mountains in what is now Colorado, New Mexico, and Wyoming

36. naked-seed plants (pines, spruces), which carry their seeds on surfaces within reproductive cones, not within the more protected parts of flowers

37. trees of late Paleozoic times with palm-like stalks and fern-like leaves, which are now extinct

38. two orogenies along the western edge of North America in late Paleozoic–early Mesozoic times

39. a major continental volcanic arc that formed along the western edge of North America during Cretaceous times and continued active into the Mesozoic

40. swimming reptiles of the Mesozoic

41. reef-building corals of the Mesozoic

42. flying reptiles

43. a shallow sea that crossed North America from the Arctic Ocean to the Gulf of Mexico during the Cretaceous period

44. an orogeny along the northwest edge of North America during late Mesozoic that involved huge thrust faults

45. an uplift of deep Precambrian rocks along deeply penetrating fault zones

_____ 46. a late Mesozoic orogeny that created the Rocky Mountains of the western United States

_____ 47. extremely large mantle plumes, some of which existed in late Mesozoic times

_____ 48. gases like carbon dioxide that contribute to atmospheric warming in a way analogous to the action of window glass

_____ 49. flowering plants that bear their seeds within the protection of flower parts

_____ 50. a solid extraterrestrial object (meteor, comet, or asteroid) that collides with Earth

_____ 51. a very heavy element found only in extraterrestrial objects

_____ 52. grains of quartz that show the signs of sudden and intense pressure

_____ 53. an impact crater on the sea floor off the Yucatan peninsula

_____ 54. the largest mountain belt, which consists of the Alps and the Himalayas, on the continents today

_____ 55. a region of rifted (stretched) continent in several western states of the United States

_____ 56. a type of plant that appeared in the Cenozoic and became the dominant plant of the plains in temperate and subtropical climates

_____ 57. the name for the last great ice age, which occurred within the last 2 million years

_____ 58. a giant star, forecast to be a step in our Sun's future development

Matching Questions

TIME-EVENTS CHART

Fill the blanks for the time interval names missing from the following chart, choosing from the list provided below. Then, review the events listed on pages 156–158 and place them in a logical sequence by writing their labels in the appropriate box in the time-events chart. For convenience, the events are loosely organized into categories. Do remember that there's often overlap of categories; moving continents and orogenies happen because of plate tectonics, climates and sea level are related, and so on.

Archean   Ordovician
Cambrian   Permian
Carboniferous   Phanerozoic
Cenozoic   Precambrian
Cretaceous   Proterozoic
Devonian   Silurian
Hadean   Triassic
Jurassic

Use the following code for each of the categories:

Climate Events (CE-1, CE-2, etc.)
The Sea (S-1, S-2, etc.)
Continents (C-1, C-2)
Orogenies (O-1, etc.)
Plate Tectonics (PT-1, etc.)
Evolution of Life (E-1, etc.)
Miscellaneous (M-1, etc.)

TIME-EVENTS CHART

_____ 545 million years ago → now		_____ era (65 million years ago → now) Events:	
		Mesozoic era, late	_____ period Events:
		Mesozoic era, early and middle	_____ and _____ periods Events:
		Paleozoic era, late	_____ and _____ periods Events:
		Paleozoic era, middle	_____ and _____ periods Events:
		Paleozoic era, early	_____ and _____ periods Events:
_____ 4.6 billion years ago → 545 million years ago	_____ eon Events:		
	_____ eon Events:		
	_____ eon Events:		

Climate Events

CE-1. For the first time since the Triassic, the climate cooled enough to allow glaciers to form in Antarctica.

CE-2. The first evidence for worldwide glaciation occurs; two short but severe episodes.

CE-3. Increased levels of carbon dioxide created extreme greenhouse conditions: much of world was tropical or subtropical, the ice sheets melted, the ocean levels were high, and the dinosaurs loved it.

CE-4. Cool conditions, combined with lack of ocean influence, caused the interior of huge, newly formed Pangaea to be desert.

CE-5. The last of the great ice ages, the Pleistocene, began 2 million years ago: the sea level fell, so the sea floor of the Bering Strait was exposed and allowed life forms to migrate from Asia to North America. Then 11,000 years ago the climate warmed and interglacial times began.

CE-6. There was cold climate on the coasts and low sea level, and the semi-arid conditions that created the red-rocks desert looked typical of the southwest United States today; sea transgression started by the end of this time.

CE-7. There was cold climate and retreating seas that left coastal swamps and river deltas; although ice sheets covered large areas by the end of this time, the coastal swamps on Laurentia had lush growth that became future coal.

The Sea

S-1. There is the first record (two episodes) of sea level high enough to flood many land areas with epicontinental seas.

S-2. Earth cooled enough that volcanic water vapor condensed; it rained *a lot*; oceans filled, and soon became salty.

S-3. Warm greenhouse conditions caused the sea level to rise and flood continents; huge reefs formed in some oceans, and great accumulations of evaporite minerals (gypsum and halite) formed in other oceans.

S-4. The climate warmed and the sea level rose; the Western Interior Seaway flooded across North America, from the Arctic Ocean to the Gulf of Mexico.

Continents

C-1. The first continental crust formed; by the end of this time interval 80% of all continental crust had formed.

C-2. The continental crust continued to form; by the midpoint of this time interval 90% of all continental crust was in existence.

Orogenies

O-1. The Acadian orogeny occurred and eventually provided the sediment for the Catskill delta of the eastern United States and the Old Red Sandstone of the United Kingdom.

O-2. The Alleghenian orogeny began and eventually produced the Appalachian Mountains and the Ouachita Mountains of eastern North America.

O-3. The Alpine-Himalayan chain was created by collisions along the southern margins of Europe and Asia.

O-4. Unrest began in the western United States, as a large island arc collided with Nevada in the Antler orogeny.

O-5. The Grenville orogeny, an early one in Earth history, occured; remnants of this orogen are still visible as part of the Appalachian Mountains.

O-6. Convergence on the U.S. West Coast began and initiated the Laramide orogeny, which formed the current Rocky Mountains.

O-7. The Laramide orogeny continued in yet another era, and the Rockies continued to be uplifted.

O-8. The Sonoma and Nevadan orogenies occured in the western United States.

O-9. The Sevier orogeny, a large thrust fault event on the west coast of North America, created the Canadian Rockies.

O-10. On the western edge of North America, the Sierran arc, a large continental volcanic arc, began.

O-11. The Sierran arc continued into this next period and created a range similar to the Andes of today.

O-12. A collision event, the Taconic orogeny, was the first stage of development of the Appalachian orogeny.

O-13. The Ural Mountains were created along the border of Europe and Asia.

Plate Tectonics

PT-1. Collision was so strong along the east coast of North America that the Midwest and West were uplifted and the Ancestral Rockies formed.

PT-2. The Americas moved west; convergence along the west coast of South America produced the Andes.

PT-3. Rifting and stretching in the western United States produced the Basin and Range Province.

PT-4. Subduction of the Juan de Fuca plate created the Cascade volcanic chain of mountains on the northwest coast of North America.

PT-5. The isthmus of Panama formed; the Atlantic and Pacific Oceans were separated. Darn! now we have to build a canal.

PT-6. The final collisions (including the Alleghenian orogeny) that completed the supercontinent Pangaea occurred.

PT-7. Pangaea began to break apart; North America and Africa separated, creating the North Atlantic Ocean.

PT-8. Pangaea continued to split; now the southern land areas started to move apart and the Southern Atlantic Ocean began.

PT-9. Pangaea was still splitting; Australia separated from Antarctica, and the North Sea was formed; Greenland separated from North America, and North America and Europe continued to part company.

PT-10. The supercontinent Rodinia rearranged itself to become Pannotia.

PT-11. The supercontinent Pannotia began to split into four major pieces: Laurentia (North America and Greenland), Gondwana (South America, Africa, Antarctica, India, and Australia), Baltica (Europe), and Siberia.

PT-12. The first supercontinent we know of, Rodinia, formed 1 billion years ago.

PT-13. Plate tectonics was happening at superrates; seafloor spreading at the mid-ocean ridges was three times its current rate, and superplumes reached the base of the lithosphere.

Evolution of Life

E-1. The first eukaryotic cells appeared; the first multicellular organisms appeared; a shell-less community known as Ediacarian fauna flourished; and photosynthetic organisms flourished and added oxygen to the atmosphere.

E-2. The first mass extinction ever known ended this time interval.

E-3. There wasn't any life.

E-4. The first vascular plants developed; swampy forests were widespread and inhabited by spiders, scorpions, insects, and crustaceans; jawed fish like sharks were in the oceans; the first amphibians emerged on land; and new species of sea creatures (trilobites, eurypterids, gastropods, crinoids, and bivalves) appeared to replace those lost in the mass extinction that ended the early Paleozoic.

E-5. Mammals diversified and dominated; huge mammals existed (giant beavers, bears, and sloths and mammoths), then died out, possibly hunted to extinction by humans; human-like primates appeared, followed by genus *Homo*, and finally our genus species, *Homo sapiens*, appeared.

E-6. The first life forms, prokaryotic cells of bacteria and cyanobacteria, appeared 3.8 billion years ago, and a little later (3.2 billion years ago), stromatolites appeared.

E-7. There was tremendous diversification of life forms, known as the Cambrian explosion of life.

E-8. Laurentia was covered with lush growth that would eventually become huge coal deposits. Huge dragonflies, ferns, club mosses, scouring rushes, insects (including big cockroaches!), gymnosperms (conifers, like pines and spruce), cycads, amphibians, and reptiles were all important life forms; the forerunners of mammals existed (pelycosaurs and therapsids). Unfortunately the era ended with major extinctions (90% of marine species disappeared).

E-9. Plesiosaurs (swimming reptiles) appeared; hexacorals built huge reefs, gymnosperms like cycads were important. Reptiles diversified; the first turtle and first flying reptile (pterosaur) appeared. The first true dinosaurs appeared and grew in size, abundance, and importance (sauropods like seismosaurus, stegasaurus, and allosaurus). The first feathered bird (*Archaeopteryx*) and first small, rat-like mammal appeared.

E-10. Modern fish (teleosts) appeared; huge swimming reptiles and giant turtles existed; cycad trees vanished; angiosperms (flowering plants) became important; dinosaurs lived almost everywhere and dominated the life scene; *Tyrannosaurus rex* existed;

pterosaurs (flying reptiles) reached huge size; birds and mammals diversified; and the era ended with the sudden extinction of most species on Earth, probably due to meteorite impact.

E-11. Common fossils of this time were conodonts, trilobites, mollusks, brachipods, echinoderms, archeocyathids (reef-forming sponges), and graptolites—all sea creatures; the earliest land plants appeared at the very end of this time interval.

Miscellaneous

M-1. Earth's atmosphere consisted of the volcanic gases nitrogen, ammonia, methane, water, carbon monoxide, carbon dioxide, and sulfur dioxide.

M-2. The atmosphere changed considerably and was similar to today's atmosphere, although there still wasn't 21% oxygen concentration which we have today; that didn't happen until the Phanerozoic.

M-3. Sedimentary sequences known as banded-iron formations (BIF) were deposited; these were interbedded layers of hematite, magnetite, and chert.

M-4. This was really a hellish world.

M-5. Earth had no permanent crust; its surface was covered with magma ocean.

M-6. The Moon was closer to Earth than it is today, so tides were huge.

M-7. Earth developed a permanent solid crust; the rigid surface became segmented into plates.

M-8. Earth began its existence as an accumulation of planetesimals.

M-9. Shallow-marine sediments (limestone and quartz sandstone) became common; the amount of limestone increased as calcite-secreting organisms (bacteria and algae) appeared and prospered.

Short-Answer Questions

Answer with a few words, numbers, or short sentences, or choose from the words in parentheses.

1. State two reasons Earth got hotter, not cooler, during the Hadean. _____
_____

2. Why doesn't the process of plate tectonics happen on the Moon? _____

3. What is the origin of the oldest Earth rock? _____
_____

4. List the five principle rock types of Archean cratons.
_____
_____

5. Why are Archean shallow-water sediments rare?
_____

6. How old is the ocean? _____

7. What extraterrestrial bodies are thought to have contributed gases to Earth's early atmosphere?
_____

8. What were the first forms of life, and how old are they? _____
_____

9. "Proterozoic" comes from Greek for "first life." Why was the time interval named this, when it *wasn't* the time interval in which the first life forms were found?
_____

10. During Archean times Earth was a world of tiny, fast-moving plates with relatively small continents and an oxygen-poor atmosphere. Describe Proterozoic times in terms of these same characteristics. _____
_____

11. Most large cratons in existence today formed how long ago? _____

12. a. How did the development of oxygen-based metabolism influence the size of organisms? ____
_____

    b. How did the increase of oxygen in the atmosphere influence radiation levels on Earth? _____
_____

13. Why did banded-iron formations (BIF) form? ____
_____

14. What reproductive advantage did reptiles gain by their ability to produce eggs with shells? _____
_____

15. State two ways dinosaurs differ from other reptiles (one has to do with leg structure, one with the blood system). _____
_____

16. Generally speaking, if the climate warms, what happens to the sea level and why? _____
_____
_____

17. What is the origin of the Sierra Nevada? _____
_____

18. The faulting styles of the Sevier and the Laramide orogenies were quite different and resulted in mountain ranges (the Canadian Rockies and the U.S. Rocky Mountains) that, despite the similarity of their names,

are structurally quite different. What is a suggested cause for these differences? _____

19. How do teleost fish differ from primitive fish? _____

20. What reproductive advantages do angiosperms (flowering plants) have over gymnosperms (naked-seed plants, like pines)? _____

21. For how long did dinosaurs exist and dominate life on Earth? _____

22. Scientists believe a bolide struck Earth at the end of the Mesozoic.
    a. How big was it? _____
    b. Where did it hit? _____
    c. What is the name of the crater it created? _____

23. According to the fossil record, how long ago did the following steps in human development occur?
    a. the appearance of the first human-like primate _____
    b. the appearance of the first member of our genus, *Homo* _____
    c. the appearance of the first *Homo erectus* (tool maker) _____
    d. the divergence of the line that lead to *Homo neanderthalenis* (Neanderthal Man) from the line that led to us, *Homo sapiens* _____
    e. the appearance of the modern human, *Homo sapiens* _____

Short-Essay Questions

1. Historians who chronicle human history gather data from written accounts, relics, monuments, recordings and videos, then outline human history in terms of daily life, wars, economics, government leaders, inventions, and explorations. What type of information are geologists looking for, and how do they collect these data to produce a history of the Earth?

2. Describe how the first continental crust formed.

3. How did geologists use data about banded-iron formations (BIF) to interpret oxygen content of past atmospheres?

4. Summarize the story of the K-T boundary event. Include bolide, iridium, shocked quartz, Walter and Luiz Alvarez, and the Chicxulub crater.

5. What does the future hold in store for Earth, in the geologic short term (5,000 years) and in terms of Earth's eventual destiny?

Practice Test

1. Which of the following is NOT useful in identifying mountain belts that have eroded away and therefore are not obvious topographic features?
    a. folds and faults
    b. unconformities
    c. shocked quartz
    d. foreland sedimentary basins
    e. basement uplifts

2. Which of the following statements is FALSE?
    a. Conglomerates containing land plants probably formed in an alluvial fan or a stream channel.
    b. Marine limestone overlying alluvial-fan conglomerate shows the sea level rose at that site.
    c. Limestone containing coral fossils probably developed in a shallow sea.
    d. Anorthosite is the common dark-colored rock of foreland sedimentary basins.
    e. Igneous and metamorphic rocks are commonly found in mountain belts.

3. The ratio of isotopes of what element in fossil shells is an indicator of past temperature?
    a. oxygen
    b. carbon
    c. nitrogen
    d. potassium
    e. calcium

4. The lack of rock record for Earth's first 600 million years
    a. indicates the surface then was one huge ocean basin filled with hot water.
    b. has lead to the name Archean ("ancient waters") for this time interval.
    c. hints the planetesimals were not accreted to each other yet.
    d. indicates all surface rock was too hot to allow the radiometric clocks to start ticking.
    e. The statement is false: There is Earth rock as old as 4.5 billion years.

5. Which geologic time interval saw the first plates, the first continents, the first oceans, and the first life?
    a. Hadean
    b. Phanerozoic
    c. Proterozoic
    d. Paleozoic
    e. Archean

6. Which of the following has nothing to do with fossils?
    a. stromatolites
    b. Rodinia
    c. cyanobacteria

d. Ediacaran fauna
   e. conodonts

7. Which of the following statements is FALSE? The first life forms
   a. probably occurred in oxygen-rich, shallow, warm ocean waters.
   b. may have been simple cells of bacteria and cyanobacteria (blue-green algae).
   c. have been found in the Archean eon.
   d. date back 3.8 billion years.
   e. were prokaryotic cells.

8. Which of the following is NOT one of the smaller continents formed by the break up of Pannotia?
   a. Laurentia
   b. Siberia
   c. Laramide
   d. Baltica
   e. Gondwana

9. Which of the following are fossil plants?
   a. trilobites
   b. gymnosperms and angiosperms
   c. graptolites and archeocyathids
   d. pterosaurs and plesiosaurs
   e. bolides

10. Which of the following has nothing to do with the K-T (Cretaceous-Tertiary) boundary?
    a. iridium
    b. shocked quartz
    c. Chicxulub
    d. bolide
    e. Thomas Huxley

11. Which of the following statements is FALSE?
    a. Human ancestors first appeared in the late Mesozoic.
    b. The human genus *Homo* first appeared more than 2 million years ago.
    c. Fossils of modern humans (*Homo sapiens*) date back to about 150,000 years ago.
    d. The Cenozoic is known as the Age of Mammals.
    e. Huge mammals like giant beavers, bears, and sloths and mammoths lived during the Cenozoic but are now extinct.

12. Our Sun will
    a. run out of nuclear fuel in about 5 billion years.
    b. first collapse, then expand after its fuel is gone.
    c. form a red-giant star with a diameter about the size of Earth's orbit.
    d. engulf and vaporize the first three planets, which include Earth.
    e. All of the above are forecast by scientists as the future of the Sun.

13. There is no marine magnetic anomaly data for pre-Jurassic time because all ocean floor older than Jurassic has subducted.
    a. True
    b. False

14. Cratering of the Moon is most intensive in the basaltic mare, indicating most of it happened between 2 and 3 billion years ago.
    a. True
    b. False

15. Rodinia, Pannotia, Grenville, and Pangaea are all names of supercontinents.
    a. True
    b. False

16. Banded-iron formations (BIFs) are forming today along mid-ocean ridges.
    a. True
    b. False

17. Gymnosperms are naked-seed plants like conifers; angiosperms are flowering plants.
    a. True
    b. False

18. *Archaeopteryx* was the first feathered bird; it appeared in the middle Paleozoic.
    a. True
    b. False

19. There is debate among geologists as to whether we currently live in an interglacial time and glaciers will return soon or human-induced greenhouse warming will cause the melting of ice sheets and flooding of the land.
    a. True
    b. False

GEOLOGIC TIME CHART

1. Write the time intervals you used to label your Time-Events Chart; do it from memory. Check your work against the answers given for the Time-Events Chart.

2. Take the group of answers for each of the time intervals on the Time-Events Chart, cover up the names of the time intervals, and identify the times by the collection of events that took place.

(Working with the collection of events should make this an easy test activity; if you can do it by individual events, you really know the material. A reminder: It's between you and your instructor which specific details should be committed to memory.)

YOU ARE THERE!

Imagine you have achieved what others have only dreamed of—time travel. For each of the following scenarios, name the time interval you are in. Be as specific as you can. Occasionally more than one answer could be correct.

1. The Rocky Mountains seem to be where they belong, but they're awfully small and there are dinosaurs in the mountain valleys. _____

2. The Moon is very big, and the tides are very high. _____

3. You could swear that's your Uncle Ed over there, but he seems a bit more primitive than you remember him. _____

4. Ouch! That's a trilobite nibbling on your toes. _____

5. You are traveling in a heat-resistant canoe; there's no solid land on Earth for you to land on. Don't fall in! _____

6. You're in north-central California all right, and you're in the mountains all right, and Yosemite ought to be around here somewhere, but something isn't quite right. These mountains aren't supposed to be volcanoes. _____

7. Thank goodness you had a space suit handy, because when you tried to take a breath, it was terrible. This air could kill you! _____

8. It's raining and raining and raining and you forgot your umbrella. At least you remembered your swim suit, and the water's not too salty yet. _____

9. Those strange looking mounds must be the stromatolites you learned about in geology, and a drop of water under your microscope shows some bacteria and cyanobacteria. Aside from that, this is not a very lively place. _____

10. You're drifting around in your scuba gear, and all those trilobites and graptolites keep getting in the way. _____

11. You came down to Mexico for a relaxing vacation, but you picked the wrong time. That asteroid's going to hit you if you don't watch out. _____

12. You're suspended above the planet for a long geologic time, and you watch one huge continent split off a few bits and pieces of itself and reattach them around on the other side. _____

13. You're not too sure you want to stay here. The dragonflies are big and scary, and there are cockroaches all over the place. The big frogs are kind of cute, but all those reptiles aren't. There's not even one dinosaur among them. You're lonesome; there's not another mammal on Earth, but a couple of things out there look like they just might develop into one any day now. _____

14. Hey, that big continent is splitting apart. There's the North Atlantic Ocean opening up, now the South Atlantic, but you still can't figure out what's going to be Antarctica, Australia, or Greenland. _____

15. You'd mow the lawn if only that mammoth would move out of the way. _____

16. Just an eon ago it was hardly crowded in the ocean. Now there seems to be a veritable explosion of life around you, and so many of the new life forms have nice hard shells. _____

17. You're drifting in your scuba gear, and there's lots of solid pieces appearing out of the ocean water and settling down all around you. Iron perhaps? _____

18. You think you'll climb the Himalayas today, before they get too big. _____

19. Watch out! That's a *Tyrannosaurus rex* coming up behind you (and the answer is not Jurassic). _____

20. You board a boat in the Gulf of Mexico and float all the way up to the Arctic Ocean, and you never have to get out and portage across dry land. _____

21. Your bird-watching finally gets some results. Just last era there wasn't one in sight, but you're sure that thing that just flew by had feathers. _____

22. Finally there are enough flowers to pick yourself a bouquet. _____

23. You finally caught a fish that looks good enough to eat; it reminds you of the one that got away the last time you went fishing back in your own time. _____

24. You're pretty sure you're looking at the first mammal ever; too bad it reminds you so much of a rat. _____

25. You're in your scuba gear again, checking out the mid-ocean ridges. There seem to be more than the usual number of hot spots to watch out for, and you could swear you can actually see that sea floor spreading. _____

26. Don't look now, but there's something flying over you that's awfully big. It's a bird (no, no feathers), it's a plane (no, hasn't been invented yet), it's Superman (no, this is a serious science fantasy), it's a huge pterosaur! _____

27. Brrrrrrrr, it's so cold! And nobody around here has invited you to share a cave. _____

ANSWERS

Completion

1. foreland sedimentary basins
2. magma oceans
3. protocontinents
4. cratons
5. prokaryotic cells
6. cyanobacteria
7. organic carbon
8. stromatolites
9. shield
10. platform
11. accretionary orogenies
12. Rodinia
13. Pannotia
14. Grenville orogen
15. passive-margin basin
16. paleogeography
17. photosynthetic organisms
18. eukaryotic cells
19. Ediacaran fauna
20. Laurentia, Gondwana, Baltica, and Siberia
21. epicontinental seas
22. Taconic orogeny
23. Cambrian explosion of life
24. conodonts
25. trilobites
26. anomalocaris
27. archeocyathid
28. graptolites
29. Acadian orogeny
30. Antler orogeny
31. vascular plants
32. Alleghenian orogeny
33. Appalachian fold-thrust belt
34. thin-skinned deformation
35. Ancestral Rockies
36. gymnosperms
37. cycads
38. Sonoma and Nevadan orogenies
39. Sierran arc
40. plesiosaurs
41. hexacorals
42. pterosaurs
43. Western Interior Seaway
44. Sevier orogeny
45. basement uplifts
46. Laramide orogeny
47. super plumes
48. greenhouse gas
49. angiosperms
50. bolide
51. iridium
52. shocked quartz
53. Chicxulub crater
54. Alpine-Himalayan chain
55. Basin and Range Province
56. grasses
57. Pleistocene ice age
58. red giant

Matching Questions

TIME-EVENTS CHART

TIME-EVENTS CHART

Phanerozoic 545 million years ago → now		Cenozoic era (65 million years ago → now) Events: CE-1; CE-5; O-3; O-7; PT-2; PT-3; PT-4; PT-5; PT-9; E-5	
		Mesozoic era, late	Cretaceous period Events: CE-3; S-4; O-6; O-9; O-11; PT-8; PT-13; E-9
		Mesozoic era, early and middle	Triassic and Jurassic periods Events: CE-6; O-8; O-10; PT-7; E-10
		Paleozoic era, late	Carboniferous and Permian periods Events: CE-4; CE-7; O-2; O-13; PT-1; PT-6; E-8
		Paleozoic era, middle	Silurian and Devonian periods Events: S-3; O-1; O-4; E-4
		Paleozoic era, early	Cambrian and Ordovician periods Events: S-1; O-12; PT-11; E-2; E-7; E-11; M-6
Precambrian 4.6 billion years ago → 545 million years ago	Proterozoic eon Events: CE-2; C-2; O-5; PT-10; PT-12; E-1; M-2; M-3; M-9		
	Archean eon Events: S-2; C-1; E-6; M-7		
	Hadean eon Events: E-3; M-1; M-4; M-5; M-8		

Short-Answer Questions

1. Kinetic energy of impacts changed to heat energy; radioactive decay produced heat.

2. It's small size allowed it to cool to greater depth, so its lithosphere is 60% of its thickness; this is too thick to break into moving plates.

3. It comes from the first continental crust, which was a combination of large piles of volcanic rock above hot-spot volcanoes and volcanic arcs adjacent to subduction zones.

4. gneiss, greenstone, granite, graywacke, and chert

5. Continents were so small little depositional environment existed, and erosion has had lots of time to destroy most of the few sedimentary deposits that did form.

6. 3.8 billion years old

7. comets

8. simple procaryotic cells (bacteria and blue-green algae called cyanobacteria); 3.8 billion years old

9. When the Proterozoic was named, scientists thought it was the time interval during which life began. Only later were bacteria found in the older Archean rock, and thus made the nomenclature misleading.

10. relatively few, large, slow-moving plates; large continents; and oxygen-rich atmosphere

11. 1.8 billion years ago

12. a. Oxygen-based metabolism produces energy more efficiently than oxygen-free metabolism does, so organisms could grow larger.
    b. Oxygen is the basis for ozone, which screens out some ultraviolet radiation from sunlight and thus protects skin, retinas, and genes.

13. In the presence of abundant oxygen iron became less soluble in seawater and precipitated out.

14. It freed them from the need to return to water to lay their eggs, since shelled eggs wouldn't dry out if laid on land.

15. Legs were positioned under their bodies rather than off to the sides, and it's likely (but not proven) they were warm blooded.

16. It rises because ice sheets melt.

17. They're the plutonic roots (a huge batholith) of the volcanic arc (the Sierran arc) of the late Mesozoic.

18. During the Laramide orogeny the subducting plate entered the mantle at a very shallow angle and scraped along the continent's base; the angle of subduction was much greater during the Sevier orogeny.

19. Teleost (modern) fish have shorter jaws, rounded scales, symmetrical tails, and specialized fins; primitive fish did not.

20. Angiosperms can produce seeds more rapidly and can attract insects to help with pollination.

21. more than 150 million years

22. a. 10 km (about 6 mi) in diameter
    b. the Yucatan peninsula in Mexico   c. Chicxulub

23. a. 4 million years ago   b. 2.4 million years ago
    c. 1.6 million years ago   d. 500,000 years ago
    e. 150,000 years ago

Short-Essay Questions

1. Geologists collect data by examining rocks, geologic structures, fossils, sediment, ice cores, and even tree rings. Using these data they outline Earth's history in terms of identifying mountain building events (orogenies); the growth of continents; recognizing past environments; recognizing past changes in sea level; recognizing past positions of continents; recognizing past climates; and recognizing the evolution of life.

2. Earth's first solid crust formed about 4 billion years ago, became segmented, and began plate tectonics motion. Plates that subducted partially melted, yielding mafic- to intermediate-composition magmas that rose and erupted as volcanic arc mountains. Immense hot-spot volcanoes over mantle superplumes also produced huge piles of intermediate composition rock, all of which was too buoyant to subduct. These masses collided, stuck, and became the first continental crust.

3. Banded-iron formations (BIFs; deposits of hematite, magnetite, and chert) form when abundant oxygen causes diminished solubility of iron in ocean water. Only sedimentary sequences of early Proterozoic have BIF deposits, so the atmosphere of that time had to have contained enough oxygen to cause this. There are no significant quantities of BIF after this because by the end of the Proterozoic there wasn't much iron left in solution to precipitate out.

4. Using modern dating techniques, geologists concluded there was almost instantaneous extinction of most species on Earth at the end of the Cretaceous period, the so-called K-T boundary—Cretaceous-Tertiary (first part of the Cenozoic) boundary. The reason was a mystery until geologist Walter Alvarez and his physicist father Luiz Alvarez found (in Italy) and

analyzed a clay layer at the K-T boundary. The layer contained a high concentration of iridium (a heavy element found only in extraterrestrial bodies), and glass spheres formed by flash freezing of a melt, wood ash, and shocked quartz (an effect caused by sudden intense pressure). Further investigations found similar clay layers in numerous locations worldwide. Geologists concluded the iridium came from fragments of a bolide (an asteroid), the wood ash from forests burned in worldwide fires, and the shocked quartz from the shock wave caused by the impact. Probably giant tsumamis (2 km high) were generated by the event, dark winter-like conditions occurred, and photosynthesis was temporarily halted. Geologists believe they've found the site of the asteroid impact, a 1,100 km-wide, 16-km deep crater named Chicxulub off the northwest coast of the Yucatan Peninsula, Mexico. There is general agreement that devastating impacts like this one have happened more than once in Earth's history and will no doubt happen again.

5. Before the topic of human-induced greenhouse warming became an important issue, most geologists agreed we live in an interglacial time, and the great ice sheets will return in about 5,000 years. The greenhouse issue does complicate the scenario. Perhaps we're headed into super-interglacial times, when so much ice will melt coastal regions will be seriously flooded and New York, Paris, London, Rome, and Tokyo will be under water. As far as the more distant future is concerned, if humankind doesn't exterminate itself with nuclear weapons or irrevocably pollute the planet or get wiped out by a life-destroying asteroid impact, our species may live for 5 billion more years, and then witness the end of the planet itself. By then the Sun will have run out of nuclear fuel (hydrogen), collapse inward due to gravitational pull, heat up tremendously when this happens, and swell to become a red-giant star. This huge star will expand to swallow up and vaporize the first three planets (Mercury, Venus, and us). The red giant will eventually collapse to become a glowing white-dwarf star and in the process eject some gas (a gaseous nebula) outward into space. Part of the gas may even be atoms that now are part of us, and possibly this material will end up in a new solar system, 5 billion years plus into the future. In the meantime, as this course is emphasizing, Earth is a dynamic planet, and we can count on lots of change, always.

Practice Test

1. c This is the result of impact, not orogeny.

2. d Anorthosite is the light-colored rock making up the Moon's highlands.

3. a The ratio of $^{18}O : ^{16}O$ in fossil shells is dependent on temperature.

4. d Temperature must fall below the blocking temperature before radioactive decay can begin.

5. e By the Archean, Earth had cooled enough from Hadean times to allow its surface to freeze and form continents and plates, and contain ocean waters. Proterozoic means "first life," but it's a misnomer; life began in Archean times.

6. b This is the name of the first supercontinent known, back in the Proterozoic eon.

7. a Early life needed an oxygen-free environment because reaction with oxygen broke down the amino acids; first life probably occurred in black smokers along mid-ocean ridges.

8. c This is the late Mesozoic-early Cenozoic orogeny that resulted in the Rocky Mountains.

9. b Trilobites, graptolites, and archeocyathids are Paleozoic animals; pterosaurs and plesiosaurs are Mesozoic reptiles; bolides are extraterrestrial bodies.

10. e Huxley was a nineteenth-century geologist who realized Earth has changed immensely over time; Walter and Luiz Alvarez were the scientists who analyzed the K-T boundary layer clay.

11. a Even ape-like primates didn't appear until well into the Cenozoic.

12. e Unfortunately, statements a to d are all true, although not due to happen in our lifetimes.

13. True Ocean floor subducted by the time it was 200 million years old; pre-Jurassic times are older than this.

14. False There's little cratering in the mare, probably because most debris had been trapped by large planets before the basalt filled the mare between 3.1 and 3.9 billion years ago.

15. False Grenville is the name of a Proterozoic age orogen; the rest are supercontinents.

16. False There hasn't been enough iron in solution in the ocean since Proterozoic times to precipitate out as BIF.

17. True Plants are classified as gymnosperms if they release their seeds from cones, and angiosperms if they retain them within the protective shelter of flower parts.

18. False *Archaeopteryx* was a feathered bird but didn't appear until the Mesozoic.

19. True Only time will tell whether human-induced greenhouse warming will keep the glaciers from returning within a few thousand years.

YOU ARE THERE!

1. Mesozoic, late (Cretaceous) (Note: It might have been a description of the early Cenozoic, except for the dinosaurs; they became extinct at the end of the Mesozoic.)
2. Paleozoic, early
3. Cenozoic
4. Paleozoic
5. Hadean
6. Mesozoic, late (Cretaceous)
7. Hadean or Archean
8. Archean
9. Archean
10. Paleozoic, early
11. K-T boundary, between late Mesozoic (Cretaceous) and early Cenozoic (Tertiary) times
12. Proterozoic
13. Paleozoic, late
14. Mesozoic
15. Cenozoic
16. Paleozoic, early (Cambrian period)
17. Proterozoic
18. Cenozoic
19. Mesozoic, late (Cretaceous)
20. Mesozoic, late (Cretaceous)
21. Mesozoic, early and middle
22. Mesozoic, late (Cretaceous)
23. Mesozoic, late (Cretaceous)
24. Mesozoic, early and middle
25. Mesozoic, late (Cretaceous)
26. Mesozoic, late (Cretaceous)
27. Cenozoic, Pleistocene Ice Age

CHAPTER 14 | Squeezing Power from a Stone: Energy Resources

GUIDE TO READING

Rocks are the natural energy resources for the essentials of modern life: electricity and fuel. The types of resources vary; this chapter presents an overview of several. Since you've had thirteen chapters' worth of geology background, it's assumed you already know the basic rock types, geologic structures, geologic time, and Earth history. Therefore this chapter needs to introduce few new concepts, but there are many new vocabulary words and a resulting long list of terms to be familiar with.

Your author begins with the energy resource modern society relies on most, fossil fuels (petroleum, natural gas, and coal). Foremost of these is petroleum (oil). To understand its diverse makeup, you learn some basic organic chemistry and the significance of the molecular size of hydrocarbons.

How does petroleum form and accumulate? You read about plankton; source rock and reservoir rock; the oil window; oil reserves, traps, and seals; and the relationship between natural gas and oil.

How do you find petroleum, get it out of the ground, and process it? Topics include seismic search methods, drilling wells on land and offshore, simple pumping methods (drill bits and drilling mud), secondary recovery techniques, and what happens to crude oil in the distillation column of a refinery.

A little history of the petroleum industry is included in this section. It's colorful stuff, ranging from the days of Egyptian mummies to pioneer days in the United States. You are reminded that oil and money seem to go together. Rockefeller got and stayed rich because of oil, while many wildcatters have gotten rich but not always stayed rich because of the same stuff. Edwin Drake earned his place in history by drilling the first oil well ever, and Spindletop Hill in Texas and Red Adair are legendary names in the oil industry.

Coal is another fossil fuel and the next-most-used energy resource. You read of its formation from swamp plant material and its sequence of development (peat, lignite, bituminous coal, and anthracite coal). How much does the world depend on coal for energy, can this situation change, and how wise is it to become more dependent on yet another fossil fuel? Topics included are economic coal seams, coal reserves, strip mining versus underground mining, the hazards of coal mining, and the potential environmental impact.

A few decades ago, when the world started to become concerned about the future availability of fossil fuels, nuclear power seemed to many people to be the answer. What happened that it hasn't become the world's chief energy source? You start with learning the difference between fission and fusion and which one is used in nuclear power plants. What are pitchblende, nuclear reactors, fuel rods, control rods, and chain reactions? Just how dangerous is nuclear energy? You read about critical mass and meltdown, and two well-known nuclear accidents that were totally different from each other, Chernobyl, in Russia, and Three Mile Island in the United States. Whether to continue using nuclear power has become a touchy societal issue. One of the chief concerns is "What do we do with the nuclear wastes?"

The last part of the chapter is based on an opinion held by most experts in both science and industry: We have been living in a unique and very limited time in human history, the Oil Age. We are quickly running out of commercial quantities of oil and the end is in sight. Despite the fact oil (and coal, too) are forming in today's world, this happens so slowly that from the human viewpoint they are nonrenewable resources. Your author gives you figures of current oil consumption, oil reserves, scientific estimates of when we'll run out, and what our options are as that time approaches. There are basically two options: give up and accept a lower standard of living, or switch to alternative energy resources.

Several of these alternative energy Earth resources are already being used, but not extensively. You read about geothermal energy, hydroelectric power, tar sands, oil shale, and solar energy.

There seems to be no perfect answer; each energy option has advantages and disadvantages in terms of efficiency, cost, political consequences, and environmental impact. But it does seem as though some choices must be made and some steps taken soon so that the Arabian saying doesn't become reality: "My father rode a camel. I drive a car. My son rides in a jet airplane. His son will ride a camel."

LEARNING ACTIVITIES

Completion

Test your recall of new vocabulary terms.

1. the capacity to do work
2. anything that can be used to produce heat, power muscles, make electricity, or move cars
3. fuels that have stored energy over long periods of geologic time, such as oil, gas, and coal
4. the process green plants use to make food: carbon dioxide plus water, in the presence of chlorophyll and light, produce sugar, water, and oxygen
5. energy produced when water, heated naturally in Earth, changes to steam and drives electrical turbines
6. compounds that are composed of carbon and hydrogen atoms arranged in chain or ring-like molecular structures
7. chemicals that make up living organisms or chemicals that are very similar to them, like oil and natural gas
8. a measure of the ability of a substance to flow
9. a measure of how easily a substance will evaporate
10. tiny (about 0.5 mm in diameter) plants and animals that float in lake or ocean waters
11. the rock composed of clay-sized particles mixed with dead plankton remains and formed by the lithification of organic-rich, muddy ooze on lake or ocean bottoms
12. any rock that contains the raw materials from which hydrocarbons eventually form, of which black organic shale is a typical example
13. the material in rock that is the waxy precursor to tar, oil, and gas
14. shale that is rich in kerogen
15. the relatively narrow temperature range in which organic material in rock can change to oil
16. a region where there's a significant amount of accessible oil underground
17. the amount of oil known to exist underground in an oil field
18. any rock that is both porous and permeable, and therefore has the potential to easily release any oil that it might contain, of which sandstone is a good example
19. a measure of the amount of open space (microscopic size, between rock grains) in a rock
20. a measure of how well fluid will flow through a rock, which is dependent on how well the pore spaces connect up with each other
21. a deep hole drilled into the ground for the purpose of withdrawing oil from the surrounding rock
22. a geologic configuration that keeps oil or gas from migrating on through reservoir rock, and instead confines it to a limited area
23. a rock that is quite impermeable and therefore stops oil or gas from moving on through, of which shale, salt, and unfractured limestone are examples
24. places where oil naturally flows onto the surface of the ground
25. a mixture of various lengths of hydrocarbon chains and rings, another name for rock oil
26. people who search for oil in areas where none has yet been found
27. a map that shows the rock formations and geologic structures of an area
28. a diagram that shows the geometric configuration of rock layers and structures as they would appear if you could expose them in vertical profile

29. a vertical cross section down through rock layers of an area, created by the interpretation of the reflections of artificially generated seismic waves

30. shock waves that move through the earth, and may be created by natural vibrations of the Earth (earthquakes) or by vibrating trucks or dynamite explosions

31. the end piece on a length of pipe that rotates its way down through rock and is made of metal studded with industrial diamonds for cutting efficiency

32. a slurry of water and clay that's forced down the center of a drill pipe and leaks out through holes in the drill bit, in order to cool the bit, flush out rock cuttings, and weight down any oil there and prevent gushers

33. fountains of oil explosively forced upward out of the oil well by underground pressures

34. a huge construction that serves as a base for a derrick and any other structures necessary for the operation of an oil well that's being drilled on the ocean bottom

35. any methods used to get more oil out of the ground than ordinary pumping can bring up

36. petroleum as it comes from the ground, before any processing

37. the building where crude oil is processed and separated into its component parts

38. a large pipe that sits upright in a refinery and guides the crude oil vapors upward, drawing off different weight molecules at different levels along the pipe

39. a black sedimentary rock, composed of carbon and minor amounts of organic chemicals and clay, that burns

40. partially decayed vegetation found in swampy areas, which is the first material in coal formation and is already about 50% carbon

41. a soft, dark-brown coal, formed from peat subjected to higher temperature

42. a dull black coal known as soft coal, formed from lignite between 100 and 200°C

43. a shiny black "hard" coal formed from soft coal between 200 and 300°C

44. the situation in which the sea level rises and the shoreline moves inland

45. the situation in which the sea level falls and the shoreline moves back off the land toward deeper water

46. a classification scheme for coal based on its carbon content

47. the layers of coal that are interspersed with layers of other sedimentary rock

48. the amount of coal that has been discovered

49. the layers of coal that exist in great enough quantity to be worth mining

50. a method of mining coal that removes the overlying burden of soil and sedimentary rock to expose the coal seams

51. a huge piece of equipment that scrapes off the soil and rock above coal seams

52. the only feasible method for mining coal that lies deeper than about 100 m

53. a disease that clogs the lungs and may lead to pneumonia, caused by the inhalation of coal dust

54. the containment vessel in a nuclear power plant where the nuclear reaction takes place

55. pellets of radioactive material, often uranium oxide, used as fuel in a certain type of power plant

56. metal tubes packed with fuel pellets, then inserted into reactors in nuclear plants

57. the nuclear fission that keeps happening within a mass of radioactive material because it gets triggered by free neutrons

58. the minimum amount of radioactive material necessary to cause a chain reaction to proceed at an explosive pace

59. the process that concentrates the U^{235} in a quantity of natural uranium by a factor of 2 or 3, so it will be suitable fuel for a nuclear power plant

60. the mineral name for uranium oxide (UO_2)

61. steel tubes inserted into nuclear reactors to absorb free neutrons and thus slow the rate of nuclear fission

_____ 62. the result of uncontrolled fission in a nuclear reactor, in which conditions may become hot enough to melt through pipes and concrete

_____ 63. spent fuel and radioactively contaminated water and equipment that results from the operation of a nuclear power plant

_____ 64. nuclear waste that is more than 1 million times as radioactive as the accepted safety level

_____ 65. nuclear waste that contains between 1,000 and 1 million times as much radioactivity as is considered safe

_____ 66. nuclear waste that contains less than 1,000 times the safe level of radioactivity

_____ 67. the rate of temperature increase as you go deeper in the Earth—15 to 50°C/km in the upper crust

_____ 68. the energy produced when water flows downslope and drives turbines and generators

_____ 69. the energy resources that Earth creates by natural processes in a short time span, even by human standards (months or several years at most)

_____ 70. the energy resources that require hundreds to millions of years to develop naturally

_____ 71. a name analogous to stone age or bronze age that could be applied to current times because of the nature of our energy dependency

_____ 72. a special type of nuclear power plant whose operation literally produces its own nuclear fuel

_____ 73. the energy that does not create pollution problems as a side effect

_____ 74. the water that becomes polluted with sulfuric acid as it filters through piles of waste rock that contains sulfides such as pyrite

_____ 75. incidents in which oil leaks from pipelines, trucks, or ships and contaminates groundwater, ocean water, or shorelines

_____ 76. the coal that has a lower sulfur content than most, for example, coal from the western United States

_____ 77. the phenomenon in which gases like carbon dioxide trap atmospheric heat close to Earth's surface

_____ 78. the global increase in average atmospheric temperature, severe enough to change climatic belts

_____ 79. an atmospheric condition in which a layer of warmer air high above Earth's surface traps cooler air below, preventing the normal pattern in which warm air would rise and disperse any pollution present at the surface

_____ 80. smoke and fog together, a term coined in London a century ago

_____ 81. the pollution formed when auto exhaust reacts with air in the presence of sunlight to form ozone, nitrogen dioxide, and other hydrocarbons

Short-Answer Questions

Answer with a few words, numbers, or short sentences, or choose from the words in parentheses.

1. a. According to your text, how much more energy does the average American use than a prehistoric hunter-gatherer did? _____
   b. What are the two most common sources for this energy? _____
   c. What are six other less common energy sources? _____

2. What are the five fundamental sources of energy on Earth? _____

3. a. When plant material combines with oxygen (burns), what are the resultant products? _____
   b. When fossil fuels combine with oxygen (burn), what are the resultant products? _____

4. What physical characteristic of a hydrocarbon influences both its viscosity and its volatility? _____

5. Short-chain hydrocarbons are usually _____ (more, less) viscous and _____ (more, less) volatile than long-chain hydrocarbons.

6. Very long chain hydrocarbon molecules exist as _____ (gases, liquids, solids), midlength-chain hydrocarbons exist as _____ (gases, liquids, solids), and short-chain hydrocarbons exist as _____ (gases, liquids, solids).

7. All of the following are components of crude oil. List them in order of increasing chain length, shortest chain first:

   gasoline   heating oil   kerosene   lubricating oil
   natural gas   tar
   _____

8. What is the primary source of the organic material that develops into oil and gas? _____

9. Why do dead plankton settle out only where clay settles out? _____

10. The first step in either oil or coal formation is partial decay in an oxygen-poor environment. What happens if there's too much oxygen? _____

11. a. At the lower cooler end of the oil window (about 100°C), what waxy hydrocarbon forms? _____

    b. Between 100 and 150°C, what types of hydrocarbons form, solids, liquids, or gases? _____

    c. Between 150 and 250°C, what type of hydrocarbon forms, solid, liquid, or gas? _____

    d. What mineral product forms if the temperature gets beyond the oil window (past 250°C)? _____

12. a. In regions that have low geothermal gradients, at what depth is the oil window found? _____

    b. At what depth is the oil window found in regions that have high geothermal gradients? _____

    c. In summary, in what percent of the crust is there any chance at all that oil might form? _____

13. What geographical area has the largest oil fields (supergiant fields) and the most oil reserves? _____

14. What are the four basic geologic requirements for an oil field to develop? _____

15. What is the basic reason energy companies are not currently processing oil shale to produce oil? _____

16. a. What is typical porosity of a shale? _____
    b. of a sandstone? _____

17. Why does oil migrate from source rock to reservoir rock, and how long does the process take? _____

18. If oil, gas, and water are all present at some particular location in a rock structure, how will they be layered (i.e., which will be on top, in the middle, and on the bottom)? _____

19. What two components are necessary to have an oil trap? _____

20. List four types of oil traps. _____

21. a. Name a few ways very early civilizations used oil. _____

    b. Name a few ways people used oil in preautomobile days in the United States. _____

22. How did John D. Rockefeller become the richest man in North America? _____

23. How did Exxon, Chevron, Mobil, Sohio, Amoco, Arco, and Conoco come into existence? _____

24. What is Spindletop Hill? _____

25. What rock type (igneous, sedimentary, metamorphic) must geologists explore to find oil? _____

26. In today's economy, what might be the cost of drilling a deep well on land? _____

27. Simple pumping of an oil well can extract about what percent of the oil present? _____

28. Secondary recovery techniques can increase the percent of recoverable oil at a site to about what amount? _____

29. Briefly describe two secondary recovery techniques. _____

30. What country is the world's largest consumer of oil, and how much oil does it use per day? _____

31. a. What percentage of total world oil reserves are in the United States? _____
    b. How dependent is the United States on imported oil? _____

32. Name four varieties of natural gas. _____

33. Why do some fields yield oil and gas, and other fields yield only gas? _____

34. What is an environmental advantage of burning gas rather than burning oil? _____

35. What is an economic disadvantage associated with the transportation of natural gas? _____

36. a. Which does the current world population use more of, oil or natural gas? _____
    b. If both are found together in a field, which is sometimes treated as a waste material and burned off at the well? _____

37. What is the difference in source material between coal and oil? _____

38. What is the earliest geologic time period when significant coal deposits could possibly have formed? Why couldn't they have formed earlier? _____

39. a. What two geologic time periods are responsible for the world's most extensive coal deposits? _____
    b. Where in the United States are these different age coal deposits found? _____

40. a. What is the chief use of coal in the United States today? _____
    b. About what percentage of current U.S. energy needs does coal satisfy? _____

41. Arrange these terms to show the sequence of development of coal, from lowest to highest rank:

    anthracite   bituminous coal   lignite   peat

    _____

42. a. What percentage of a rock must be carbon before it can be called coal? _____
    b. As the coal-developing process proceeds, what materials continue to be driven off? _____

43. Why is anthracite coal always associated with mountain-building processes but never specifically with metamorphic rocks? _____

44. What is the minimum thickness for an economically valuable coal seam? _____

45. What is the maximum depth to which the strip mining of coal is economically feasible? _____

46. Explain how canaries were used as a safety device in coal mines. _____

47. How do firefighters extinguish fires in underground coal mines? _____

48. What is the difference between nuclear fusion and nuclear fission? _____

49. Where does the energy come from in both fusion and fission? _____

50. State in a very short and simple way why uranium gets concentrated into magma and stays in solution there longer than most materials, rather than being used up in common minerals early in the magma-freezing process. _____

51. What instrument is a uranium detector? (It is either carried across the land or flown across it.) _____

52. Name the worst nuclear power plant accident in the United States. Give the location, date, cause, and damages. _____

53. List five possible types of geologic sites where sealed containers of radioactive nuclear wastes might be safely disposed of. _____

54. Where geographically is the U.S. government currently proposing to store nuclear waste materials, and geologically, what type of environment is this? _____

55. Name two ways geothermal energy can be used to produce heat and electricity. _____

56. What is one of the practical problems encountered in creating energy in geothermal situations? _____

57. a. Name two areas where geothermal energy is used extensively. _____
    b. Geologically, what do these areas have in common? _____

58. How can dams and reservoirs produce electricity? _____

59. List some positive aspects of hydroelectric power generation and some negative aspects. _____

60. How can ocean waters provide hydroelectric power? _____

61. What was the average price for oil
    a. when it was first drilled in 1859? _____
    b. in the 1960s? _____

c. in the early 1970s? _____
d. in 1979? _____

62. What is OPEC and what does it regulate? _____
_____

63. a. What is the current annual rate of oil consumption worldwide? _____
b. How many barrels of obtainable oil might there logically be in the world today? _____
c. On the basis of these figures, when will the world run out of commercial quantities of oil? _____
_____

64. If the above estimates are reasonable, when society runs out of commercial quantities of oil, about how many centuries will humans have lived in the Oil Age? _____

65. Name two locations of huge quantities of tar sand. _____

66. Name four nonnuclear energy resources that may be turned to as quantities of oil diminish. _____
_____

67. Why have the following methods of producing energy not been developed more extensively?
a. nuclear power _____
b. hydroelectric power _____
c. solar energy _____
d. wind power _____
e. fusion power _____

Short-Essay Questions

1. Could there possibly be a very porous rock that is not very permeable? Would such a rock be a good reservoir rock for oil?

2. Once natural processes create oil in the Earth, does it last forever? Is any oil forming today to add to or replace the oil already in existence?

3. Outline the story behind the first U.S. oil well, drilled in 1859.

4. Who is Red Adair and why is he legendary?

5. What do successive transgression and regression events of the sea and a sinking sedimentary basin have to do with the quantity of coal produced at a particular site?

6. What is routinely done these days to combat the potential environmental damage from strip mining?

7. How does uranium get concentrated in veins in igneous rock and end up in stream bed deposits and concentrated in sedimentary rock?

8. Explain why a nuclear generating plant is really just a fancy way to use steam to generate electricity.

9. What was the worst nuclear power plant disaster ever to occur? State what happened and why, where, and when it happened.

10. Discuss environmental problems associated with the following procedures involved in energy production: drilling for oil, transporting oil, mining for coal or uranium, and burning fossil fuels.

Practice Test

1. Which of the pairs below shows a highly volatile petroleum product as the first item and a very viscous petroleum product as the second item?
   a. tar (40-carbon chain) and butane (4-carbon chain)
   b. kerosene (12-carbon chain) and gasoline (5-carbon chain)
   c. kerosene (12-carbon chain) and heating oil (15-carbon chain)
   d. propane (3-carbon chain) and kerosene lubricating oil (26-carbon chain)
   e. methane (1-carbon molecule) and ethane (2-carbon chain)

2. A typical source rock of oil, which started out as mud in which dead organic matter settled, is
   a. shale.
   b. granite.
   c. conglomerate.
   d. sandstone.
   e. slate.

3. The oil window
   a. is the late Cenozoic, when all current oil was formed; any older oil has already decomposed.
   b. means the depth in Earth at which oil forms, roughly 40–50 km.
   c. is the relatively narrow range of temperatures in which oil can form, 100–250°C.
   d. is the holes in the drill bit through which oil enters.
   e. is the time of year at which OPEC allows new oil fields to be drilled in its member countries.

4. A typical reservoir rock, into which oil has migrated and collected, is
   a. shale.
   b. granite.

c. slate.
   d. basalt.
   e. sandstone.

5. Oil formed because
   a. leaves, twigs, and tree trunks decomposed under conditions of heat, pressure, and low oxygen.
   b. tiny plankton decomposed under conditions of heat, pressure, and low oxygen.
   c. the decomposition of organic material remained aerobic (lots of oxygen) for a long time.
   d. large sauropod-type dinosaurs decayed in swamps.
   e. the temperature reached a minimum of 1,000°C.

6. Which of the following is a typical oil trap?
   a. anticline
   b. syncline
   c. neck of an ancient volcano
   d. an igneous dike
   e. All of the above would make good oil traps.

7. A seal rock
   a. is a necessary ingredient of an oil trap.
   b. typically is fractured limestone or poorly cemented sandstone.
   c. must lie beneath the reservoir rock in a trap to keep oil from filtering downward.
   d. has high porosity and high permeability.
   e. All of the above are true statements.

8. Which of the following has never been a use for oil?
   a. a waterproof sealant
   b. a preservative to embalm mummies
   c. an ingredient in patent medicines of the 1800s
   d. lamp oil
   e. Oil has been used at some time in history as all of the above.

9. The first oil well was drilled
   a. 3,000 years ago by the Babylonians in what is now Iran.
   b. by John D. Rockefeller in 1901 in Texas.
   c. by Edwin Drake in 1859 at Titusville, Pennsylvania.
   d. to a depth just short of 1 mi.
   e. Both c and d are true statements.

10. A seismic-reflection profile
    a. is created by interpreting reflected seismic waves.
    b. shows the shapes but not the depths of underground sedimentary layers.
    c. is used even by amateurs looking for oil because it's relatively easy and cheap.
    d. is essentially the same thing as a geologic map.
    e. All of the above are true statements.

11. Drilling mud
    a. is a slurry of water and clay.
    b. is pumped down the center of an oil drilling pipe.
    c. is used to cool the drill bit and to flush rock cuttings up and out of the drill hole.
    d. weights down oil in the drill hole and thus helps prevent gushers.
    e. All of the above are true statements.

12. Which of the following statements is FALSE? Secondary recovery techniques
    a. are used to coax more oil out of a drilled hole.
    b. include the use of steam to make oil less viscous.
    c. include the use of steam pressure to help push oil up and out of the drill hole.
    d. include creating fractures in the rock through which oil can more readily flow.
    e. usually enable drillers to get 90% of the oil out of the ground.

13. Natural gas
    a. consists of a mixture of petroleum products with carbon chains 10 to 20 carbons long.
    b. is used more extensively than oil is.
    c. is found layered between oil and water in a drill hole.
    d. burns more cleanly than oil does.
    e. All of the above are true statements.

14. Coal develops in the following order:
    a. lignite → peat → bituminous coal → anthracite
    b. peat → lignite → anthracite → bituminous coal
    c. bituminous coal → peat → lignite → anthracite
    d. peat → lignite → bituminous coal → anthracite
    e. plankton → lignite → bituminous coal → anthracite

15. Which of the following statements is FALSE?
    a. Fossil fuels are the hydrocarbon remains of past life forms.
    b. Oil is usually several millions of years old.
    c. Most of the coal mined today in the United States is used in electric power plants.
    d. Coal is considered a renewable resource because it's currently forming in swamps.
    e. Coal rank is based on the carbon content of the coal.

16. Macroscopic organic material (leaves, stems, trunks) of swampy areas undergoes heat and pressure in an oxygen-poor situation over a geologically long time. The result is
    a. oil shale.
    b. kerogen.
    c. tar sands.
    d. petroleum.
    e. coal.

17. Currently the United States
    a. recovers 90% of the oil from each well drilled.
    b. knows it can get a minimum of 100-years' worth of energy needs out of Alaskan oil.
    c. is finding huge new oil resources in new geologic areas, like igneous deposits.
    d. uses more oil than natural gas or coal to supply its energy needs.
    e. All of the above are true statements.

18. Which of the following statements is FALSE?
    a. Coal couldn't form until vascular land plants became abundant.
    b. Anthracite coal develops at depth along mountain belt borders where temperatures reach 300°C.
    c. Underground coal mining is dangerous because of possible tunnel collapse and possible methane gas explosions.
    d. Coal in the western United States is of Carboniferous age and has higher sulfur content than Cretaceous age coal of the Midwest.
    e. Strip mining can be done economically down to a depth of about 100 m.

19. Reasonable conservation of fossil fuels probably
    a. will allow the world to have fossil fuel energy for as long as humankind exists.
    b. will help by buying time to develop other energy resources.
    c. is unnecessary, because oil shortages are not real; they're simply the result of political manipulations.
    d. will keep us going until the fossil fuel developing naturally today will be usable.
    e. will allow us to be oil dependent for a minimum of 500 more years.

20. From the rock layers that are drawn in the diagram below, it is logical to say

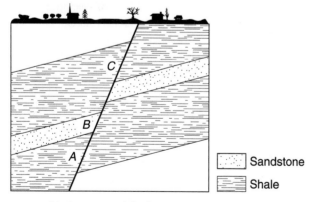

    a. this is a normal fault.
    b. this is a reverse fault.
    c. oil would most likely be found at C.
    d. both a and c are true.
    e. both b and c are true.

21. On the diagram below,

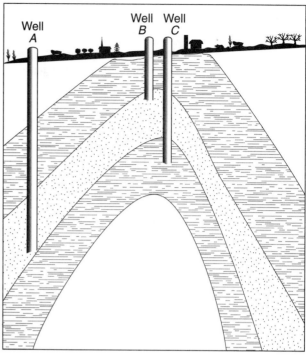

    a. oil well A would yield the most oil because it goes the deepest into permeable reservoir rock.
    b. oil well B would yield the most oil because it's on an anticlinal crest in permeable reservoir rock.
    c. oil well C would yield the most oil because it's drilled directly into source rock.
    d. All three wells would yield about the same amount of oil because all three reasons offered are true.
    e. No well could yield oil because no reservoir rock is shown.

22. Which statement is TRUE?
    a. If the diagram in question 21 showed a syncline, the oil would be found in the synclinal bottom.
    b. At the anticlinal crest in the diagram, it would be normal to have gas sitting on top of oil sitting on top of water.
    c. Anticline traps make good diagrams for test questions, but they're rare in nature.
    d. There's no way any of these wells could turn into gushers because the top shale layer would prevent that.
    e. All of the above are true statements.

23. In a nuclear power plant,
    a. fusion is happening.
    b. there must be enough nuclear fuel to equal a critical mass.

c. control rods regulate the rate of the reaction by absorbing excess neutrons.
  d. pitchblende (U^{238}) is the most common fuel.
  e. All of the above are true statements.

24. Which statement is TRUE?
    a. Hydroelectric power has many advantages and no disadvantages, provided the dam is properly constructed.
    b. Geothermal energy is commonly used in Japan and Mexico.
    c. The Organization for Petroleum Environmental Concerns (OPEC) establishes rules concerning where oil well rigs can be placed.
    d. A barrel of oil is 42 gal of oil.
    e. Tar sands in western Canada are especially valuable because the oil is so close to the surface that wells need to be only about 200 ft deep.

25. Large oil fields contain such great quantities of oil because they've developed from the largest dinosaurs.
    a. True
    b. False

26. Lignite is described as a low-rank coal, bituminous coal is midrank, anthracite is high rank, and peat isn't ranked at all.
    a. True
    b. False

27. Green plants produce food through photosynthesis, which is the combination of oxygen and chlorophyll to yield carbon dioxide and water.
    a. True
    b. False

28. The largest oil reserves in the world are in countries around the Persian Gulf.
    a. True
    b. False

29. All oil ever formed on Earth is still in existence, trapped beneath seal rock.
    a. True
    b. False

30. Spindletop Hill was the site of a tremendous gusher that started the Texas oil boom.
    a. True
    b. False

31. Offshore drilling platforms are floated above mid-ocean ridges to drill down to oil coming from black smokers.
    a. True
    b. False

32. The United States is the largest oil consumer (roughly 25% of world consumption) but has only about 4% of the world's total reserves.
    a. True
    b. False

33. Red Adair and other oil well firefighters usually have to extinguish oil fires with dynamite.
    a. True
    b. False

34. Drag lines are used in strip mining operations both to remove overlying soil and rock from the coal seams and to replace top soil after the mining is done.
    a. True
    b. False

35. Black-lung disease is becoming more common with coal miners because shallow coal has all been mined and underground mining has had to go deeper than ever before.
    a. True
    b. False

36. Fission and fusion both yield energy because a little matter is lost in either reaction and changed to energy.
    a. True
    b. False

37. Geiger counters are used to prospect for the uranium ore pitchblende.
    a. True
    b. False

38. There have been two meltdown events at nuclear power plants; the first was at the Three Mile Island plant in Pennsylvania, the second at Chernobyl in Russia.
    a. True
    b. False

39. Experts believe it is highly probable that humans will have exhausted all usable quantities of oil by the year 2150.
    a. True
    b. False

40. A weather inversion is a situation in which smog rises higher in the atmosphere than usual.
    a. True
    b. False

41. After years of delay, radioactive wastes with low-level radiation but long half-lives are finally being placed in a facility for permanent waste disposal. The facility (the Waste Isolation Pilot Plant, WIPP) outside of Carlsbad, New Mexico, places wastes at a depth of more than 2,000 ft in rooms carved in late Paleozoic age salt beds that are 3,000 ft thick. Comment on the wisdom of this decision. Personal opinion is allowed, even encouraged, in your answer, as long as it's backed up by scientific fact.

ANSWERS

Completion

1. energy
2. energy resource
3. fossil fuels
4. photosynthesis
5. geothermal energy
6. hydrocarbons
7. organic chemicals
8. viscosity
9. volatility
10. plankton
11. black organic shale
12. source rock
13. kerogen
14. oil shale
15. oil window
16. oil field
17. oil reserve
18. reservoir rock
19. porosity
20. permeability
21. oil well
22. oil trap
23. seal rock
24. seeps
25. petroleum
26. wildcatters
27. geologic map
28. cross section
29. seismic-reflection profile
30. seismic waves
31. drill bit
32. drilling mud
33. gushers
34. offshore drilling platform
35. secondary recovery techniques
36. crude oil
37. refinery
38. distillation column
39. coal
40. peat
41. lignite
42. bituminous coal
43. anthracite
44. transgression
45. regression
46. coal rank
47. seams
48. coal reserves
49. economic seams
50. strip mining
51. drag line
52. underground mining
53. black-lung disease
54. nuclear reactor
55. nuclear fuel
56. fuel rods
57. chain reaction
58. critical mass
59. enrichment
60. pitchblende
61. control rods
62. meltdown
63. nuclear waste
64. high-level waste
65. intermediate-level waste
66. low-level waste
67. geothermal gradient
68. hydroelectric power
69. renewable resources
70. nonrenewable resources
71. oil age
72. breeder reactor
73. clean energy
74. acid mine runoff
75. oil spills
76. low-sulfur coal
77. greenhouse effect
78. global warming
79. weather inversion
80. smog
81. photochemical smog

Short-Answer Questions

1. a. a. 110 times as much   b. oil and gas   c. coal, wind, flowing water, nuclear energy, geothermal energy, and solar energy
2. energy created by nuclear fusion in the Sun and transferred to Earth as electromagnetic radiation
   energy created by the pull of gravity
   energy created in nuclear fission reactions
   energy that has been stored in Earth's interior since Earth's beginnings
   energy stored in chemical bonds of compounds
3. a. carbon dioxide plus water plus carbon (soot) plus other gases plus heat
   b. the same as a
4. the size of its molecules
5. less; more
6. solids; liquids; gases
7. natural gas, gasoline, kerosene, heating oil, lubricating oil, tar
8. plankton
9. They're both tiny pieces and will settle out only in quiet waters.
10. An oxygen-rich environment would support aerobic bacteria that would eat/decompose the plankton's or swamp plants' remains, leaving nothing to develop into oil or coal.
11. a. kerogen   b. a mixture of solids, liquids, and gases   c. gas   d. graphite
12. a. between 4 and 6.5 km for oil; possibly down to 9 km for gas
    b. 2.5–4.5 km
    c. top 15–20% of the crust
13. area bordering the Persian Gulf
14. source rock, reservoir rock, oil trap, and an oil seal
15. Too expensive; it's cheaper to drill for already existing oil.
16. a. 10%   b. 35%
17. Buoyancy drives it upward; it takes millions of years.
18. The gas will be on top of the oil which will be on top of the water.
19. a seal rock and a rock setting (geometrical arrangement) that causes the oil to collect in a limited area
20. anticline trap, fault trap, salt dome trap, stratigraphic trap
21. a. as cement, as a waterproof sealer (Noah's arc was caulked with bitumin, a solid form of petroleum), and as a preservative to embalm mummies
    b. to grease wagon wheels, to make patent medicines, and to burn in kerosene lamp
22. Started Standard Oil Company.

23. They were the companies formed when the supreme court mandated Standard Oil must split because it was violating antimonopoly laws.

24. the location of a fantastic gusher in 1901 that started the Texas oil boom

25. sedimentary rock

26. over $10 million

27. 30%

28. 50%

29. Steam pressure heats the oil, lessens its viscosity, and helps push it into the well.
Underground explosions cause rock fractures through which oil can flow more readily.

30. the United States; 7 million barrels per day, which is 25% of world consumption

31. a. 4%   b. The United States imports more than half of the oil it uses.

32. methane, ethane, propane, and butane

33. Subsurface temperatures sometimes are hot enough to have caused all oil molecules in the area to break into small enough chains to be gas.

34. Gas burns cleaner, producing only carbon dioxide and water; burning oil produces those and also complex organic pollutants.

35. It requires high pressure pipe lines or container ships, so it's expensive.

36. a. oil   b. gas

37. Oil forms from microscopic marine (usually) plankton; coal forms from macroscopic pieces (wood, stems, and leaves) of land plants from swamps, wetlands, or rain forest.

38. after the late Silurian period (mid Paleozoic), about 420 million years ago; because vascular land plants didn't appear until then

39. a. 178Carboniferous (late Paleozoic), 290–360 million years ago, and Cretaceous (late Mesozoic), 65–145 million years ago
   b. Carboniferous age coal is in the Midwest; Cretaceous coal is in the Rocky Mountain regions of both the United States and Canada.

40. a. fuel for electric power plants   b. 23%

41. peat, lignite, bituminous coal, anthracite

42. a. more than 70%   b. hydrogen, nitrogen, and sulfur

43. Anthracite requires the high temperatures that develop only along the borders of mountain belts, but temperatures high enough to produce metamorphic rocks arrange carbon atoms into the metamorphic mineral graphite, rather than into anthracite.

44. 1–3 m

45. 100 m below ground surface

46. Miners carried canaries with them to detect the deadly gas methane; presence of even a tiny amount would kill the bird.

47. They can't extinguish them.

48. Nuclear fusion means small atoms are melted together into a bigger atom by extreme heat; nuclear fission means a large atom splits into smaller pieces.

49. In either reaction, a tiny bit of matter disappears and is changed to energy according to Einstein's equation $E = mc^2$.

50. Uranium atoms are too big to fit into the crystal structure of most common minerals.

51. a Geiger counter

52. Three Mile Island power plant near Harrisburg, Pennsylvania, in 1979; faulty pumps allowed the reactor to overheat; there were very little escape of radioactive material and no known injuries.

53. underground tunnels in mountain rock in an area not prone to quakes or volcanic activity
interior of impermeable salt domes
landfills surrounded by impermeable clay
landfills where groundwater can react with radioactive atoms and convert them to nonmovable materials
deep-ocean floor near subduction zones

54. Yucca Mountain in the Nevada desert; in a rhyolite mountain

55. pumping the naturally hot water out of the ground and running it through heating pipes in buildings; converting the hot water to steam which is used to drive turbines and generate electricity

56. running out of water if the hot waters are being pumped out of the area faster than they can be replaced

57. a. Iceland and New Zealand   b. They're volcanic areas.

58. The potential energy of the water positioned high in reservoirs changes to kinetic energy as it flows through pipes down to the bases of dams where it turns fly wheels in electric-generating turbines.

59. Positive aspects: Produces no smoke or radioactive waste; reservoirs can provide flood control, irrigation waters, and recreation areas.

Negative aspects: Any dam and reservoir change the environment, and the changes may be undesirable for the landscape and ecology.

60. Flood gates impound water at an inlet as the tide rises; when the tide is low, released water flows downward to operate electricity-generating turbines.

61. a. $20 a barrel   b. $1.80 a barrel   c. $18 a barrel
    d. over $30 a barrel

62. The Organization or Petroleum Exporting Countries (middle East countries) regulates much of the oil industry.

63. a. 25 billion barrels per year
    b. between 1,100 and 3,000 billion barrels
    c. sometime between the years 2050 and 2150

64. less than three centuries

65. Athabasca tar sands of western Canada and deposits in Venezuela

66. tar sands, oil shale, coal, and natural gas

67. a. Concern has grown over radiation itself, accidents, and waste storage.
    b. Most appropriate river locations are already in use, and in today's world, building new dams and reservoirs is usually not a popular proposal due to environmental concerns.
    c. currently too expensive
    d. Areas where the wind is strong enough are limited; putting huge windmills all over the landscape lacks appeal to residents.
    e. Scientists don't know how to control fusion; currently they can only use it in bombs.

Short-Essay Questions

1. A rock could have lots of microscopic spaces between rock grains and therefore be very porous, but if the spaces didn't connect up with each other, the rock would not be very permeable. Such a rock would not serve as a good practical reservoir rock because even if somehow oil had gotten into the many spaces available, there would be no flow through the rock when it was drilled and the oil could not be extracted.

2. No oil reserves last forever. Some get destroyed by microbes that eat (metabolize) oil. No rock is absolutely impermeable, so eventually oil will leak past the seal rock of a trap and be dispersed, possibly to Earth's surface where it may partially disappear by evaporation of its more volatile components. Oil is forming today in appropriate environments (as uniformitarianism predicts), but it forms so slowly it cannot do current society any good. Oil is definitely a nonrenewable resource.

3. In 1854 George Bissel realized petroleum could be used as a replacement for animal fat in oil lamps, and it might have many other uses, all of which could make him rich. He and a group of investors backed "Colonel" (a phony title) Edwin Drake, who drilled the first oil well ever on August 27, 1859, in Titusville, Pennsylvania. The well was only 21 m deep but still filled with oil and yielded 10–35 barrels per day. It was not the best-organized operation; originally the only containers available for the crude oil were whiskey barrels. It certainly made good money; $20 a barrel back in 1859 was very impressive.

4. Red Adair became famous in the oil industry for his success in putting out oil well fires that accompanied gushers. He became famous with the general public when John Wayne played him in the movie *Hellfighters*. You can't put oil fires out with water or foam; they are literally blown out by dynamite explosion. Red led an exciting and dangerous life, one which was certainly worthy of a John Wayne movie.

5. The more times the sea transgresses, then recedes from the land, the more sediment gets deposited, the more the surface sinks, and the deeper the sedimentary basin becomes. The sediment that accumulates is the future coal, and the sinking basin and continuing cover-up of deposits improves conditions necessary for coal formation—compaction, heat, and squeezing out the water.

6. As huge drag lines remove soil and rock to expose coal seams, they separate out and preserve the top soil. After the coal has been removed the hole is filled with rock, covered with the top soil, and replanted.

7. Uranium gets in veins because hot water circulating through entire plutons dissolves out uranium and concentrates it by re-deposition in cracks. Uranium can be weathered and eroded out of plutons and carried into streams where, due its high density, it sinks in calmer waters. Uranium can be concentrated in sedimentary rock because it gets dissolved out of large areas of sedimentary rock by circulating groundwater and later deposited in pore spaces in more-limited areas of sedimentary rock.

8. The fission reaction in the nuclear reactor liberates heat, which is transferred to water circulating in pipes around the reactor. The heated water is converted to high-pressure steam, which is used to rotate the fan blades of a turbine, which in turn drive a dynamo that creates electricity. The steam is returned to cooling towers, condensed, and recycled back into the system.

9. Meltdown almost occurred at the Chernobyl nuclear power plant in Russia in April 1986. While experimenting with cooling procedures for the reactor, operators of the plant shut down safety devices. The

equipment got hot enough to cause a steam explosion that literally raised the roof of the reactor and spread fragments of reactor and fuel around the plant grounds. Some entered the atmosphere and dispersed over eastern Europe and Scandinavia. Some people died immediately of radiation sickness, and many others have suffered serious health problems since.

10. Drilling for oil involves huge equipment that can stress and greatly alter and damage the land. This is true also when coal and uranium mines are developed. Transporting oil via pipelines, tanker trucks, or tanker ships has too often resulted in oil spills and severe environmental damage. Coal and uranium mines have the potential for acid mine runoff. Burning fossil fuels can produce the pollutants carbon monoxide, sulfur dioxide, nitrous oxide, and unburned hydrocarbons. Not only are they pollutants as such, but they can lead to smog, acid rain, and increased greenhouse effect with resultant global warming.

Practice Test

1. d The shorter the chain, the more volatile and less viscous the substance.

2. a Igneous granite and metamorphic slate are not possible because of the intense heat associated with their formation; the other two don't represent the oil-forming environment necessary, which is mud at the bottom of standing water.

3. c Choices a, d, and e are just nonsense; oil forms no deeper than 6.5 km; gas no deeper than 9 km.

4. e Shale is impermeable and would not permit oil to move into or through it. The others are igneous or metamorphic rocks, and the heat involved in those processes destroys any material that might develop into oil.

5. b Choice a describes coal formation; oxygen would lead to total decay; dinosaurs had nothing to do with oil formation; and temperature can't get above 250°C.

6. a Oil rises because of buoyancy and would not settle at the bottom of a syncline; the heat of igneous activity would destroy any oil formed.

7. a Choice b would be good reservoir rocks; oil doesn't leak downwards; and the porosity and permeability must be low.

8. e Statements a to d are all correct.

9. c The well was a little over 20 m deep (60+ ft); Rockefeller started the Standard Oil Company in the early 1900s.

10. a The profile is created by experts and shows shapes and depths of rock layers underground. A geologic map shows rock layers and geologic structures on Earth's surface.

11. e Statements a to d are all true.

12. e Best recovery is usually about 50%.

13. d Gases have very short carbon chains; gas is found above oil and water in the drill hole.

14. d If you chose e, note that plankton bodies may eventually form oil, not coal.

15. d "Renewable" means on a human time scale, not a geologic one; oil and coal are both forming in today's world, but too slowly to replace themselves as we use them up

16. e Choices a to d name energy resources that began as microscopic marine organisms.

17. d Best recovery is about 50%; the total U.S. oil reserve will last only a few years; and the heat of igneous activity would destroy oil.

18. d Coal in the western United States is of Cretaceous age and has lower sulfur content than the Carboniferous coal of the Midwest.

19. b Experts believe oil supplies will last a maximum of 200 years, and other fossil fuels will also be exhausted faster than they're forming. Alternative energy sources must be developed if our standard of living is to be maintained.

20. a The fault is normal because the hanging wall did go down; oil would be found at B, not C, because C (shale) is the seal rock, keeping the oil in B, the reservoir rock.

21. b Oil flows to the top of an anticline and stays there (anticline trap). Source rock is not permeable, so it prohibits flow into the well.

22. b Gas is less dense than oil or water so it would sit above them; anticline traps are common; and great enough gas pressure causes gushers no matter what rock is present.

23. c Fission is happening; critical mass is the amount that would result in an explosion; and U^{235} is the most common fuel.

24. d Dams and reservoirs always change the environment, not always in a good way; geothermal energy is common in Iceland and New Zealand; tar sands are mined, not drilled; and OPEC is the Organization of Petroleum Exporting Countries.

25. False Oil develops from tiny plankton, not from dinosaurs.

26. True   The varieties of coal are ranked on the basis of their fuel efficiency; lignite is the lowest and anthracite is the highest.

27. False   Photosynthesis is the combination of carbon dioxide and water, in the presence of chlorophyll and light, to yield sugar and oxygen and water.

28. True   The largest oil reserves are found in the Persian Gulf area.

29. False   Oil eventually decomposes by bacterial action or leaks through seal rock.

30. True   The discovery of oil on Spindletop Hill did start the Texas oil boom.

31. False   Offshore drilling platforms either float or are anchored along coast lines in order to drill into strata on the continental shelves.

32. True   Sobering facts, aren't they?

33. True   Water doesn't work because oil floats on it; foam is ineffective on very high-energy fires; and explosions drive oxygen away from the fires and therefore extinguish them.

34. True   Drag lines are used to strip away the overburden and uncover the coal and to replace the overburden after the coal is removed.

35. False   More stringent regulations have helped control this and other mining problems.

36. True   Nuclear energy, whether fission or fusion, results from the conversion of matter into energy.

37. True   Geiger counters detect radiation from uranium ores.

38. False   The accident at Chernobyl was a steam explosion that released radioactive debris, not a meltdown; the Three Mile Island event was a much less serious accidental release of some radioactive material.

39. True   Most experts agree that the world will have exhausted its commercial quantities of oil by the year 2150, if not sooner.

40. False   A weather inversion is a situation where warm air lies above a colder air layer, so ordinary upward air circulation doesn't occur and pollution gets trapped close to Earth's surface.

41. There are many comments that could be made, including:
The disposal of radioactive waste in impermeable salt deposits has been studied and approved by experts.
The issue is not whether it is or isn't safe and morally correct to continue nuclear activity; the issue is that radioactive wastes exists and should be disposed of in the safest way possible.
Salt is impermeable and flows slowly. If left alone, it should seal off these rooms.
Salt of course is soluble, but this particular salt has been around for more than 200 million years and it's not likely to dissolve soon.
The depth of 2,000+ ft takes the waste out of the realms of groundwater flow and of surface disasters like fires, tornadoes, floods, or explosions.
The alternative to moving the waste to a permanent disposal site like this is to leave it in a temporary surface facility; that no doubt would be riskier.
Any decision this important should be based on much more data than was given in this question (which it definitely was).

CHAPTER 15 | Riches in Rock: Mineral Resources

GUIDE TO READING

This chapter deals with many of the geologic resources that give us the raw materials for our very comfortable lifestyles in our industrial society. It discusses the geologic reasons for their existence, methods of extracting them from the Earth, products made from them, concerns about future supplies, and the environmental impact caused by their extraction.

The raw materials are both metallic and nonmetallic mineral resources. The metals are generally less abundant and more expensive than the nonmetals. Some are not only useful to humankind but are highly prized by society. The native metal, gold, influenced the course of history in our own West and in other places worldwide. The author discusses gold and many more mundane metals (including copper, tin, iron, aluminum, lead, zinc, nickel, and manganese). Topics of discussion are:

- What are metallic characteristics and how do you work with metals? You read about metallic bonds, metallurgy, malleability, tempering, cold working, smelting, slag, alloys like bronze and steel, precious metals, and base metals.
- How do you extract metals from the Earth? You read about underground mining, open pit mines, ores, ore minerals, grade of an ore, shows, assays, adits, and rock bursts.
- How can you classify ore deposits? On the basis of their origins: magmatic (including massive-sulfide deposits), hydrothermal (including disseminated, vein, and porphyry deposits, black smokers, and ophiolites), secondary-enrichment deposits (including banded-iron formations and manganese nodules), residual mineral deposits (what's left after leaching by groundwater), and placer deposits (of dense materials).
- Where do you look for ore deposits? Metallic minerals aren't evenly distributed worldwide. Why are deposits found in some places and not in others? Plate tectonics, of course! You get details of ores in the Andes of Peru and Chile, in the western United States, and in southern Africa. You've gotten so used to plate tectonics' being the reason for everything, you may be surprised to learn plate tectonics had no direct bearing on the formation of banded-iron formations (BIFs) or on residual ore deposits like bauxite.

In all those discussion of metals, and throughout the rest of the chapter, the author has had to make choices of what examples he'll use. While it's never a bad idea to learn as much detail as is reasonable, it's most important you understand the concepts being explained by use of the examples. Your teacher may use local examples of minerals to illustrate concepts.

Nonmetallic mineral resources don't have the glamour of gold, but they've been equally useful to humankind. Centuries ago people began to quarry stone to construct impressive and enduring buildings, and as time has passed, society has continued to use building stone and numerous other nonmetallic resources. The chapter mentions many of these resources and products made from them, including cement, concrete, gravestones, flagstones, bricks, crushed rock, window glass, asbestos, landscaping rock, salt, and gypsum.

As Chapter 14 pointed out, society has become very concerned about the diminishing amounts of energy reserves; it is also starting to be concerned about diminishing amounts of mineral reserves. As with energy resources, mineral resources are nonrenewable and unevenly distributed around the world. This causes economics and politics to play important roles in any scenario that concerns mineral reserves. How long a resource lasts depends in part on how badly people want it and what they're willing to pay for it. Many of these materials have been fought over in the past; the United States stockpiles some of its strategic metals to hopefully avoid some battles in the future.

The chapter ends with the sensitive topic, mining and the environment. Even if things run smoothly in the political sense, extracting minerals on a large scale can be costly to the environmental quality on Earth. We want all the good things of life Earth can supply. Balancing our material desires with minimal environmental impact is becoming a growing challenge.

LEARNING ACTIVITIES

Completion

Test your recall of new vocabulary terms.

_____ 1. metallic minerals extracted from Earth that are useful to civilization (things like gold, copper, aluminum, and iron)

_____ 2. nonmetallic metals extracted from Earth that are useful to civilization (things like building stone, gravel, sand, gypsum, phosphate, and salt)

_____ 3. the group of opaque, shiny, smooth solids that because of their chemical bonds, can conduct heat and electricity well

_____ 4. chemical bonds that allow outer electrons to flow freely from atom to atom and thus are responsible for good electric conductivity

_____ 5. being easily bent and shaped

_____ 6. alternately heating and cooling a metal in order to affect its properties

_____ 7. shaping a metal after it has cooled

_____ 8. metals that occur in rock not as compounds but as single elements (examples: copper, silver, and gold)

_____ 9. heating rock to high temperature in order to separate its metal components from its nonmetallic components

_____ 10. the nonmetallic components of rock that separate from the metallic components during smelting

_____ 11. a mixture of two or more metals (or a metal and nonmetal) melted together to produce a material with different properties than either metal alone

_____ 12. a time span of roughly 500 years when people used a copper-tin alloy to make swords, battle axes, plows, and spades (roughly 4,000+ years ago, from 2500 to 2000 B.C.E.)

_____ 13. an alloy of iron and carbon that is the most widely used metal today

_____ 14. an alloy of iron, carbon, and chrome that is a corrosion-resistant metal

_____ 15. metals that are exceptionally expensive, like gold, silver, and platinum

_____ 16. metals that aren't as expensive as gold, silver, and platinum (examples: copper, lead, zinc, and tin)

_____ 17. minerals that are composed of high concentrations of metal that can be easily extracted (example: galena, which yields lead)

_____ 18. the measurement of the concentration of useful metal within an ore, which is higher the higher the concentration

_____ 19. a body of rock that is composed of a high enough concentration of economic minerals to make it worth mining

_____ 20. the type of deposit formed when heavy minerals crystallize early during the cooling process, sink to the bottom of the magma chamber, and accumulate there

_____ 21. accumulations of sulfide minerals at the base of a solidified magma chamber

_____ 22. any deposit formed when metallic ions dissolved in hot water associated with magma or igneous intrusions precipitate in fractures and in pores

_____ 23. a hydrothermal deposit in which the precipitated ore mineral is finely and broadly scattered throughout an intrusion

_____ 24. a hydrothermal deposit in which the ore mineral is precipitated in cracks in preexisting rocks

_____ 25. a hydrothermal deposit in which copper is precipitated out in igneous intrusions that are composed of large clasts of copper within a finer matrix material

_____ 26. fountains of hot water along mid-ocean ridges that are rich in metallic sulfides

_____ 27. chunks of sea floor pushed up onto continents during mountain-building events

_____ 28. an ore deposit formed when metals are dissolved out of a preexisting ore deposit and re-deposited at a new location

_____ 29. ore formed when very hot groundwater at depth dissolves metals, rises, and precipitates in cooler near-surface rocks (named for a location where it's common)

_____ 30. iron-rich sedimentary layers that precipitated out of seawater because oxygen concentration increased

_____ 31. lumpy deposits of manganese oxide chemically deposited on the sea floor

_____ 32. high concentrations of metals like iron or aluminum left in the soil after other elements have been leached out

_____ 33. name of the soil that is an ore deposit because it's so rich in residual aluminum

_____ 34. deposits of grains of heavy metals in places in streams where water slows down, looses energy, and can continue to carry only lighter material

_____ 35. exposures of ore minerals at ground surface

_____ 36. testing rock to determine how much metal can be economically extracted from it

_____ 37. a network of tunnels into an ore body, created by drilling and blasting

_____ 38. sudden spontaneous explosions of rock off ceilings or walls of tunnels

_____ 39. rock when it's used as a construction material (architects' usage)

_____ 40. entire slabs or blocks (not just small pieces) of granite or marble removed whole to serve as construction stone

_____ 41. any location from which construction stone is removed

_____ 42. any location from which ore is removed

_____ 43. a tool used to cut rock, consisting of a loop of braided wire moving between two pulleys, augmented with abrasive grains and water during its use

_____ 44. a tool used to cut rock, consisting of a blowtorch surrounded by cold-water nozzles, so the combination of extreme heat and cold acting simultaneously on the rock breaks it into fine pieces

_____ 45. a construction material that starts as a mixture of water, lime, silica, aluminum oxide, and iron oxide but is intended to lose water and harden

_____ 46. a specialized variety of cement, named by its English developer because of its resemblance to a local rock

_____ 47. the artificial conglomerate that cement mixed with sand or gravel hardens into

_____ 48. tabular units which are a common construction material and which are made by pouring wet clay into molds and then baking it

_____ 49. metals such as manganese, platinum, chromium, and cobalt that are alloyed with iron to make special-purpose steels needed in the aerospace industry

_____ 50. huge hills of waste rock sitting around mines

_____ 51. the precipitation that has become acid by filtering through tailings piles or exposed ore, and so it can do environmental damage

Matching Questions

MINERAL AND ROCK RESOURCES

Match the resource or product with its description.

a. azurite and malachite
b. bauxite
c. brick
d. cement
e. concrete
f. copper (gets used twice)
g. galena
h. gold
i. granite
j. gypsum
k. halite
l. hematite and magnetite
m. iron
n. lead and zinc
o. manganese
p. marble (gets used three times)
q. massive-sulfide deposits
r. native gold in quartz veins in granite
s. oil
t. quartz sand
u. stainless steel

_____ 1. a metal commonly found in porphyry deposits

_____ 2. copper carbonate minerals that are found as secondary-enrichment deposits

_____ 3. two metals found in dolomite beds in the Mississippi Valley region

_____ 4. a native metal found in placer deposits at Sutter's Mill, California

_____ 5. two minerals that are iron oxides

_____ 6. an alloy of iron, carbon, and chrome

_____ 7. the chief ore mineral of lead

_____ 8. the category of mineral deposits around vents of black smokers

_____ 9. the metallic oxide in which nodules on the sea floor are particularly rich

_____ 10. the mineral ore of aluminum, which occurs as a residual mineral deposit

_____ 11. geologic description of the mother lode of the Sierra Nevadas

_____ 12. any polished carbonate stone (architects' name)

_____ 13. any rocks that contain feldspar and quartz (architects' name)

_____ 14. a popular choice of stone with sculptors because it's soft and easy to carve

_____ 15. not a good choice to make for your tombstone if you want people to read about you for centuries to come

_____ 16. the hard material that results when a slurry of water, lime, silica, aluminum oxide, and iron oxide dries out

_____ 17. cement plus sand plus gravel

_____ 18. tablets of baked clay used in construction

_____ 19. the nonmetallic Earth resource for drywall (wallboard)

_____ 20. the basic ingredient of window glass

_____ 21. what electric wiring is usually made of

_____ 22. the main ingredient in nails

_____ 23. the common inorganic Earth resource for plastics

_____ 24. the mineral used to fill salt shakers

Short-Answer Questions

Answer with a few words, numbers, or short sentences, or choose from the words in parentheses.

1. What two characteristics of the molecules of a metal influence its behavior? _____

2. What is the name for the science of working with metals? _____

3. Iron is not found as a native metal in Earth's crust, but there are occasional natural chunks of native iron on the surface. Where do they come from? _____

4. a. Of the three principal metals in use today, copper, iron, and aluminum, why was copper the first to be used by society? _____

   b. Why did it get replaced by bronze? _____

   c. What must be added to copper to make bronze? _____

5. Iron is stronger, harder, and more abundant than copper or bronze. Give some reasons it took society so much longer to start using it. _____

6. Aluminum is a strong, light-weight metal and extremely abundant in crustal rocks. Why didn't early humans make use of it? _____

7. a. Name the two main ore minerals of iron. _____

   b. Chemically, what type of compounds are they? _____

8. What metal is obtained from each the following ore minerals? (Table 15.1 in the text) a. azurite and malachite _____   b. cassiterite _____
   c. cinnabar _____   d. corundum and kaolinite _____   e. galena _____
   f. rutile _____   g. sphalerite _____

9. Why did copper-bearing rocks require at least 3% copper to be considered economic ore in 1880, yet in 1970 rock containing only 0.3% copper was considered economic ore? _____

10. What mineral is commonly associated with gold in veins? _____

11. a. Why do geologists check the strength of Earth's magnetic field and the pull of gravity in regions they think may have ore deposits? _____

    b. Why do they analyze plants in these same areas? _____

12. Give two practical matters that must be considered when deciding whether to mine an ore deposit. _____

13. Open pit mines can get huge, because on average how much waste rock must be removed for every ton of ore recovered? _____

14. What metal is recovered at the huge open pit mine in Bingham Canyon near Salt Lake City, Utah? _____

15. If you were working in the deepest mine on Earth, how deep could you go and how hot would it be? _____

16. Give two reasons architects have for centuries built with stone. _____

17. From information given in the text and from your own experience, give several uses for each of the following nonmetallic mineral resources:
    a. granite _____
    b. limestone and sandstone blocks _____
    _____
    c. slate _____
    d. crushed stone _____

18. What two steps are involved in producing crushed stone from solid masses of stone? _____
    _____
    _____

19. Quantify what voracious consumers we are.
    a. How many metric tons of geologic resources does the population of the United States consume each year? _____
    b. To supply these resources, how many metric tons of geologic material must be mined, quarried, or pumped out of the Earth each year? _____
    c. As a comparison, how many metric tons of sediment does the Mississippi River transport into the Gulf of Mexico each year? _____
    d. Compare the amount of material mined/quarried/pumped to the amount of material carried by the river. Which is greater, and by a factor of how much? _____
    _____

20. Mineral and rock resources are being formed in the world today, as the principle of uniformitarianism does correctly suggest, yet they are considered nonrenewable resources. Why? _____
    _____

21. Use Table 15.4 in the text to find:
    a. the strong, light-weight metal that the United States uses extensively but has only a 2-year supply of (luckily the world has more than a three century supply) _____
    b. a precious metal that the United States has about 2/3 of the world supply _____
    c. a precious metal that, if left on her own, the United States would run out of in about 1 year _____
    d. the metal for which the United States owns about 33% of the world supply, which is nice, since it's an ingredient of the alloy steel _____

22. Name the country that owns all of the mineral resources it needs. _____

23. As a defense precaution, what does the United States do about its supplies of strategic metals, such as manganese, platinum, chromium, and cobalt?
    _____

Short-Essay Questions

1. What is the origin of the nickel deposits at Sudbury, Ontario?

2. Explain briefly how totally different plate tectonics situations produced valuable metal deposits in Peru and in southern Africa.

3. Using both text material and your own experience, list many items found in a typical home that are made from geologic resources (metallic and nonmetallic).

4. Why might calculated reserve figures of some mineral resources change in the future?

5. Present a brief discussion of the values of mining versus environmental concerns.

Practice Test

1. Which of the following statements is FALSE? Metals
    a. shine because their outer electrons are so firmly fixed in place they have a high reflectivity index.
    b. may be malleable.
    c. are opaque, shiny, smooth solids.
    d. are termed native when they're found in Earth as the element alone, not as a compound.
    e. are termed strategic if they're alloyed with iron to make special purpose steels for the aerospace industries.

2. Which of the following statements is FALSE? Bronze
    a. is an alloy of copper and tin.
    b. is harder than copper.
    c. retains a sharper cutting edge than does copper.
    d. was used as early as 4000 B.C.E. to replace less desirable iron tools.
    e. was so influential its name was applied to several hundreds of years of human history.

3. Copper
    a. alloyed with chrome becomes bronze.
    b. is used in electric wiring.
    c. is not easily smelted because extremely high temperature is required.
    d. is usually found in placer deposits.
    e. All of the above are true statements.

4. Iron
    a. is forming today in the Great Lakes region as banded-iron formations (BIFs).
    b. is stronger and weighs less than aluminum.
    c. was the first of the common metals (iron, aluminum, and copper) to be used by humankind.
    d. is alloyed with carbon and chrome to make stainless steel.
    e. All of the above are true statements.

5. Pick out the metal which is both a precious metal and a strategic metal.
   a. gold
   b. silver
   c. platinum
   d. copper
   e. zinc

6. Miners who search for gold in the mother lode
   a. need waterproof boots and a pan.
   b. are wasting their time; only silver and platinum are found in this type of deposit.
   c. geologically speaking are looking for a hydrothermal disseminated deposit.
   d. should look carefully at milky-white quartz veins.
   e. Both c and d are true.

7. What is NOT a logical place to look for gold?
   a. floating along in a stream
   b. in stream gravels
   c. in quartz veins
   d. in exposed granitic plutons
   e. You could find gold in all of the above situations.

8. If you're prospecting for metallic ores, you should be paying attention to
   a. assay reports.
   b. magnetic surveys.
   c. gravity surveys.
   d. shows (geologic, not the TV kind).
   e. All of the above.

9. Open pit mines are
   a. developed in areas where there's high-grade ore.
   b. developed only when the ore is several hundred meters deep.
   c. used only for gold minerals and coal.
   d. well illustrated by the large one at Bingham Canyon, Utah.
   e. All of the above are true.

10. Which of the following could NOT be an ingredient of cement?
    a. lime
    b. silica
    c. aluminum oxide
    d. iron oxide
    e. They could all be ingredients.

11. Which is NOT a hazard or problem associated with mining?
    a. acid mine runoff
    b. tailings piles
    c. rock bursts
    d. air pollution
    e. They are all potential hazards or problems.

12. Tempering is heating a mineral to high enough temperature to separate its metallic component from any nonmetallic residue called slag.
    a. True
    b. False

13. Malleability is a measure of how easy it is to bend and shape a metal.
    a. True
    b. False

14. Malachite and azurite are two native metals that are also classified as strategic metals.
    a. True
    b. False

15. Metallic iron is released from iron-oxide minerals when they're heated in the presence of carbon dioxide.
    a. True
    b. False

16. The nickel deposits of Sudbury, Ontario, are massive-sulfide deposits.
    a. True
    b. False

17. Sulfide deposits in ophiolites are accessible to humans; those associated with black smokers are not.
    a. True
    b. False

18. Metallic deposits in south Africa owe their existence to rifting, which produced decompression, which in turn allowed magma to form and rise along rift axes.
    a. True
    b. False

19. A green or red stain of milky-white quartz in a vein is bad news for prospectors; it shows where the gold had been before it got weathered out.
    a. True
    b. False

20. Analyzing plants in an area can detect traces of metal present in the ground.
    a. True
    b. False

21. Adits are the shafts that go deep into a mine to reach to the depth at which ore is found.
    a. True
    b. False

22. Portland cement is a type of concrete named that because it was used so extensively in Portland, Oregon.
    a. True
    b. False

23. The Rosendale Formation of Michigan is the most famous banded-iron formation.
    a. True
    b. False

METALS

Match the metal, rock, or mineral with its description. Some terms may be used more than once.

a. aluminum
b. clay
c. copper
d. gypsum
e. halite
f. iron
g. lead
h. manganese
i. marble
j. nickel
k. quartz
l. sulfides
m. zinc

_____ 1. found as a native metal in meteorites
_____ 2. the three principal metals in use today
_____ 3. galena is the chief ore of this metal
_____ 4. the origin of its largest deposit was a meteor impact 2 billion years ago that triggered igneous activity
_____ 5. often found in disseminated porphyry deposits
_____ 6. of the three metals so widely used today, this was the last to be used by society, because it took a long time to figure out how to separate the metal from its ore
_____ 7. ores found around black-smoker vents
_____ 8. found in sedimentary layers as oxide minerals, interbedded with layers of chert
_____ 9. two metals, found in dolomite in the Midwest, that are typical MVT ores
_____ 10. the metal that's very abundant in nodules found on the sea floor
_____ 11. the metal whose ore, bauxite, is a typical residual mineral deposit
_____ 12. a dimension stone and a sculptor's delight
_____ 13. the basic component of wallboard (also called Sheetrock)
_____ 14. table salt
_____ 15. the basic ingredient of window glass
_____ 16. bricks
_____ 17. the metal the United States uses so much of, but without imports, would have a mere 2-year supply of

ANSWERS

Completion

1. metallic mineral resources
2. nonmetallic mineral resources
3. metals
4. metallic bonds
5. malleability
6. tempering
7. cold working
8. native metals
9. smelting
10. slag
11. alloy
12. Bronze Age
13. steel
14. stainless steel
15. precious metals
16. base metals
17. ore minerals (or economic minerals)
18. grade of an ore
19. ore deposit
20. magmatic deposit
21. massive-sulfide deposits
22. hydrothermal deposits
23. disseminated deposit
24. vein deposit
25. porphyry copper deposits
26. black smokers
27. ophiolites
28. secondary-enrichment deposit
29. Mississippi Valley type (MVT) ores
30. banded-iron formations (BIFs)
31. manganese nodules
32. residual mineral deposits
33. bauxite
34. placer deposits
35. shows
36. assay
37. adits
38. rock bursts
39. stone
40. dimension stone
41. quarry
42. mine
43. wireline saw
44. cutting jet
45. cement
46. Portland cement
47. concrete
48. bricks
49. strategic metals
50. tailings piles
51. acid mine runoff

Matching Questions

MINERAL AND ROCK RESOURCES

1. f
2. a
3. n
4. h
5. l
6. u
7. g
8. q
9. o
10. b
11. r
12. p
13. i
14. p
15. p
16. d
17. e
18. c
19. j
20. t
21. f
22. m
23. s
24. k

Short-Answer Questions

1. strength of its chemical bonds and the shape and dimensions of its crystalline structure
2. metallurgy
3. They've fallen to Earth as meteors.

4. a. because copper smelting was so easy b. bronze is stronger and keeps a sharp edge better, so it made better cutting tools. c. tin

5. Iron has a very high melting temperature, difficult to achieve, so it can't be easily worked; its common forms are oxides, which don't even look metallic; and it requires a chemical reaction in the presence of carbon monoxide to separate out the metallic iron.

6. Separating aluminum metal from rocks is a very complex procedure; it couldn't have happened accidentally, as no doubt the first smelting of copper ores did.

7. a. hematite and magnetite b. oxides

8. a. copper b. tin c. mercury d. aluminum e. lead f. titanium g. zinc

9. New technology for mining and processing ores made it possible to make money developing lower-grade ore.

10. milky (white) quartz

11. a. Ore minerals tend to be dense and so exert greater than average gravity pull, and they're more magnetic than ordinary rock.
    b. Plants can absorb metals through their roots.

12. whether the ore can be mined at a profit and whether environmental concerns can be accommodated

13. 1,200 tons of waste rock for every ton of ore

14. copper

15. 3 km deep (a little less than 2 mi) and more than 55°C (131°F)

16. its durability and its visual appeal

17. a. cobblestones, curbstones, tombstones, stairs, countertops (in expensive homes today), statues
    b. foundation stones, fireplaces, building stones, limestone fenceposts in pioneer days in Kansas, flagstones, and other landscaping stones
    c. roofing shingles (but hopefully not in earthquake-prone areas), gravestones, pool table beds, floor tiles, and in the past, blackboards
    d. substrate of highways and railroads, material for manufacturing concrete and asphalt, landscaping material

18. blasting with explosives and then putting the blasted-out stones through a jaw crusher

19. a. 4 billion metric tons b. 18 billion metric tons c. 190 metric tons d. The amount of material mined/quarried/pumped is 94,736,842 times as much as the amount of material the Mississippi River transports.

20. They're forming slowly, on a geologic time scale; renewable and nonrenewable are decided on a human time frame.

21. a. aluminum b. gold c. platinum d. iron

22. There is no such country.

23. It stockpiles them.

Short-Essay Questions

1. A huge meteor hit Earth 2 billion years ago and produced a crater so deep some of the mantle there melted because of the decompression from above. Magma rose to the crust, and as it cooled, nickel sulfide sank to the bottom of the magma chamber (an example of a massive sulfide deposit).

2. The Andes Mountains are the result of compression and volcanic activity as the Pacific Ocean floor subducts under the South American Plate. Erosion has exposed granite plutons formed during mountain building, which contain gold, copper, and silver. Miners find the metals as disseminated deposits within the plutons, as vein deposits, or weathered-out deposits in placers. Secondary enrichment has occurred to produce deposits that are geologically newer than those formed during the original mountain building.

    Rifting is the plate tectonics story in Africa. It has produced magma dikes formed by the partial melting of the mantle along the axis of the rift. Some of the magma contained metal ions which have accumulated in magmatic or hydrothermal deposits, the most valuable of which are Precambrian deposits in southern Africa.

3. Your lists will vary, but may include: concrete driveways and foundations, bricks, windows, mirrors, anything else that is glass, Sheetrock (also called drywall or wallboard), copper electric wiring, anything made of steel (beams, nails, etc.), aluminum window frames, anything that's plastic, plumbing pipes (whether they're copper or plastic), marble or granite desktops or counter-tops, gold (or brass) plumbing or light fixtures, barbecue grills and the lava rock in them, garden tools and snow shovels, salt on the table, landscaping rock, cans for soda in the refrigerator, potash and phosphate fertilizers for your garden—just look around; the list is endless!

4. "Reserve" is a practical term, meant to convey amounts of a material that can be *economically* extracted from Earth. Current figures of mineral reserves are based on today's prices, rates of consumption, and known deposits, all of which will change in the future. New deposits will be found, prices will change, population increases will tend to increase consumption, and con-

servation and recycling may decrease consumption. The changes will influence the economics of recovery, which in turn will influence what deposits are considered "reserves."

5. As the answer to essay question 3 points out, Earth's resources supply much of the material necessary for our comfortable lifestyles, which most of us would not care to give up. Also, mining does provide jobs for many people. But mining is not gentle to Earth; even if done carefully, great physical changes are made to Earth. The minimum effect is the removal of huge amounts of Earth material which cannot be replaced on the human time scale. If done carelessly, mining and processing ores can result in ugly scars on the land, dangerous tailings piles, acid mine runoff, polluted waters, damaged vegetation and wildlife, and various types of air pollution. Newer mining techniques and stricter regulation of procedures are intended to make mining more environmentally friendly, but no perfect solution has been found to the problem of how to mine the Earth and produce absolutely no negative impact.

Practice Test

1. a   "Shine" may sound good, but it's nonsense; in fact the outer electrons move quite freely between atoms, which accounts for metals' good conduction of electricity.

2. d   Copper was used that long ago, not bronze; iron tools were not less desirable, they just hadn't been made yet.

3. b   Copper is alloyed with tin; it's easily smelted; it's not dense enough to be found in placer deposits.

4. d   The BIFs occurred in the Precambrian, iron is heavier than aluminum, and it was the second of the common metals to be used; copper was the first.

5. c   Gold and silver are precious metals but not strategic metals (those alloyed with iron to make special-purpose steel); copper and zinc are base metals.

6. d   Choice a describes gold panning, in which gold has already been weathered out of the mother lode deposit. Choice c is correct except for the disseminated part; it's a hydrothermal vein deposit.

7. a   Gold is too dense to float along.

8. e   Choices a to d are all true.

9. d   Open pit mines are used for low-grade ore fairly near the surface.

10. e   Choices a to d are all true.

11. e   Choices a to d are all potential hazards or problems.

12. False   This is a definition of smelting; tempering is alternate heating and cooling to influence the metal's properties.

13. True   Metals that can be bent and shaped easily are described as malleable.

14. False   They're minerals that are copper-carbonate compounds, and they're not alloyed with iron to produce special steels.

15. False   They must be heated with carbon monoxide, not carbon dioxide (sounds similar, but they have totally different properties).

16. True   The nickel sulfide deposits were formed almost 2 billion years ago as the result of a huge meteor impact.

17. True   Both have ocean origins, but ophiolites are ocean-bottom rock that has been thrust up onto continents; black smokers remain deep along mid-ocean ridges.

18. True

19. False   It shows that metal-containing minerals are present, because the stain is caused by their partial oxidation.

20. True   Plants can absorb metals through their roots.

21. False   They're a maze of tunnels into an ore body in an underground mine.

22. False   It's a type of cement named by its developer Isaac Johnson because of its resemblance to a natural rock near Portland, England.

23. False   It's a formation of the Hudson River area, New York, which contains the right mix of lime and silica to make an excellent cement, which is widely used in New York City.

METALS

1. f	7. l	13. d
2. a, c, f	8. f	14. e
3. g	9. g, m	15. k
4. j	10. h	16. b
5. c	11. a	17. a
6. a	12. i	

CHAPTER 16 | Unsafe Ground: Landslides and Other Mass Movements

GUIDE TO READING

Put simply, this chapter is about gravity. To elaborate a bit, the chapter discusses mass movement (or mass wasting), which is simply the movement of rock, regolith, or snow and ice downslope.

Naturally there are details. You learn there are different types of mass movement, classified on four factors: type of material (rock, regolith, or snow and ice); velocity of movement (fast, intermediate, or slow); character of the moving mass (chaotic cloud, slurry, or coherent body); and environment of the event (subaerial or submarine). You will study the causes of mass movement, its consequences to Earth and to humans, and the ways to protect ourselves and our structures from its damaging effects.

One difficulty with the subject matter is it's almost too common; words like "landslide" are used so casually they can mean almost anything. As you read, concentrate on the exact meaning of even the simple words. For example, a fall implies a vertical drop, whereas a slide means slipping as a coherent mass along an inclined slope, and a flow means turbulent, tumbling motion in which fluid (gas or liquid) is involved. You may still have to remind yourself it's not all repetition as you read about mudflows, debris flows, landslides, rock slides, debris slides, snow avalanches, debris avalanches, rock falls, debris falls, and submarine debris falls, and you may even welcome more-sophisticated terms like "creep" (well, they're not all sophisticated), the cold-climate variation of creep called solifluction, and other new terminology such as rock glaciers, slumping (and its components the slump, glide horizon, and head scarp), lahars, submarine slumps, and turbidity currents.

To remind you this is serious science, the author relates some of the classic cases of mass movement. Localities to note are:

Yungay, Peru, 1970
Pacific Palisades, California, 1958
Rio de Janeiro, Brazil, 1988
Armero, Colombia, 1985
Vaiont Dam, Italy, 1963
Austrian Alps, 1999
Yosemite National Park, California, 1996
Elm, Switzerland, 1881
Madison Canyon, Montana (near Yellowstone), 1959
Turnagain Heights, Alaska, 1964
Lituya Bay, Alaska, 1958
Gros Ventre River Valley (near Jackson Hole), Wyoming, 1925
La Conchita, southern California, 1994–95
Olympus Mons, Mars
Portuguese Bend Slump of the Palos Verdes area, California, 1956–85
Los Angeles, a mobile society (tongue-in-cheek)

Why does *anything* fall down? Gravity brings things down, but how do things get up in the first place? More scientifically worded, what causes the relief of the area? Possible causes are:

- convergence at plate boundaries and either resultant subduction or continental collision
- faulting in rift areas
- epeirogenic movements
- pile up of extrusive rocks at volcanic vents
- buildup of sediments
- human activity that builds things up
- human activity that tears things down
- weathering and erosion that tear things down
- lowering of sea level

Why do just *some* things fall down? Why isn't everything that's up in the process of falling down? You'll read about

- fragmentation and weathering that weaken the surface
- slope stability (stable and unstable slopes, slope failure, downslope forces, and resistance forces)
- angle of repose
- weak surfaces that act as glide horizons

What triggers a mass movement event? Here you read about shocking events and vibrations (earthquakes), and special cases that involve quick clay.

Whether a slope moves is partially dependent on inherent characteristics of the slope itself. You read about changing slope angles, slope loads, slope support, and changing slope strength (due to weathering, vegetation, and water content).

Getting down to the very basic reason for all this instability and resultant falling down, what else could any basic reason in this book be but plate tectonics. You're given a short reminder that plate tectonics is the basic cause of volcanic eruptions, faulting, and earthquakes, which in turn are basic causes of the topography that produces mass movements.

What can we do about all of this? We can't turn off gravity. Ideally no one should build, work, or play in areas prone to mass wasting. But that covers so much territory, and the world is too crowded to allow everyone to avoid the danger zones. Also, any area with significant relief is a potential danger zone and also a scenic place. Many people choose not to stay out of such areas. So what is the answer? The goal of geology is to ascertain just what the risks are in potentially dangerous areas so each person can make an informed decision about whether to be there. To do this, geologists determine how long rock surfaces have been stable by use of cosmogenic dating. They assess risk factors by looking for features like pressure ridges and by analyzing factors like slope steepness; strength of substrate; degree of water saturation; dip of bedding, jointing, and foliation relative to slope; vegetation cover, climate, undercutting, and seismicity. They compile all this information into landslide-potential maps which are made available to government agencies and to the general public. It's up to them to decide what to do with the information.

What can be done to stop mass movement? In one sense that's a silly question, because we all agree you can't turn gravity off. But there are some factors that can be controlled and steps that can be taken to reduce the risk that mass wasting will occur. These include revegetation and regrading of slopes, reducing subsurface water, preventing undercutting, using proper construction practices, and even doing controlled blasting of unstable slopes to make mass wasting occur when you want it to happen, not when it chooses to happen.

In summary, this chapter reminds us that Earth is dynamic, and even when it's not doing something dramatic, like spewing out lava or quaking and shaking, it can do things that upset and even endanger our lives. Life is a gamble. Living in areas of high relief makes life an even bigger gamble. How big a gambler are you? Are you going to take a mountain vacation soon?

LEARNING ACTIVITIES

Completion

Test your recall of new vocabulary terms.

1. _____ land that's at risk of moving downslope
2. _____ soil, sediment, and debris sitting on Earth's surface
3. _____ the movement of unconsolidated material downslope due to gravity
4. _____ a natural feature of the environment that could cause damage to people or their property
5. _____ the gradual movement of a slope back toward the hill crest
6. _____ the very slow movement, measured in centimeters per year, of regolith downslope
7. _____ the ground at high altitudes or high latitudes which remains permanently frozen
8. _____ especially fast creep that happens in areas of permafrost because the ground above it tends to get soggy and weak
9. _____ huge accumulations of rock interspersed with ice that move slowly down a slope or valley
10. _____ a type of mass-wasting event (and the mass itself) in which regolith slips downward along a sliding surface that is concave up, giving the impression that the mass has been scooped out and some of it still sits there but out of place
11. _____ any surface on which a mass slides
12. _____ the slightly curved little cliff that sits along the top of a slump site
13. _____ a mass of mud flowing downhill (Yes, it's as simple as you think.)
14. _____ a mass of mud and rock flowing downhill
15. _____ mudflows associated with volcanic events
16. _____ the sudden movement of rock alone, rock and debris, or regolith alone down a nonvertical slope
17. _____ turbulent clouds of regolith or snow and ice mixed with air that travel down steep hillsides at great speed
18. _____ snow and air, or rock and dust and air, traveling down a steep hillside at great speed

_____ 19. pathways that are used over and over again by turbulent downhill flows of either snow or debris

_____ 20. the free fall of rock or debris (Don't try to make these hard; they're obvious.)

_____ 21. slump events that happen underwater

_____ 22. slump blocks that get buried and preserved on the ocean bottom

_____ 23. debris flows that happen underwater

_____ 24. turbulent clouds of sediment suspended in water that avalanche down submarine slopes and settle out as graded beds

_____ 25. the difference in elevation between the highest high point and the lowest low spot of an area

_____ 26. an inclined (tilted) surface

_____ 27. a phenomenon that exists because the polarity of water molecules causes them to bond to mineral surfaces and to attract each other to form a tenuous "skin" that allows water drops to form, sand castles to hold together, and very light objects to "sit" on water surfaces

_____ 28. two types of slopes, on the first of which sliding is unlikely, on the second sliding is likely

_____ 29. gravity has won and material is moving down an unstable slope

_____ 30. two forces that counteract each other, of which the first tends to pull material downslope due to gravity and the second to inhibit sliding

_____ 31. the steepest angle at which unconsolidated material can sit without sliding

_____ 32. a type of clay that behaves like a solid when it sits still but turns into a slurry and flows like a fluid when it is shaken

_____ 33. a hillside of sedimentary rock in which the bedding planes parallel the face of the hill

_____ 34. the process in which weathering attacks a vertical rock face from below and causes an overhang which will eventually collapse

_____ 35. a type of clay whose mineral structure allows it to absorb water at the molecular level, between its layers of silica tetrahedra, and thus swell to several times its original size

_____ 36. a dating technique that allows geologists to determine how long a boulder has been stable by measuring the concentration of isotopes formed by the impact of cosmic rays on minerals in the rock surface

_____ 37. an upward bulge along the toe of a slump

_____ 38. maps that are created as the result of studies that assess the potential hazards from mass movement in areas

_____ 39. a special type of cement that is sprayed on road cuts to prevent water infiltration, subsequent freezing and thawing, and eventual disintegration

_____ 40. "roofs" constructed above highways at points where they run along the bases of avalanche chutes, to serve as artificial ground surfaces and keep snow off the highways proper

Matching Questions

HISTORIC MASS MOVEMENT EVENTS

Match the location of the event with its description.

a. Armero, Colombia (1985)
b. Austrian Alps (1999)
c. Elm, Switzerland (1881)
d. Gros Ventre River Valley (near Jackson Hole) Wyoming (1925)
e. Madison Canyon, Montana (near Yellowstone) (1959)
f. Pacific Palisades, California (1958)
g. Portuguese Bend Slump of the Palos Verde area, California (1956–85)
h. Rio de Janeiro, Brazil (1988)
i. Turnagain Heights, Alaska (1964)
j. Vaiont Dam, Italy (1963)
k. Yosemite National Park, California (1996)
l. Yungay, Peru (1970)

_____ 1. An earthquake triggered an ice fall in the Andes, which started a debris flow that buried a town and killed thousands of people

_____ 2. A kilometer-long section of cliff slumped down onto Highway 1 and closed the road for weeks. Luckily this time nobody was injured, just inconvenienced.

_____ 3. Heavy rains on steep hillsides covered with thick soil started mudflows and debris flows that buried entire communities overnight.

_____ 4. Experts' warnings were ignored, and a lahar in the Andes buried a town and killed 20,000 people.

5. The building of a dam and reservoir was the last straw in an already unstable area. A huge chunk of Mt. Toc slid into the reservoir, and the resulting splash drowned people downstream.

6. A recent example of a too common situation: snow blanketed the ground in European mountains, temperatures rose, partial melting occurred, temperatures fell, surface snow froze, a blizzard occurred, and skiers were happy, until serious avalanches killed many.

7. A rock fall occurred in a national park because a huge slab loosened along an exfoliation joint.

8. A 600-m-high crag of slate, undermined by quarrying, fell to the valley and buried an entire village.

9. An earthquake triggered a debris slide along foliation planes in metamorphic rock on a mountainside. Tragically the slide buried a campground below.

10. Shaking due to an earthquake caused quick clay to liquefy and slump, dumping a cliff-side subdivision into the sea.

11. The undercutting of a weak shale layer by a stream produced the largest observed landslide in U.S. history and dammed a river in the process.

12. Intense efforts have finally, after three decades, stopped the slumping of some coastline in southern California that was triggered by suburban development.

MASS MOVEMENT: CAUSES AND PREVENTION

Label the following statements with P if they are actions which would help Prevent mass wasting or damages from it, or with T if they would be likely to Trigger mass wasting.

1. blasting in construction areas
2. building avalanche sheds over portions of highways
3. blasting unstable rock loose to bring it down on purpose
4. bolting loose slabs of rock to the rock beneath
5. covering road cuts with chain-link fencing
6. draining excess subsurface water by using of drainage ditches or pipes or by drilling and pumping
7. excavating wide, deep shoulders along highways to catch falling rock
8. earthquakes
9. regrading slopes to bring them safely below their angle of repose
10. building retaining walls along highway embankments
11. revegetating areas of bare ground
12. using riprap or building offshore breakwaters along coastlines
13. removing, by solution, salt that helps hold quick clay
14. spraying road cuts with shotcrete
15. storms that produce heavy precipitation and winds
16. cutting terraces on a slope to create flat surfaces for roads or building foundations
17. triggering avalanches on purpose with explosions
18. the passing of large trucks through the area
19. undercutting by streams or waves of weak horizontal rock layers that lie under more resistant rock layers
20. adding large quantities of water to hillsides
21. adding just enough water to dampen dry sediment on hillsides
22. waves' crashing against the base of a slope or sea cliff
23. the chemical and physical weatherings of rock
24. adding significant weight to the top of a slope

Short-Answer Questions

Answer with a few words, numbers, or short sentences, or choose from the words in parentheses.

1. What are the four characteristics that define which type of mass movement event has occurred? List them, and give a brief description of each. _____

2. Slumping causes continual retreat of sea cliffs along the southern coast of California. What is the average amount of retreat annually? _____

3. a. How fast do slumps move? _____
   b. Do they retain their shape or fall apart as they move? _____

4. What two factors influence the velocity of either a mudflow or a debris flow? _____

5. What are possible sources of lahar material? _____

6. Slides happen when bedrock or regolith slides down on a glide horizon that is _____ (dipping into, parallel to) the slope surface.

7. a. What is maximum velocity for slides? _____
   b. Under what circumstance is a slide likely to move exceptionally fast? _____

194 | Chapter 16

8. What weather sequence is likely to cause snow avalanches? _____

9. Why is there usually a better geologic record of submarine mass movements than of subaerial mass movements? _____

10. What is the name of the equipment that was used to discover a slump block the size of Scotland that had slid into the North Sea 7,000 years ago? _____

11. In order for mass movement to take place in an area,
    a. what must happen to the topography there? _____
    b. what must happen to the strength of the surface rock? _____

12. How does steepness of slope affect velocity of mass movement? _____

13. Name three processes that can weaken surface rock and thus allow mass movement to happen. _____

14. State three ways rock is held together. _____

15. What is a typical angle of repose for dry, unconsolidated materials like dry sand? _____

16. Why can talus slopes sit at angles of repose of up to about 45°, which is significantly steeper than normal? _____

17. List four examples of particularly weak surfaces that are prone to become glide horizons. _____

18. Joints can act as glide surfaces. Name two different processes that can pry loose a rock layer along a joint and thus set it up to slide. _____

19. Which potential glide surface is more likely to fail, one with a dip that parallels the slope surface or one that dips into the slope? _____

20. List several types of triggers that can initiate slope failure. _____

21. What are some possible sources of shocks and vibrations that might initiate slope failure? _____

22. How does southern California's climate contribute to mass movement problems there? _____

23. How does housing development contribute to mass movement problems in southern California? _____

24. What is the only real way for people to avoid being killed or loosing their property to mass movement events? (Unfortunately, it's often a very unrealistic solution.) _____

25. List eight factors that are evaluated by geologists (usually with the aid of computer programs) as part of hazard assessment studies and used to produce landslide-potential maps. _____

Making Use of Diagrams

Refer to the diagram at the bottom of the page.

1. Label each force as *H* (holds block in place) or *S* (increases tendency of block to slide)
   _____ a. downslope force
   _____ b. normal force
   _____ c. pull of gravity
   _____ d. resistance force

2. a. On a gentle slope, which force predominates, downslope or normal? _____
   b. On a steep slope, which force predominates, downslope or normal? _____

Gentle slope: Resistance force, $F_N > F_D$, a. Downslope force, b. Normal force, c. Pull of gravity

Steep slope: d. Resistance force, $F_D > F_N$

Short-Essay Questions

1. Name the mountain that was nicknamed "the mountain that walks" and discuss the mass movement event associated with it.

2. What is GLORIA? Discuss some of the discoveries GLORIA has made around Hawaii.

3. List eight ways the relief of an area can be increased.

4. Why does shaking cause quick clay to act like a slurry instead of a solid?

5. Briefly explain why plate tectonics is—again—the ultimate reason for a geologic phenomenon, that of mass movement.

Practice Test

1. Which of the following statements is FALSE? Mass movement
   a. simply explained means gravity exists so things move downward.
   b. was the basic cause of the tragedies of Yungay, Peru, and Armero, Colombia.
   c. is more likely to happen under wet conditions than under dry conditions.
   d. happens when the slope of a hill gets steeper than the angle of repose.
   e. can't happen under water because the buoyancy force of water is too great.

2. Landslides are likely to happen when
   a. the angle of repose of a sandy slope gets greater than 9°.
   b. the downslope force becomes greater than the resistance force.
   c. weak surfaces dip into the slope.
   d. weight is added at the bottom of a slope.
   e. All of the above are true statements.

3. Slides with a spoon-shaped (concave upward) surface are called
   a. block glides.
   b. debris flows.
   c. avalanches.
   d. slump.
   e. creep.

4. Larry, Curly, and Mo are buried in a cemetery on a hillside. Their headstones were placed upright at their funerals, but now they all tilt downhill as shown in the diagram that follows. Larry's stone tilts 20° down from the vertical, Curly's tilts 37°, and Mo's tilts 6°. Who died first?
   a. Larry
   b. Curly

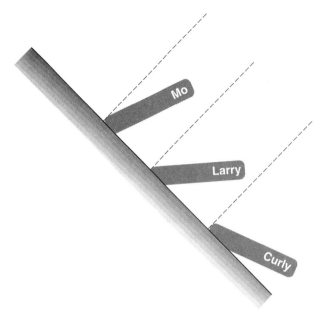

   c. Mo
   d. There's not enough difference between the tilt figures to judge.
   e. Tilt is not related to how long the headstones have been in the ground.

5. What process most logically explains the different tilts of grave stones in a hillside cemetery?
   a. slump
   b. ghosts
   c. mudflow
   d. creep
   e. liquefaction

6. The Vaiont Dam disaster
   a. occurred in the early 1800s before engineers understood how to build safe dams.
   b. was the worst-ever dam disaster in the United States.
   c. came as a total surprise, with no warning signs.
   d. happened because a very fine dam was built in a geologically unstable location.
   e. resulted in the dam's being completely destroyed.

7. Which of the following would increase the risk of having a landslide?
   a. allowing water to infiltrate the slope
   b. removing weight at the top of the slope
   c. adding support to the toe of the slope
   d. recontouring the slope to reduce its slope angle
   e. planting vegetation that has deep roots

8. The tragedy in 1970 at Yungay, Peru
   a. was surprising, because there was no evidence of previous landslides there.
   b. started with an earthquake that triggered an icefall that triggered a debris flow.

c. is not considered mass wasting because it happened quickly, and by definition mass wasting is slow movement of material downslope.
d. unfortunately could happen again; the government rebuilt the town in the same place.
e. All of the above are true statements.

9. Which of the following statements is FALSE? Avalanches
   a. are high-velocity mass movement events.
   b. can happen when frozen snow layers get buried by new snow.
   c. can be triggered by explosions, people, or even just new snow.
   d. may be turbulent clouds of debris mixed with air.
   e. never happen twice in the same place; therefore the pathway an avalanche creates is a safe place to build.

10. Solifluction
    a. means the water-table level fluctuates with precipitation.
    b. is a faster-than-usual kind of slump in wetlands.
    c. is a faster-than-usual kind of creep on slopes that are underlain by permafrost.
    d. is the technical name for a rock glacier.
    e. is the proper term for alternate expanding and contracting of swelling clays.

11. Which of the following statements is FALSE?
    a. Lahars are mudflows that have been triggered by earthquake shaking.
    b. Permafrost is permanently frozen ground in arctic regions or at high elevations.
    c. Slow movement of unconsolidated material downslope is called creep.
    d. The angle of repose is the steepest angle at which unconsolidated sediments can sit without slipping downhill.
    e. Sediments carried by turbidity currents settle out in sequence from coarse to fine, creating graded beds.

12. Which of the following statements is FALSE? Classification of mass movement events is based on the
    a. type of material involved (rock, regolith, snow, and ice).
    b. age of the material involved (historic, thousands of years old, or millions of years old).
    c. velocity of the movement (fast, intermediate, or slow).
    d. character of the moving mass (chaotic cloud, slurry, or coherent body).
    e. environment (subaerial or submarine).

13. Which of the following statements is FALSE? Mass movements can happen
    a. only in areas where there is relief.
    b. on slopes.
    c. along joints and faults.
    d. when strong, intact rock gets weathered.
    e. when downslope force equals resistance force.

14. Plate tectonics
    a. is often the ultimate cause of mass movement because it generates relief in an area.
    b. can cause volcanic eruptions that provide gases and water that weather and weaken crustal rock.
    c. may cause earthquakes that trigger mass movement.
    d. is responsible for the San Andreas Fault, which causes lots of mass movement events in California.
    e. All of the above are true statements.

15. The Vaiont Dam area was vulnerable to slide because limestone beds there dipped parallel to the mountain slope and were interlayered with weak shale.
    a. True
    b. False

16. Snow avalanches no longer are triggered on purpose by explosions because the results are too unpredictable.
    a. True
    b. False

17. Highways can be protected from snow avalanches by building avalanche chutes above them.
    a. True
    b. False

18. The angle of repose of dry sand is typically between 30 and 37°.
    a. True
    b. False

19. Talus slopes typically have higher angles of repose than hills of dry sand because the irregular rocks of talus interlock and hold each other in place.
    a. True
    b. False

20. To stabilize steep hillsides, it is wise practice to weight them down with buildings.
    a. True
    b. False

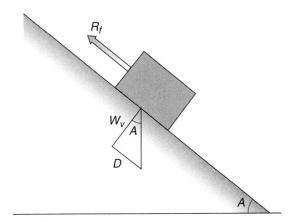

21. In the diagram above, if angle *A* is increased, the slope of the hill becomes steeper.
    a. True
    b. False

22. In the diagram above, if angle *A* is increased, the downslope force (*D*) gets larger.
    a. True
    b. False

23. In the diagram above, if the hillside gets wet, the resisting force (R_f) gets greater
    a. True
    b. False

24. In the diagram above, if angle *A* is increased, the normal force (W_v) becomes relatively smaller in relation to the driving force, represented by line *D*, and the block is more likely to slide.
    a. True
    b. False

25. Landslides happen when the angle of repose is somehow exceeded and the slope readjusts itself.
    a. True
    b. False

26. A rock glacier is a glacier that is so thickly covered with surface rock it is pushed firmly to the ground and can no longer move.
    a. True
    b. False

27. Potential glide horizons are weak layers of rock or sediment below ground level that parallel the slope surface.
    a. True
    b. False

28. GLORIA is a special type of sonar that has discovered evidence of huge slumps along the west coast of South America.
    a. True
    b. False

29. A slope of damp sand is more likely to slide than a comparable slope of dry sand.
    a. True
    b. False

30. Quick clay may change from solid state to slurry state when salt between the clay grains is dissolved by heavy rains or when the clay is shaken by an earthquake.
    a. True
    b. False

31. Landslide-potential maps designate areas in which construction is illegal because of the high risk of mass movement there.
    a. True
    b. False

32. Write the letters of the phrases below on the line under the proper title.

    a. careful inventory and mapping to determine dangers
    b. weathering of minerals to produce clay
    c. controlled blasting of unstable slopes
    d. water drainage
    e. retaining walls along highway embankments
    f. spraying shotcrete on road cuts
    g. removing support at the toe of the slope
    h. adding weight at the top of the slope

    Factors that Lead to Slide   Ways to Reduce Losses
    _____   _____

For questions 33 to 37, match the description of the mass-wasting event with its location.

   a. Armero, Colombia (1985)
   b. Turnagain Heights, Alaska (1964)
   c. Vaiont Dam (1963)
   d. Yosemite National Park (1996)
   e. Yungay, Peru (1970)

_____ 33. An earthquake triggered an icefall which triggered a debris flow which buried thousands of people.

_____ 34. Part of a mountain slid into a reservoir and caused a splash that drowned residents downstream.

35. A layer of rock broke loose along an exfoliation joint and caused damage in a famous recreation area.

36. Despite warnings from geologists, a lahar in the Andes buried this town and its inhabitants.

37. Earthquake shaking caused quick clay to liquefy and slump, ruining a cliff-side subdivision.

38. And the question you've all been waiting for: The cause of all mass wasting is the reason things fall down. Why do things fall down? _____ (one word)

ANSWERS

Completion

1. unstable ground
2. regolith
3. mass movement (or mass wasting)
4. natural hazard
5. slope retreat
6. creep
7. permafrost
8. solifluction
9. rock glacier
10. slumping/slump
11. glide horizon
12. head scarp
13. mudflow
14. debris flow
15. lahars
16. rockslide; landslide; debris slide
17. avalanches
18. snow avalanche or debris avalanche
19. avalanche chutes
20. rock falls or debris falls
21. submarine slumps
22. olistostromes
23. submarine debris flows
24. turbidity currents
25. relief
26. slope
27. surface tension
28. stable slopes; unstable slopes
29. slope failure
30. downslope force; resistance force
31. angle of repose
32. quick clay
33. dip slope
34. undercutting
35. swelling clay
36. cosmogenic dating
37. pressure ridge
38. landslide-potential maps
39. shotcrete
40. avalanche sheds

Matching Questions

HISTORIC MASS MOVEMENT EVENTS

1. l
2. f
3. h
4. a
5. j
6. b
7. k
8. c
9. e
10. i
11. d
12. g

MASS MOVEMENTS: CAUSES AND PREVENTION

1. T
2. P
3. P
4. P
5. P
6. P
7. P
8. T
9. P
10. P
11. P
12. P
13. T
14. P
15. T
16. T
17. P
18. T
19. T
20. T
21. P
22. T
23. T
24. T

Short-Answer Questions

1. type of material involved (rock, regolith, or snow and ice)
   velocity of movement (fast, intermediate, slow)
   character of the moving mass (chaotic cloud, slurry, or coherent body)
   environment of occurrence (subaerial or submarine)

2. a few meters

3. a. anything from a few millimeters per day to tens of meters per minute   b. fall apart

4. angle of slope and water content

5. volcanic ash, regolith on the mountain slopes, melt water from snow and ice heated by volcanic activity, and rain

6. parallel to

7. a. 300 km/h   b. when a cushion of air is trapped under the moving material, so friction is reduced

8. snowfall, warming and partial melting, cooling and freezing of the top of the snow layer, and more snowfall (and strong wind at this stage can make matters worse)

9. The piles of debris resulting from submarine mass-wasting events tend to be buried by younger sediments and preserved; similar piles of debris left by subaerial events tend to weather and erode away.

10. GLORIA (a special type of sonar)

11. a. Relief must be established (i.e., elevation differences must be created on the ground surface).
    b. The surface material must be weakened, possibly by fracturing and weathering.

12. The steeper the slope, the faster the movement.

13. jointing, faulting, and weathering

14. strong chemical bonds within mineral crystals, mineral cement, and the interlocking of grains

15. between 30 and 37°

16. Talus is composed of large, irregularly shaped pieces of rock that can interlock with each other.

17. wet clay layers in regolith, joints in massive rocks, bedding planes in sedimentary rock, and foliation planes of metamorphic rocks

18. alternate freezing and thawing of water in the joint, or wedging by tree roots

19. one with a dip parallel to the slope surface

20. shocks and vibrations and changes in slope angles, slope loads, slope support, and slope strength

21. earthquakes, storms, passing of large trucks, and construction blasting

22. The normal hot, dry weather supports only sparse semi-desert flora and encourages brush fires that decrease even this cover; heavy winter rains add weight to the slopes and infiltrate the land.

23. Developers may oversteepen slopes, overload slopes, and change levels of groundwater in areas that are already borderline stable.

24. Don't build, live, or work in an area where mass movement is likely.

25. slope steepness; strength of substrate; degree of water saturation; dip of bedding, joints, or foliation relative to slope; vegetation cover; climate; undercutting; seismicity

Making Use of Diagrams

1. a. *S*   b. *H*   c. *S*   d. *H*
2. a. normal   b. downslope

Short-Essay Questions

1. As the reservoir behind Vaiont Dam in the Italian Alps began to fill with water in the early 1960s, residents began to call Monte Toc *la montagna che cammina* (the mountain that walks) because its slopes began to creak, shake, and rumble as creep down the hillside increased due to the infiltration of water. The geology of the valley encouraged mass movement; limestone beds interlayered with weak shale dipped parallel to the surface of the mountain and curved under the reservoir. Heavy rains caused so much concern engineers began to lower the reservoir level, but it was too late. In October 1963 a huge chunk of Mt. Toc slid into the reservoir and created a huge splash that went over the dam and drowned 1,500 people downstream. The dam held; there was nothing wrong with its engineering. It was simply a geologically bad place to put a dam.

2. GLORIA, which stands for Geologic Long-Range Inclined Asdic, is a special type of sonar. Because it can see sideways, it has been used to map areas where the sea floor slopes, such as along margins of islands, along the walls of trenches, and along continental slopes. It has found huge slumps along the margins of Hawaii that over time have altered the shape of the island and created an area of hummocky sea floor around it. Major slumping events seem to recur every 100,000 years and probably generate huge tsunamis in the Pacific Ocean basin. A large crack is currently developing along Hawaii's coastline and may be the forerunner of another huge slump.

3. The following are ways relief in an area can increase:

    Either subduction or continental collision at convergent plate boundaries thickens and elevates the crust, forming mountains.
    Faulting associated with rifting drops blocks of rock down between other masses of rock.
    Changes in mantle temperature cause the overlying continent to either gradually rise or gradually subside; these are called epeirogenic movements.
    Extrusive rock can pile up mountain-high around volcanic vents.
    Sediment deposition, both subaerial and submarine, can produce hills of material with sloping sides.
    Human activity can build hills (examples: create mounds to build on, pile garbage and sediment in landfills, and pile slag around mines).
    Human activity or stream action can dig pits or troughs.
    A drop in sea level, such as during an ice age, results in shorelines that slope down to the sea.

4. When the quick clay is not being shaken, the surface tension associated with the clay's normal water content holds the clay flakes together. Shaking separates the flakes, destroys the cohesive force of surface tension, and causes the material to act like a water suspension of fine particles, not as the solid it seemed to be before the shaking.

5. Most naturally unstable ground is that way because it has been raised up and then weakened. Plate tectonics is the cause of most major uplifting, and thus generates most of the significant relief that exists in the world. If this uplifted land gets weakened, it's likely to fall, slide, or flow. Enter typical plate tectonics activities: volcanic eruptions provide gases and water that encourage weathering and erosion, and faulting and earthquakes fracture rock, allowing infiltration of water.

Practice Test

1. e   There's abundant evidence of submarine mass movements, including submarine slumps, debris flows, and turbidity currents.

2. b   The angle of repose of sandy slopes is 30–37°; sliding happens on weak surfaces that parallel the slope; and adding weight to the tops of slopes may trigger movement.

3. d   The other choices are other types of mass movement that have nothing to do with concave surfaces.

4. b   Creep of the ground surface causes vertical structure in it to gradually tilt downhill. The longer the time in the ground the greater the tilt.

5. d   Creep is slow movement of the ground surface downhill.

6. d   The disaster happened in the 1960s in Italy; there was so much slope movement before the slope failure the locals called Mt. Toc "the mountain that walks"; the dam survived the disaster.

7. a   All other choices would help prevent a landslide.

8. b   Previous slides had occurred; mass movement can be slow or fast; and the government has forbidden rebuilding in the area.

9. e   Pathways created by avalanches are called avalanche chutes, and avalanches typically run down the same chute many times.

10. c   The term has nothing to do with water tables, wetlands, rock glaciers, or swelling clays.

11. a   Lahars are mudflows associated with volcanic activity.

12. b   The age of the material has nothing to do with the classification of mass movement.

13. e   Downslope force must exceed resistance force before there's movement.

14. e   Choices a to d are all true.

15. True

16. False   It's standard procedure to do this because it works so well and avoids unexpected and therefore more dangerous avalanches.

17. False   Build avalanche sheds, not chutes; chutes are the pathways created and followed by avalanches that happen often in the same location.

18. True   The angle at which unconsolidated material can sit without sliding is its angle of repose; for dry sand this is between 30 and 37°.

19. True   Angular pieces of talus are more stable than rounded pieces of sediment or rock.

20. False   Added weight is more likely to trigger than to prevent mass movement.

21. True   Increasing the size of angle A on the diagram produces a steeper hillside.

22. True   If you sketch a larger angle A on the diagram, you must increase the length of line D to complete the triangle.

23. False   The resisting force gets smaller because wet surfaces offer less frictional resistance.

24. True   If you sketch a larger angle A on the diagram, thus increasing the length of line D, it is apparent line W_v, representing a force that helps keep the block in place, has become smaller in relation to the force causing the block to move downhill, represented by line D.

25. True   Material moves downhill when its angle of repose is exceeded, and it doesn't stop until a new equilibrium is reached.

26. False   A rock glacier is a mix of rock fragments and ice that does slowly move.

27. True   Weak layers of rock or sediment under the ground surface and parallel to it are typically where movement occurs.

28. False   GLORIA is a sonar, but the slumps are along the coast of Hawaii; the west coast of South America is bordered by trenches caused by subducting ocean plate.

29. False   The surface tension created by minimum amounts of water tends to hold small rock pieces together.

30. True   Quick clay may become fluid when it gets wet or is shaken.

31. False   The maps show the high-risk areas as a guideline for intelligent construction choices, not as legal orders.

32. Factors that Lead to Slide   Ways to Reduce Losses
          b, g, h   a, c, d, e, f

33. e

34. c

35. d

36. a

37. b

38. gravity

CHAPTER 17 | Streams and Floods: The Geology of Running Water

GUIDE TO READING

Water is so common on Earth we rarely pause to appreciate how rare it is in our solar system and how unique it makes our planet. Water has literally shaped Earth's surface and enabled life to exist here.

The chapter begins by presenting a concept you've probably encountered before; water cycles through all of Earth's "spheres"—atmosphere, hydrosphere, lithosphere, and biosphere. That's a lot of territory to cover, so this chapter focuses on water only while it flows on land, not while it's in the oceans (that's Chapter 18) or under the land (that's Chapter 19), or in the atmosphere (that's Chapter 20).

For very common events, geologists use lots of very common terms, such as surface water, runoff, sheetwash, infiltration, and downcutting. They use a mix of common and specialized vocabulary to discuss the origin of surface water (meteoric water) and to describe every part and aspect of a stream and it system: bed, mouth, headwaters, stream gradient, tributaries, distributaries, trunk stream, drainage network, meanders, braided stream, reach of a stream, drainage basins, drainage divides, incised meanders, wetted perimeters, and drainage networks (dendritic, radial, rectangular, and trellis).

Once stream parts and patterns are defined, the author points out that characteristics of a stream are usually related to its geologic surroundings. Folds, fractures, and faults serve as structural controls of network patterns, except in special cases where the timing of events allows streams to ignore such controls and become superposed and antecedent streams. Streams with beds above the water table are ephemeral streams; those with beds below the water table are permanent streams. Not only are streams *affected* by their geologic environment, streams in turn *affect* their geologic environment. The nature and magnitude of the effect are influenced by the amount of water (discharge) and character of flow (thalweg, turbulent, and laminar flows).

The work that streams do can be organized into three categories:

- Transporting materials: You read about types of sediment load (dissolved, suspended, and bed load), the size of the sediment particle (competence), and the quantity of load (capacity).
- Eroding the land by scouring, abrasion, breaking, lifting, and dissolution: Erosion can produce varied and spectacular results, such as the badlands of South Dakota and the goosenecks of the San Juan River in Utah.
- Depositing materials (generically called alluvium or fluvial deposits): Materials can be deposited as bars in a stream's channel, as natural levees along its banks, as alluvial fans where it quickly loses gradient on land, and as deltas where it quickly loses gradient in quiet waters.

Like the Earth itself, a stream evolves (changes over time). It experiences youth, maturity, and old age, and its features change as the stages progress. Its longitudinal profile develops a gentler gradient as it downcuts lower and gets closer to local base levels and to its ultimate base level (an ocean). It may achieve a condition known as graded (no net erosion or deposition), create and eventually lose waterfalls and rapids, create and rework valleys or canyons, and develop meanders on a floodplain. Meanders are themselves a complex story involving cut banks, point bars, neck cutoffs, oxbow lakes, backswamps, and yazoo streams.

You shouldn't be surprised to find, toward the end of the chapter, the usual reminder, with examples, that plate tectonics is the ultimate reason for many geologic happenings. The uplift of broad areas due to plate tectonics events caused complete drainage reversal of the Amazon River; ancient rifts provided channels for the Amazon, Mississippi, and Congo Rivers; and Africa's continental drifting changed its latitude and caused the once lush Sahara area to become the desert we know today.

The chapter ends with a discussion of human interaction with rivers. Early civilizations needed to be close to rivers because they provided so many essentials of life, like food, drinking water, and transportation. Today rivers provide us these same things, plus electric power and recreation sites. But rivers don't always behave as we want them to; sometimes they get out of control and kill us. You read about several great historic floods (Johnstown, Pennsylvania; Big Thompson River, Colorado; Yellow and Yangtze Rivers, China; Bangladesh; and the Mississippi River) and a humongous prehistoric flood 11,000 years ago (Great Missoula Flood) that created the channeled scablands of Washington State.

Can humans ever totally control stream flow and prevent flooding? Probably not. A stream seems ultimately to win every battle to confine it or to direct its waters against its will. But society hasn't given up the fight. Scientists analyze potentially dangerous situations, determine recurrence intervals of various size floods, and recommend the construction of appropriate dams, reservoirs, levees, and concrete flood walls to protect humans and their property.

Society has come to realize that while sometimes rivers harm us, our growing human population is doing increasing amounts of damage to rivers. The mere building of a city (urbanization) can increase the likelihood of flooding of an area. Humans have carelessly polluted streams, and this has resulted in severe damage to the physical and biological environment. Our increasing population has caused us to possibly go overboard on dam and reservoir construction and has resulted in serious overuse of stream waters (as in the Aral Sea region of Central Asia and along the Colorado River in the southwestern United States). The quality and quantity of a water supply anywhere can change. Mars once had surface water, but it no longer does. We currently have usable surface water, but this condition is not assured forever.

In summary, there are a lot of science facts in this chapter, plus the message that life as we know it is intimately connected with the water flowing across Earth's surface. We shouldn't take this water system for granted, and we should be careful with the changes we're making to it.

Completion

Test your recall of new vocabulary terms.

_____ 1. an event during which the stream volume gets so great, the stream overflows its banks and covers land that's usually above water

_____ 2. ribbons of water that flow downslope in channels

_____ 3. troughs on Earth's surface which were cut by stream erosion and which contain stream waters and direct their flow downslope

_____ 4. the precipitation that falls to Earth when atmospheric water vapor condenses

_____ 5. seep into the Earth

_____ 6. liquid water that sits on the Earth and doesn't infiltrate it

_____ 7. surface water that flows downslope in response to gravity

_____ 8. the film of water, only millimeters thick, that covers the ground after a rain or snowfall and slips downslope as a broad thin sheet

_____ 9. a stream eroding its channel and making it deeper

_____ 10. the process in which a stream lengthens in an upslope direction, which happens because the flow is more intense at the entry to the channel (upslope) than in the surrounding downslope sheetwashed areas

_____ 11. smaller streams that flow into a single larger channel

_____ 12. the single larger stream into which smaller streams flow

_____ 13. an array of interconnecting streams that collect excess water from an area and lead it away downslope

_____ 14. the fairly large area that is drained by an array of interconnecting streams

_____ 15. the strip of high ground that separates one watershed from another

_____ 16. a very major drainage divide, which separates a drainage area whose waters flows into one ocean from an area whose waters drain into another ocean

_____ 17. the pattern resembling tree branches achieved by a drainage network that flows over fairly uniform substrate with fairly uniform initial slope

_____ 18. the drainage network pattern that forms on a cone-shaped surface, as streams radiate out from the center like spokes of a wheel

_____ 19. the pattern achieved by a drainage network that flows across ground fractured with vertical joints, causing streams to join each other at right angles

_____ 20. the pattern achieved by a drainage network that develops across an area of parallel valleys and ridges, where the major flow is down the valleys and minor trunk streams cut across ridges

_____ 21. the cut created where a trunk stream that is part of a trellis network cuts across a resistant ridge

_____ 22. a drainage network that develops under the influence of geologic structures such as folds, fractures, and faults

_____ 23. streams that carved an initial flow pattern into flat-lying strata, then kept this established pattern as they cut downward into lower strata, even when the geometry of the stream pattern seemed at odds with what would be expected in this lower strata

_____ 24. streams that established their flow patterns in accordance with surface features in an area, then kept these patterns as the ground rose below them, even when the patterns seemed at odds with the new rock features the streams became exposed to

_____ 25. meanders that lie at the bottom of steep-walled canyons, because their stream was either a superposed or an antecedent one

_____ 26. the level in the ground that marks the top of the zone in which all pores between rock grains are filled with water

_____ 27. a stream that flows year-round because its bed lies below the water table and it therefore is replenished not only by surface waters but also by groundwater seepage

_____ 28. a stream whose bed lies above the water table, so it may dry up when the amount of water it loses to normal infiltration exceeds the amount of replacement water from surface sources

_____ 29. a dry ephemeral stream bed

_____ 30. lands that are dry, unvegetated, and dissected by dendritic patterns of gullies with sharp-crested ridges between

_____ 31. the volume of water passing any particular point on a stream bank in 1 s

_____ 32. the fastest moving part of a stream flow, generally near the surface in the center of the channel

_____ 33. all of the stream channel wall that is touched by water

_____ 34. the twisting, swirling motion of stream water that may keep it going around in circles or heading upstream, instead of going in the more normal downstream direction

_____ 35. the flow pattern in which all parcels of water follow parallel paths and flow downstream

_____ 36. the process by which running water removes loose fragments of sediment

_____ 37. the process by which running water grinds and rasps away at the stream channel

_____ 38. a bowl-shaped depression caused by abrasion by sand or gravel in a whirlpool

_____ 39. the sediment carried by a stream

_____ 40. the minimum velocity of stream flow which will keep sediment suspended in the water and whose value depends on the specific gravity, size, and shape of the sediment grains

_____ 41. a type of bed-load movement in which the sediment pieces along the bottom "skip," are lifted up and fall down along a curved trajectory, over and over again

_____ 42. a measure of the maximum particle size a stream can carry

_____ 43. a measure of the total quantity of sediment a stream can carry

_____ 44. any deposits that have been laid down by flowing water in the stream channel or along its banks

_____ 45. the sorted, stream-washed sediments that have been laid down by stream water

_____ 46. a lens-shaped deposit of alluvium in a stream channel

_____ 47. the segregating of different sizes of sediment because of differences in the velocities of stream flow

_____ 48. a measure of the steepness of the downstream slope of a stream

_____ 49. areas of steep, bouldery stream bed where the stream flow is extremely turbulent

_____ 50. places where stream waters plunge over escarpments

51. a cross-sectional view along the length of a stream that shows its changing elevation as it progresses downslope

52. the lowest elevation the floor of a stream channel ever gets (i.e., the lowest elevation along a stream's longitudinal profile); a temporary version of this, located upstream of a drainage network's mouth; and the final, lowest level possible for trunk stream erosion (sea level)

53. an equilibrium condition of a stream in which it is neither eroding nor depositing sediment but instead just carrying along all the sediment that has been supplied it

54. a trough created by a stream in which the walls slope gently

55. a trough created by a stream in which the walls slope steeply

56. a very large valley occupied by a small stream

57. a feature of turbulence, formed above submerged boulders, in which the crest and trough of a wave remain in place and individual water particles move through the wave

58. a depression scoured out by falling water at the base of a waterfall

59. a small side valley whose floor lies higher than the floor of the main valley it enters and which was created by a small tributary glacier that flowed into a large glacier

60. delta-shaped, gently sloping aprons of sediment deposited along mountain fronts where streams abruptly lose their steep gradients

61. a stream composed of a network of entwined channels

62. a series of wide curves of a stream snaking its way across a broad floodplain

63. the outside wall of a stream meander, where the water travels the fastest and does the most erosion

64. the wedge-shaped deposit of sediment on the inside wall of a stream meander, where the water is traveling the slowest and therefore has the least energy

65. the narrow strip of land between the cut banks at both ends of a stream meander

66. a straight reach that develops when a stream cuts through a meander neck

67. the horseshoe-shaped lake that may result when a stream no longer flows through the meander curve but instead follows the cutoff

68. the dried out channel of a meander after the stream has taken the cutoff

69. a long cliff (escarpment) that borders a flood plain

70. low ridges of sediment on both sides of a stream channel, deposited when the stream overflowed its banks, immediately lost energy, and had to drop part of its load

71. part of a stream's floodplain that sits between natural levees and bluffs, which gets marshy because flood waters can't flow back into the stream channel

72. tributaries that are blocked by natural levees from joining the trunk stream and therefore must run in the floodplain parallel to the stream and its levees

73. numerous small streams that a trunk stream divides into when it flows over broad areas of relatively soft alluvial deposits

74. any wedge of sediment formed at the mouth of a river, of which some are triangular but many are not

75. a delta that consists of many elongate lobes that extend into the sea

76. the situation where a river overflows a natural levee and begins to flow in a new direction

77. low swampland that originated as a delta and has distributaries flowing slowly across it

78. areas that started as mountains, eroded to hills, and become almost flat plains

79. a natural event in which one stream's headward erosion causes it to intersect the channel of another stream and divert its waters

_____ 80. a feature formed when stream capture diverts stream flow and leaves behind a dry water gap

_____ 81. the renewed vigor in a stream's ability to downcut, due to a drop in its base level, an increase in its discharge, or an uplift of the land it's flowing through

_____ 82. the remnants of an old floodplain left on both sides of a stream when rejuvenation allows it to resume downcutting and create a new floodplain at lower elevation

_____ 83. the change of direction of stream flow throughout an entire drainage network due to plate tectonics uplift

_____ 84. the condition of a stream when the volume of water flowing through it exceeds the volume of the stream channel

_____ 85. the floods that happen at a certain time every year because of weather conditions typical for that time

_____ 86. the floods that happen quickly and unexpectedly due to violent storms or broken dams

_____ 87. a high-altitude wind current that influences weather patterns

_____ 88. a region in eastern Washington with unusual topography that resulted from huge floods 11,000 years ago

Matching Questions

ANATOMY OF A STREAM

Match the term with its definition.

a. bed
b. braided stream
c. course
d. floodplain
e. headwaters (source)
f. meandering stream
g. meanders
h. mouth
i. reach
j. stream gradient

_____ 1. the place a stream begins
_____ 2. where a stream empties into another stream, lake, or ocean
_____ 3. flat land on either side of a stream that is covered only when water leaves its channel
_____ 4. the path a stream follows
_____ 5. the slope of the channel downhill
_____ 6. the floor of a stream channel
_____ 7. any specific segment of a stream
_____ 8. wide curves in the course of a stream
_____ 9. a stream reach that consists of entwined subchannels instead of one main channel
_____ 10. a stream reach that contains many sweeping curves

Short-Answer Questions

Answer with a few words, numbers, or short sentences, or choose from the words in parentheses.

1. Name several things that streams do that help
   a. the land _____
   b. human society _____
   c. life in general _____

2. What do all standing bodies of freshwater have that salty lakes don't have? _____

3. List and describe three major drainage divides on the North American continent. _____

4. What stream feature is illustrated by the goosenecks of the San Juan River in southern Utah? _____

5. What national park offers mammal fossils and scenery that is dramatic but none the less bare, dry, unvegetated, and gullied? _____

6. State whether the discharge of a given stream will *increase* or *decrease* in the following conditions and *explain why*:
   a. in a temperate region, going farther downstream
   b. in an arid region, going farther downstream
   c. in the spring _____
   d. during a flood _____

7. Give two reasons why all parcels of water in the same stream don't move at the same speed. _____

8. What two factors determine how much friction will slow the flow of stream water? _____

9. Other factors being equal, which is more turbulent?
   a. faster water or slower water _____
   b. deep stream or shallow stream _____
   c. stream flowing in a smooth, clay-lined channel or one flowing in a rough, gravelly channel _____

10. What energy transformation occurs in the downslope flow of a stream? _____

11. What is the average elevation of continents above sea level, in meters and in feet (1 m = 3.28 ft)? _____

12. List four ways running water causes erosion. _____

13. What factors influence the efficiency of stream erosion? _____

14. Name and briefly describe the three components of the sediment load. _____

15. Which component of the sediment load of the Mississippi River is by far the largest, and what percent of the total sediment load does it constitute? _____

16. If the velocity of a stream doubles, what happens to its competence? _____

17. If all other factors are equal, which stream in each pair has greater competence?
    a. stream with low density due to sediment load or stream with high-density, high-sediment load _____
    b. clear water or muddy water _____
    c. saltwater or freshwater _____

18. What three factors determine a stream's capacity? _____

19. When a stream's velocity decreases, what happens?
    a. to its competence _____
    b. to its sediment load _____

20. Name four situations in which water in a stream would slow down. _____

21. Why does different-sized sediment accumulate in different locations in and along a stream? _____

22. Briefly describe what a longitudinal profile of a stream looks like. _____

23. A steep mountain stream looks like it's racing along, but its average downstream velocity is actually less than that of a large, flat, muddy river. Why is this so? _____

24. Name a few places that can act as local base levels along a stream. _____

25. Why don't local base levels last for geologically long times? _____

26. A stream has reached the condition of being a graded stream, and then something occurs to change its sediment supply, discharge, or base level. How does the stream react? _____

27. List three situations in which a graded stream can lose the condition of equilibrium. _____

28. Explain why sometimes a small stream occupies a very large valley _____

29. In rapids, what is a typical ratio between the water depth and the diameter of the coarsest sediment in the stream bed? _____

30. List two possible pathways that water flowing over submerged clasts can take. _____

31. Why do all waterfalls eventually disappear? _____

32. Samuel Clemens, the author of *Huckleberry Finn*, was also a riverboat pilot on the Mississippi River. What connection does this have with the pen name he used? _____

33. How did the depositional structures at the mouths of rivers come to be called deltas? _____

34. For how long has the Mississippi delta been developing? _____

35. List four factors that influence the shape of a delta. _____

36. Briefly state the conditions necessary to create the following structures at a river's mouth:
    a. bird's-foot-shaped delta _____
    b. delta-shaped delta _____
    c. delta with arc-shaped lobes _____
    _____
    d. no delta _____

37. Why does downtown New Orleans lie below the level of the Mississippi River channel, and why isn't it currently underwater? _____
    _____

38. Geologically speaking, what kind of structure is the Cumberland Gap in the Appalachian Mountains, through which pioneers gained access to the Kentucky wilderness? _____
    _____

39. What plate tectonics feature provides the channels for the Amazon, Mississippi, and Congo Rivers?
    _____

40. State two ways the shifting of continents due to plate tectonics can alter climates. _____
    _____
    _____

41. What geologic feature did radar reveal below the sands of the Sahara Desert? _____
    _____

42. What geologic feature is poetically known as China's Sorrow, and why is it called this? _____
    _____
    _____

43. Give the locations for the following events:
    a. frequent seasonal floods (delta-plain or floodplain floods), particularly bad in 1887 and 1931, when a few million people died due to famine after the floods _____
    b. flooding due to monsoons in 1990 that drowned 100,000 persons on the delta plain _____
    c. an 1889 rainstorm that collapsed a dam resulting in a flash flood that completely destroyed a town and killed 2,300 residents _____
    d. an intense thunderstorm in 1976 in the mountain foothills that caused a flash flood that took out a canyon road and buildings and killed close to 150 persons _____
    e. an especially wet spring in 1993 that produced floodplain flooding in nine midwestern states in the United States _____

44. What did the jet stream have to do with flooding of the Mississippi River in 1993? _____
    _____

45. Give two reasons the lower Mississippi River basin (south of St. Louis) didn't experience flooding in the summer of 1993. _____
    _____

46. State two things the U.S. Army Corps of Engineers did, in accordance with the Flood Control Act of the 1920s, to control the Mississippi River. _____
    _____

47. Can artificial levees provide permanent flood protection for a vulnerable location? Explain your answer. ____
    _____

48. List some ways, besides just building levees and reservoirs, that can help prevent flood damages in flood-prone areas. _____
    _____

49. a. If the recurrence interval of a certain-sized flood in a particular area is 200 years, what is the average time between floods of this size in this area? _____
    b. What is the probability that a flood of this size will occur in any one year? _____
    c. If a 200-year flood occurs this year in area X, is that area safe from having a comparable flood for the next 199 years? _____
    d. Could a 200-year flood happen in the same area in 2 consecutive years? _____

50. List several reasons humankind has for centuries chosen to settle in river valleys and on floodplains.
    _____
    _____

51. a. Why does the Colorado River no longer reach its mouth (the Pacific Ocean) but instead dry up at about the Mexican border? _____
    _____
    b. Can the Colorado River meet the demands of cities and farmers along its route? _____

52. What Asian location dramatically illustrates human overconsumption of stream waters, and what has happened there? _____
    _____

53. Why can urbanization of an area increase the incidence of flooding there? _____
    _____

54. a. Why does damming a river decrease the amount of silt carried downstream? _____
    _____
    b. Why can agriculture increase the amount of sediment carried into streams? _____
    _____

Making Use of Diagrams

Label the four basic types of stream networks shown.

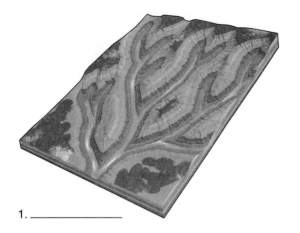

1. _____

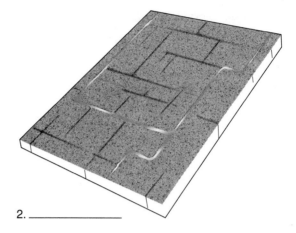

2. _____

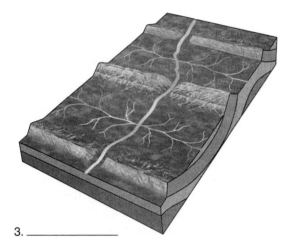

3. _____

4. _____

5. What is the discharge of a stream that has a cross-sectional area of 100 ft^2 and a velocity of 10 ft/s? _____ cfs

6. If a stream is 30 ft wide and 20 ft deep at a stream gauge site where its discharge registers 6,000 cfs, what must its velocity be at this point? _____ ft/s

Short-Essay Questions

1. Explain how sheetwash spread across an entire slope can evolve first into a stream and eventually into an entire drainage network.

2. Explain why stream erosion sometimes produces a valley and sometimes a canyon.

3. Discuss four possible situations that can produce waterfalls.

4. Discuss the possibility that New Orleans may someday not be a riverfront city.

5. Explain how distributaries form at the mouth of a stream.

6. What did the U.S. geologist William M. Davis believe about the development of streams and their associated landscapes?

7. What happened to the Colorado River as the Colorado Plateau was rising that eventually allowed the Grand Canyon to be formed?

8. What is possibly the greatest flood that we have geologic evidence of? Discuss when, where, and why it occurred. What does J. Harlan Bretz have to do with the story?

9. List and briefly describe three environmental issues that directly involve mankind's overuse or abuse of streams.

Practice Test

1. Which of the following statements about the hydrologic cycle is FALSE?
   a. There is an exchange of water among oceans, land, and atmosphere.
   b. Water that manages to infiltrate the land is lost to the cycle.

c. Solar energy evaporates water from Earth's surface, mainly from the sea.
d. Meteoric water is formed when vapor rises, cools, and condenses.
e. Surface ice is part of this cycle.

2. Which of the following statements about drainage divides is FALSE?
   a. A drainage divide is a ridge that separates one watershed from another.
   b. Precipitation that falls on the east side of the continental divide flows to the Gulf of Mexico and the Atlantic Ocean.
   c. The divide that runs along the crest of the Appalachians separates the Atlantic Ocean drainage from the Arctic Ocean drainage.
   d. The ultimate base level for streams on the west side of the continental divide is the Pacific Ocean.
   e. A divide that runs just south of the Canadian-U.S. border separates Gulf of Mexico drainage from Hudson Bay–Arctic Ocean drainage.

3. Which of the following statements is FALSE?
   a. The termination point of a stream is called its mouth.
   b. The beginnings of a stream are called its headwaters.
   c. Any specific segment of a stream is called a reach.
   d. A stream that swings back and forth in snake-like curves is called a braided stream.
   e. Flat land on either side of a stream that is usually not underwater is called a floodplain.

4. Incised meanders
   a. are meanders that lie at the bottom of a steep-walled canyon.
   b. can result from either superposed or antecedent stream activity.
   c. are demonstrating a feature of stream maturity (meanders) in a youthful setting (steep walled).
   d. are illustrated by the goosenecks of the San Juan River, Utah.
   e. All of the above are true.

5. An ephemeral stream
   a. is one whose bed lies below the water table.
   b. is replenished by both precipitation and groundwater.
   c. flows year-round.
   d. in dry climates may sometimes become a dry wash.
   e. All of the above are true statements.

6. The discharge of a stream is
   a. calculated by dividing its cross-sectional area by its velocity.
   b. constant for the length of the stream.
   c. likely to decrease downstream in arid regions and increase downstream in temperate regions.
   d. typically lower in spring than during summer.
   e. All of the above are true statements.

7. Which statement is TRUE?
   a. Stream flow is fastest in the center of the channel near the surface.
   b. The slowest-moving part of a stream is called the thalweg.
   c. Water from distributaries flows into the main trunk stream.
   d. A shallow stream is less turbulent than a deep stream.
   e. All of the above are true statements.

8. A typical longitudinal profile of a stream
   a. shows a cross section of the stream from bank to bank at one particular site.
   b. is roughly a convex-upward curve.
   c. illustrates that a stream's gradient is steeper near its headwaters than near its mouth.
   d. shows almost horizontal plains near the stream's headwaters, and deep valleys near its mouth.
   e. is used to calculate discharge: the stream length multiplied by the drop in elevation equals the discharge.

9. Which of the following statements is FALSE? Waterfalls
   a. often scour out plunge pools at their bases.
   b. may be found where a stream crosses a resistant ledge of rock.
   c. may form at the boundary between a hard pluton and softer country rock.
   d. may drop from a shallow hanging valley into a deeper valley in areas that have been carved by glaciers.
   e. last for millions of years because they are created by elevation differences in exceedingly hard bedrock.

10. Which of the following statements is FALSE? Deltas
    a. may form where a stream empties into the quiet waters of a lake or the ocean.
    b. are likely to slowly sink and become swampland called a delta plain.
    c. make poor farmland because they flood regularly and flooding leaches out the minerals.
    d. received their name because some have a triangular shape, but many don't.
    e. are places where the main trunk stream divides into many distributaries.

11. Which of the following statements is FALSE? The Mississippi River delta
    a. consists of several distinct lobes.
    b. is the product of several avulsions.
    c. developed its shape because the ocean current there was stronger than the river current.
    d. is a bird's-foot delta.
    e. is the site of the city of New Orleans.

12. A stream carrying lots of sediment empties into a human-made reservoir. Pick the most logical follow-up statement.
    a. The reservoir will have no effect on the environment of the area.
    b. If left alone, the reservoir will eventually fill with sediment and become unusable.
    c. The reservoir will increase the frequency of flooding in the area.
    d. The animal and plant population of the area will remain unchanged.
    e. Downstream of the reservoir the waters will have increased load and therefore increased abrasive power.

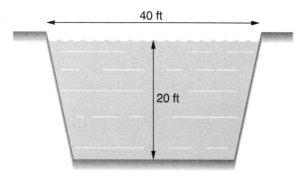

13. The stream in the diagram flows at a velocity of 6 ft/s. Its discharge is
    a. 4,800 ft³/s
    b. 240 ft³/s
    c. 120 ft³/s
    d. 66 ft³/s
    e. 800 ft³/s

14. Youthful streams are likely to have
    a. backswamps.
    b. V-shaped valleys.
    c. natural levees.
    d. oxbow lakes.
    e. All of the above.

15. Streams come down out of steep areas onto much flatter land, lose velocity, and drop material in triangular-shaped structures which are called
    a. point banks.
    b. alluvial fans.
    c. levees.
    d. sandbars.
    e. deltas.

16. For the diagram that follows, which of the statements below is TRUE?
    a. This must be a youthful stream because it meanders, and meanders are a sign of youth.
    b. Point A is called a point bar.
    c. Point A represents a place of extreme erosion.
    d. You should expect lots of waterfalls and rapids along this reach of the stream.
    e. All of the above are true statements.

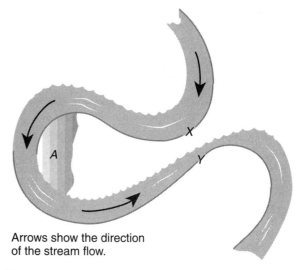

Arrows show the direction of the stream flow.

17. For the diagram above, which statement below is TRUE?
    a. If the stream flow were reversed, the locations of point bars and cut banks would also be reversed.
    b. Point A represents a cut bank.
    c. The meander shown is geologically long lived and will exist for thousands of years.
    d. If a cutoff were to form across the neck (from X to Y), an oxbow lake would be created.
    e. All of the above are true statements.

18. Which of the following was the location of a flash flood caused by failure of a dam?
    a. Bangladesh, 1990
    b. Yangtze River, China, 1931
    c. Big Thompson River, Colorado, 1976
    d. Johnstown, Pennsylvania, 1889
    e. upper Mississippi River Valley, 1993

19. Which of the following statements is FALSE? The channeled scablands
    a. are an area of unusual topography in Alaska.
    b. were created by a great flood about 11,000 years ago.
    c. were flooded because the glacial dam of Lake Missoula broke.
    d. are littered with giant boulders and hills that are giant ripples.
    e. include the now-dry Grand Coulee, which was once a giant waterfall.

20. Which of the following statements is FALSE?
    a. Early civilizations of Egypt, India, and China were established in river valleys and on floodplains.
    b. For centuries humankind has used rivers for transportation; to supply food, irrigation water, and drinking water; to generate power; and as sites for recreation and for waste disposal.
    c. Floodplains are areas of fertile soil, replenished yearly by seasonal floods.
    d. If a dam is constructed properly, it will not change the ecosystem of the area it's in.
    e. Overuse of the waters of the Colorado River has reduced it to a mere trickle near its mouth.

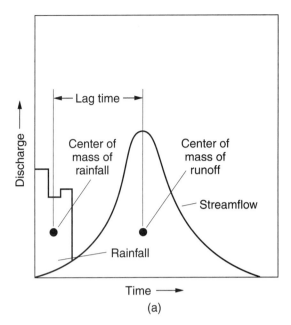
(a)

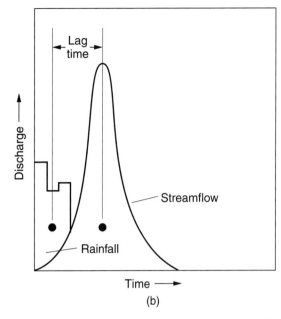
(b)

21. Identify the statement that best describes the effects of urbanization on stream flow as shown in the diagrams above.
    a. Diagram A illustrates that there is more infiltration after urbanization, and therefore less water reaches the stream channel.
    b. Diagram A shows water from the storm getting into the stream channel faster than it does in diagram B.
    c. Diagram A must show the situation after urbanization of an area.
    d. Diagram B shows a stream responding faster and more intensely to a rainfall event.
    e. Flooding potential for a stream is less after urbanization than before.

22. The largest river in the world, on the basis of discharge, is the Mississippi.
    a. True
    b. False

23. Broad sweeping bends of a stream channel are called oxbow lakes.
    a. True
    b. False

24. The terms "fluvial deposits" and "alluvium" both refer to sediment deposited by a stream.
    a. True
    b. False

25. The area in which flood waters may sit because they can't climb up over natural levees and get back into their stream channel is called a yazoo stream.
    a. True
    b. False

26. A stream is capable of erosion, transportation, and deposition of materials.
    a. True
    b. False

27. Streams carry their load in suspension, in solution, and as bed load.
    a. True
    b. False

28. The total load that a stream can carry is called its competence.
    a. True
    b. False

29. Since the Big Thompson Flood of 1976 was a 100-year flood, it's almost impossible for such a flood to happen again for another 99 years.
    a. True
    b. False

30. The Mississippi River Flood Control Act of the 1920s resulted in the construction of dams and reservoirs, levees, and flood walls by the U.S. Army Corps of Engineers.
    a. True
    b. False

31. Building gauging stations and studying an area to identify possible problems is a logical first step in flood control.
    a. True
    b. False

32. The Sahara Desert is one of the few places on Earth that has never had water running across its surface.
    a. True
    b. False

33. All standing bodies of water have an outlet through which water escapes and continues to the sea.
    a. True
    b. False

34. A drainage network is an interconnected group of streams that collects water over a large area and drains it away toward the sea.
    a. True
    b. False

35. Turbulent flow moves water downstream faster than laminar flow.
    a. True
    b. False

36. Potholes form when water rich in calcium carbonate dissolves a section of the stream bed.
    a. True
    b. False

37. Muddy saltwater can carry larger fragments than can clear freshwater.
    a. True
    b. False

38. The average downstream velocity of a steep mountain stream is greater than that of a large, flat muddy river.
    a. True
    b. False

39. Local base levels are short lived, because running water removes obstructions that create them.
    a. True
    b. False

40. A stream that is eroding more material than it is depositing is termed a graded stream.
    a. True
    b. False

41. A stair-step-shaped canyon develops where a stream downcuts through alternating layers of hard and soft rock.
    a. True
    b. False

42. Stream terraces are the result of extremely high flood levels that deposit debris far from the stream channel.
    a. True
    b. False

43. Plate tectonics events can produce mountains whose uplift can cause drainage reversal in existing river systems.
    a. True
    b. False

44. The jet stream was a contributing factor to Mississippi River flooding in 1993 because it trapped warm, moist Gulf of Mexico air over the central United States.
    a. True
    b. False

45. The diversion of river waters in central Asia for irrigation has caused the Aral Sea to shrink drastically and destroyed the fishing economy of much of its coastal area.
    a. True
    b. False

ANSWERS

Completion

1. flood
2. streams
3. channels
4. meteoric water
5. infiltrate
6. surface water
7. runoff
8. sheetwash
9. downcutting
10. headward erosion
11. tributaries
12. trunk stream
13. drainage network
14. drainage basin (or catchment or watershed)
15. drainage divide
16. continental divide
17. dendritic network
18. radial network
19. rectangular network
20. trellis network
21. water gap
22. structural control
23. superposed streams
24. antecedent streams
25. incised meanders
26. water table
27. permanent stream
28. ephemeral stream
29. dry wash
30. badlands
31. discharge
32. thalweg
33. wetted perimeter
34. turbulence (or turbulent flow)
35. laminar flow
36. scouring
37. abrasion
38. pothole
39. sediment load
40. settling velocity
41. saltation
42. competence
43. capacity
44. fluvial deposits
45. alluvium
46. bar
47. sediment sorting
48. gradient
49. rapids
50. waterfalls
51. longitudinal profile
52. base level; local base level; ultimate base level
53. graded stream
54. valley
55. canyon
56. oversized stream valley
57. standing wave
58. plunge pool
59. hanging valley
60. alluvial fan
61. braided stream
62. meanders
63. cut bank
64. point bar
65. meander neck
66. cutoff
67. oxbow lake
68. abandoned meander
69. bluff
70. natural levees
71. backswamp
72. yazoo streams
73. distributaries
74. delta
75. bird's-foot delta
76. avulsion
77. delta plain
78. peneplains
79. stream piracy (or stream capture)
80. wind gap
81. stream rejuvenation
82. terraces
83. drainage reversal
84. flood stage
85. seasonal floods
86. flash floods
87. jet stream
88. channeled scablands

Matching Questions

ANATOMY OF A STREAM

1. e
2. h
3. d
4. c
5. j
6. a
7. i
8. g
9. b
10. f

Short-Answer Questions

1. a. Drain excess water from the landscape and carry it eventually to the sea.
   b. Supply avenues of commerce, water for homes, agriculture and industry, and electric power.
   c. Transport nutrients and provide a home for various life forms.

2. an outlet that lets water escape and travel onward toward the sea

3. The continental divide along the Rocky Mountains: water that falls on its west side ends up in the Pacific Ocean, on its east side, in the Atlantic Ocean via the Gulf of Mexico
   a divide that runs along the crest of the Appalachians and separates the Atlantic Ocean drainage from the Gulf of Mexico drainage
   a divide that runs just south of the Canadian-U.S. border and separates the Gulf of Mexico drainage from the Hudson Bay (Arctic Ocean) drainage

4. incised meanders, that is, meanders that lie at the bottom of a steep-walled canyon

5. Badlands National Park in South Dakota

6. a. increase, because of added water from tributaries
   b. decrease, because of the loss of water to ground seepage and evaporation
   c. increase, because of the addition of melt water from snow and ice or heavy spring rains
   d. increase, because by definition, a flood is a time when the volume of water gets so great it covers areas outside the streams normal limits

7. a. Friction between water and channel walls slows the flow, and the channel walls are not uniform for the length of the stream.
   b. Turbulence disturbs the flow of water downstream, and the amount of turbulence varies along the stream's length.

8. the roughness of the channel walls and the shape of the stream channel

9. a. faster water   b. shallow stream   c. in a rough, gravelly channel

10. Gravitational potential energy stored in the water at higher elevation is transformed into kinetic energy as it goes to lower elevation.

11. 820 m; 2,690 ft (820 m × 3.28 ft/m)

12. scouring, breaking off chunks of rock and lifting rock fragments off the stream bed, abrasion, and dissolution (dissolving)

13. the velocity and volume of water and its sediment content

14. Dissolved load. Ions of soluble minerals (like silica, calcium carbonate, halite, or gypsum) that have dissolved in the water
    Suspended load. Small solid grains, clay or silt sized, that are carried along by the water and don't settle to the channel floor
    Bed load. Large particles of sand, pebbles, and cobbles that bounce or roll along the channel bottom

15. suspended load; 67% of the total sediment load

16. It increases by a factor of 4.

17. a. high-density, high-sediment load   b. muddy water   c. saltwater

18. its competence, discharge, and supply of sediment

19. a. Decreases.
    b. Sediment is deposited, coarse components first, finer particles when and where the water slows even more.

20. It passes from a deeper part of the channel to a shallower part.
    It spreads out over a floodplain.
    It enters a standing body of water.
    Its gradient decreases.

21. because maximum stream velocity varies with location

22. a curve that's concave up, steeper near the headwaters than the mouth, and interspersed with little plateaus and steps (lakes and waterfalls)

23. The turbulence of a mountain stream causes the water to swirl in eddies and, despite appearances, actually makes slow progress downstream.

24. lakes or reservoirs, a ledge of resistant rock, and the channel of a larger stream at the site where a tributary joins it

25. Running water erodes away the obstruction that was serving as a local base level.

26. It erodes or deposits sediments until it once again becomes graded.

27. damming the stream to create a reservoir
    faulting that drops down the lower reach of a stream
    the sudden addition of a large amount of sediment due to landslide

28. A larger stream in a past wetter climate (possibly the end of an ice age when it not only was wetter but there

was melting ice) created the valley; only a small stream remains in today's drier climate.

29. Water depth is less than about twice the diameter of the coarsest sediment.

30. It can rise over the upstream side of a clast and plunge into a depression on the downstream side.
    It can be diverted to the side and create a whirlpool on the downstream side.

31. due to headward erosion that destroys the resistant rock layer and leads to the establishment of a graded stream condition

32. His pen name, Mark Twain, was the phrase called out by boatmen to announce the water was 2 fathoms deep (about 4 m or 12 ft) and therefore the boat was not going to run aground.

33. The Greek historian Herodotus noticed that the depositional structure at the mouth of the Nile River was triangular and thus resembled the shape of the Greek letter delta; the name ended up being applied to similar deposition features at rivers' mouths, even if the structures were not particularly triangular.

34. about 9,000 years

35. the supply of sediment by the river, the current strength of the river, the degree to which ocean tides redistribute sediment, and the degree to which ocean currents carry away sediment

36. a. The strength of a river current exceeds ocean current strength and carries sediment far offshore.
    b. The strong ocean currents erode away lobes and redistribute sediment parallel to the shoreline.
    c. The tides rework the sediment extensively.
    d. The strong ocean waves and currents carry away sediment as fast as it's deposited.

37. The delta plain there is sinking. This is a totally natural occurrence; huge walls keep the river out.

38. a wind gap, created when stream piracy caused a water gap to become a waterless wind gap

39. The river channels follow the axes of ancient continental rifts.

40. Continents move to positions where atmospheric mean temperatures are different.
    Shifted land masses result in different ocean currents, which strongly influence climates.

41. an ancient drainage network (interconnected stream channels)

42. the Yellow River (Hwang River), because over the centuries it has flooded frequently and severely and has caused the deaths of millions of Chinese by drowning and by postflood famine and disease

43. a. Yellow River, plains of China
    b. Bangladesh
    c. Johnstown Flood in the Conemaugh River Valley, Pennsylvania
    d. Big Thompson Flood, north of Denver, Colorado
    e. upper Mississippi River basin, summer 1993

44. The position of the jet stream trapped warm moist air from the Gulf of Mexico and kept it stationary over the central United States, where it rose, cooled, and condensed as heavy rains for long periods of time.

45. The river channel there is wider and deeper and can accommodate more water.
    The rain pattern was shifted north, which left the southern (lower) basin drier than normal, not wetter.

46. Built about 300 dams to create reservoirs to hold excess runoff.
    Built thousands of kilometers of sand and mud levees and concrete flood walls.

47. No, defensive efforts are delaying the inevitable, because it's unrealistic to construct levees high enough to handle all conceivable floods.

48. Allow floodplains to become natural wetlands, which absorb lots of excess water; map floodways, then remove or abandon buildings located there; move levees farther away from the river; discourage, rather than encourage, the rebuilding of flood-damaged structures in floodways.

49. a. 200 years   b. 1 in 200   c. no   d. yes

50. Rivers serve as avenues of transportation, places for waste disposal (unfortunately), and as sources of food, drinking water, irrigation water, power, recreation, and fertile soil from their flood plains.

51. a. Huge pipes and canals carry the water to Phoenix and Los Angeles.
    b. No, its discharge is only 90% of the discharge amount allocated to and paid for by cities and farmers along its route.

52. The Aral Sea, once the world's fourth largest freshwater lake; it has shrunk drastically because the rivers that feed it are so heavily overused by the local human population.

53. Covering the land with impermeable concrete and blacktop inhibits the infiltration of precipitation, so more surface water is left to enter stream channels, and it may be enough to cause them to overflow their banks.

54. a. Silt is deposited in the still waters of the reservoir and is therefore unavailable downstream.
    b. Agriculture may actually decrease the amount of vegetation covering the land as it allows rain to wash more soil away into streams.

Making Use of Diagrams

1. dendritic

2. rectangular

3. trellis

4. radial

5. 1,000 cfs (100 ft² × 10 ft/s = 1,000 cfs)

6. 10 ft/s (30 ft × 20 ft × x ft/s = 6,000 cfs); (600 ft²)x ft/s = 6,000 cfs, x = 10 ft/s

Short-Essay Questions

1. Ground areas are not equally flat, equally resistant to erosion, or equally vegetated. Therefore the velocities of sheetwash flowing over different areas differ, and where sheetwash is flowing slightly faster, it digs a shallow channel into the substrate. From here on, it's a self-perpetuating cycle; adjacent sheetwash flows into the slightly deeper channel, which gets deeper yet with the increased flow. Given time, the channel is deep enough to be called a stream. Because of this same process more channels keep forming in the area. Every stream channel is lengthening upslope (headward erosion) because the flow at the entry point is intense. Eventually the smaller channels merge with the larger channel because it offers the lower elevation, and gradually a system of linked small streams (tributaries) flows into a single larger trunk stream. The whole array is called a drainage network.

2. A steep-walled narrow canyon forms where a stream downcuts rapidly and the canyon's walls are strong and do not collapse. Still, the canyon eventually widens as blocks split off along joints and the stream undercuts the side walls. If the stream is downcutting at the same speed as its walls are collapsing, unconsolidated material along its sides assumes its angle of repose and a V-shaped valley results. In the special case where the walls of a stream consist of horizontal layers of alternating strong and weak rock, a stair-step effect results (the Grand Canyon look). When the channel floor approaches base level, stream energy goes into widening the valley, and the result is the stream channel proper's occupying a small portion of a broad floodplain.

3. Waterfalls occur where the gradient of the stream bed is so steep water free-falls to the lower level. This can happen where a stream crosses a resistant ledge of rock, such as hard sandstone, limestone, or gneiss, or a tabular igneous intrusion. Another possible waterfall location is at the boundary between a hard pluton and softer country rocks. A fault may serve as a waterfall site, either where a scarp has created higher ground or where a fault has moved softer rock next to harder rock. A fourth possibility is where a large glacier has carved a deep glacial valley, smaller tributary glaciers have carved more shallow valleys leading into it, and the ice has melted and been replaced with stream waters.

4. The lower reach of the Mississippi River is a giant floodplain, composed of numerous lobes of a bird's-foot delta, with many distributaries flowing on it. Geologic evidence shows that avulsion (change of direction of stream flow) has happened many times in the area. The city of New Orleans is built on this floodplain, along a current distributary of the Mississippi. Parts of the city already are lower than the river bed, because the delta land is sinking. In the future, if an upstream levee breaks, avulsion can again take place, and the distributary could flow to the Gulf of Mexico via the Atchafalaya River channel, leaving New Orleans without a riverfront.

5. Water at a stream's mouth slows as it empties into a standing body of water (lake or ocean), loses energy, and thus must deposit some sediment as a midstream bar. The presence of this bar splits the stream flow, and the resulting two smaller streams have less energy and must deposit sediment within their channels, again as midstream bars. The process continues to eventually form a braided stream with several distributaries emptying into the lake or ocean.

6. During the latter half of the nineteenth century, following Darwin's proposal of the theory of evolution, William Morris Davis proposed the idea that streams and the landscape they produce evolve as the relief, and therefore the potential energy diminishes with time. He suggested the fluvial (stream) landscape evolves from one of youth (steep gradients and energetic flow) through maturity (rounded hills and V-shaped valleys) into old age (meanders across wide floodplains or peneplains). In reality many factors, including climate and the composition of the substrate, affect the process, and peneplains rarely have time to develop before some plate tectonics event interrupts and starts the process over again. Nevertheless, the general idea that a stream and the land around it change in a predictable way over time is accepted today.

7. The Colorado River was a "victim" of stream piracy. It began as a stream on the eastern side of the local drainage divide and flowed eastward into the Gulf of Mexico. As the Colorado Plateau was gradually uplifted, a stream on the western side of the divide got renewed energy and did enough headward erosion to invade the channel of the Colorado and steal its waters. This newer, larger stream flowed westward into the Pacific, and between its increased waters and its increased energy of position high on the elevated Plateau, it downcut vigorously, forming the Grand Canyon.

8. J. Harlan Bretz studied an unusual area in eastern Washington State and proposed the idea that a great flood had occurred at the end of the last Ice Age, roughly 11,000 years ago, when the glacial dam of a large lake, Lake Missoula, finally broke. All of the lake emptied, roared across eastern Washington State, into the Columbia River valley and eventually out to the Pacific Ocean. It left a weird landscape of bare rock stripped of its soil and regolith, huge boulders, giant rock ripples, now-dry Grand Coulee (a former immense waterfall), and huge potholes. There was much opposition, but now Bretz's theory of the origin of the channeled scablands of eastern Washington and the occurrence of the Great Missoula Flood is widely accepted.

9. First, rivers have been improperly used and overused for waste disposal and today are polluted by raw sewage, storm drainage from cities, toxic chemicals from industrial sides, and excess fertilizer and animal waste from agriculture. The harmful effects are as varied as the agents of pollution. Aquatic life is poisoned; algal blooms sap oxygen from waters; and, particularly in developing countries, polluted waters cause serious health problems.

Second, the number of dams has increased from about 5,000 in 1950 to over 38,000 today. The results are both good and bad. Reservoirs provide irrigation water and hydroelectric power, help in flood control, and create recreation areas. They also destroy "wild rivers" and alter ecosystems in numerous ways. A good example is the building of the Aswan Dam on the Nile River, which has produced many unexpected changes and altered many aspects of Egyptian civilization that for thousands of years had been influenced by the natural ecology of the Nile River plains and its annual flooding.

Third, growing populations have in some places literally dried up mighty rivers and lakes, like the Colorado River in the U.S. Southwest and streams that feed the Aral Sea in central Asia. Water usage has increased threefold from 1950 to 1995, and the end is not in sight.

Practice Test

1. b Groundwater eventually resurfaces, evaporates, and thus remains part of the cycle.

2. c The divide separates Atlantic Ocean drainage from Gulf of Mexico drainage.

3. d The stream is called a meandering stream.

4. e Choices a to d are all true.

5. d The other statements are true of permanent streams, not ephemeral streams.

6. c The discharge is calculated by multiplying the area by the velocity; it varies along stream length; and it is lower in summer.

7. a The thalweg is the fastest-moving part; tributaries flow into the trunk stream; and a shallow stream is more turbulent.

8. c A profile shows the cross section along the stream's entire length; the curve is concave-up; the profile shows horizontal plains near the mouth, deep valleys near the headwaters; and the multiplication in e may sound good, but it's nonsense.

9. e Geologically speaking waterfalls are short lived because headward erosion eats back the resistant ledge.

10. c Regular flooding adds new rich soil to farmland and helps keep the land fertile.

11. c The Mississippi River delta formed because the river current was stronger than the ocean current.

12. b Reservoirs always affect the local ecosystem; reservoirs serve for flood control; and downstream waters have decreased load, since sediment settles out in the reservoir.

13. a Discharge = cross-sectional area × velocity; 40 ft × 20 ft × 6 ft/s = 4,800 ft^3/s

14. b All of the other choices are typical of maturity or old age.

15. b Deltas are analogous structures, but they form under water.

16. b Meanders are a sign of maturity and old age; waterfalls and rapids are signs of youth; and a point bar is a place of deposition, not erosion.

17. d Stream-flow reversal would not change what was the inside or outside of the curve, and this is what determines where the cut bank and point bar are; meanders are changing constantly.

18. d The Big Thompson River Flood was a flash flood, but no dam broke; there was just too much rain. Choices a, b, and e are examples respectively of seasonal, floodplain, and delta-plain floods that build up slowly and last for weeks or months.

19. a Scablands are in eastern Washington State.

20. d Building a dam and reservoir will always change the ecosystem, often in ways that are totally unexpected.

21. d Diagram A shows the situation before urbanization, when there is more infiltration, slower stream response, and less water reaching the stream channel. Diagram B shows the situation after urbanization.

22. False The Amazon River of South America is the largest.

23. False Broad sweeping bends of a stream channel are called meanders.

24. True Both terms refer to material deposited by a stream.

25. False This area is called a backswamp.

26. True Any stream causes erosion of the land it flows through, transports the eroded material downstream, and eventually deposits it.

27. True Streams carry material visibly suspended in the water, invisibly dissolved in the water, or moved along on the stream bed.

28. False The total load that a stream can carry is called its capacity; competence is a measure of how large a piece a stream can carry.

29. False Flood recurrence terms are only statistical probabilities, and thus every year the chance of a 100-year flood are 1 in 100; 100-year floods have happened in consecutive years.

30. True The Flood Control Act very wisely mandated a variety of flood-control activities, including the construction of reservoirs, levees, and flood walls.

31. True Before you can address any problem, you should understand as thoroughly as possible what the problem is. Data from gauging stations and area studies help define the potential for flooding and initiate a relevant flood control plan.

32. False Satellite imagery shows large stream channels buried under the Sahara sands.

33. False All standing bodies of freshwater have outlets.

34. True A drainage network is a collecting basin for area precipitation and a series of channels to conduct the water out of the area and toward the sea.

35. False Turbulent flow moves water downstream slower than laminar flow.

36. False Potholes form by the action of sand or gravel abrasion in a whirlpool.

37. True

38. False The turbulence of mountain streams causes water to swirl in eddies and make little progress downstream, lowering average downstream velocity significantly.

39. True Local base levels, such as ponds, lakes, and reservoirs, are eventually eroded and carried away by the waters they at first contained.

40. False In a graded stream, there's no net erosion or deposition.

41. True Erosion of soft rock layers creates slopes; erosion of hard rock layers creates cliffs; and erosion of alternating soft and hard rock layers produces stair-step-shaped canyons.

42. False Terraces form when stream rejuvenation allows a stream to renew downcutting through its existing floodplain.

43. True The flow pattern of a stream changes when elevations along it change; significant plate tectonics uplift can result in the complete reversal of the direction of a stream flow.

44. True The presence of the jet stream can block the flow of an air mass; in 1993 it blocked the flow of the warm, moist Gulf air and therefore it condensed its moisture in a very limited area and produced flooding there.

45. True Diversion of river water has reduced the size of the Aral Sea to the point it is now useless for commercial fishing.

CHAPTER 18 | Restless Realm: Oceans and Coasts

GUIDE TO READING

Chapter 17 dealt with freshwater on Earth's surface; Chapter 18 deals with the larger realm of saltwater on Earth, the oceans. It begins with the physical structure of ocean basins. Although these are difficult places to access, over the past century scientists have managed to collect and piece together information that gives quite a comprehensive picture of both the geology and geography of the ocean floor. They have gathered an impressive amount of data directly, starting with research cruises of HMS *Challenger* in the 1870s and continuing through sea-floor explorations of the deep-sea submersible *Alvin* a century later. In this section some of your reading is a review of plate tectonics activities and features (oceanic crust and lithosphere, mid-ocean ridges, fracture zones, trenches, and active and passive continental margins). There are also many new concepts concerning ocean depths and landscapes (bathymetry, continental shelves, slopes and rises, abyssal plains, submarine canyons, turbidites, submarine fans, seamounts, and guyots).

The composition and characteristics of ocean water are discussed next. You read about salinity, the halocline, the heat capacity of water, the thermocline, the pycnocline, and the human need to take saltwater and turn it into freshwater (desalinization plants).

Ocean waters exhibit numerous patterns of movement. There are surface currents, deep currents, eddies, the Coriolis effect, the Ekman spiral, gyres, upwelling zones, downwelling zones, thermohaline circulation, water masses, and weather-related oceanic events like storm surges and El Niño. Tides, of course, are also water motion, but they are treated thoroughly enough in this chapter to warrant their own subcategory. Some special advice: This is a topic that seems so elementary it's easy to skim too lightly over the section. Everybody knows tides go in and out, but unless you live on the coast, you probably don't really know much about tidal reach, mean sea level, flood tide, ebb tide, tidal flats, tidal bores, intertidal zones, the tide-generating force, neap tides, spring tides, and the effect tides have on the rotational speed of Earth.

Waves are another one of those topics that you should be careful not to skim over too lightly, for the author presents more than the average landlubber would believe there is to know about waves. You read about their causes and geometric shapes and nomenclature, including the terms "wave base," "strength and fetch of a wind," "ripples," "swells," "amplitude and wavelength of a wave," "interference and refraction of waves," "breakers and surf zones," "swash and backwash," "effects on embayments and headlands," "longshore currents," and "rip currents."

Ocean study includes a look at ocean boundaries, that is, coastal areas and shorelines. One type of shoreline is a beach. Beaches may be composed of different types of sand, including silicic sands or carbonate sands, and they have distinct areas, including a beach face, foreshore zone, intertidal zone, backshore zone, and berm. Beaches, geologically speaking, are here today, gone tomorrow, and change is constantly occurring. In connection with this you read about active and inactive sand layers, bioturbation of sediments, beach drift, sand spits, baymouth bars, barrier islands, lagoons, and sediment budgets of beaches.

A shoreline may be a rocky coast rather than a sandy one. Here wave erosion may produce wave-cut notches, cliff retreat, wave platforms (benches), honeycomb-weathering patterns, sea arches, sea stacks, and tombolos.

Coastal areas may be coastal wetlands that are flooded with shallow water but experience no wave action. The three basic types of wetlands you read about are swamps, marshes, and bogs.

Some coastal areas are flooded stream valleys called estuaries. Here a mix of fresh- and saltwater supports complex ecosystems inhabited by unique salt-tolerant organisms.

Fjords are dramatic coastlines that result from the flooding of glacial valleys.

Coral reefs are specialized communities found in shallow, warm, well-lit seawater. Their basic physical structures (limestone mounds) are created by colonial marine animals (cnidarians), that live in a symbiotic (mutually beneficial) relationship with the algae called zooxanthellae. In addition to these two creatures, the reef provides the environment for a complex community of marine organisms. Coral reefs are classified on the basis of their shapes, which are determined by their origins (fringing reefs, barrier reefs, and atolls).

Over time the ocean level rises and falls, and these changes affect coastal areas. You read about glacial rebound of land after large masses of ice melt, eustatic (global) sea-level changes, emergent and submergent coasts, and erosional (losing area) and accretionary (gaining area) coasts.

Coastal areas have always experienced change due to natural events; today the human population is large enough to add its influence. The author concludes the chapter with a discussion of some human-induced changes and the problems they have created. For example, people build groins, jetties, breakwaters, and seawalls to fight coastal erosion, and when this doesn't work well enough to suit them, they bring in their own sand (beach nourishment). The results of such actions are often unpredictable; quite often they benefit one area of a beach and harm an adjacent area. Thoughtless human activity has destroyed huge amounts of coastal wetlands and endangered coral reefs. As in Chapter 17 the author is reminding us that as large and powerful as the water world is, the human presence is affecting it and not always in a desirable fashion.

LEARNING ACTIVITIES

Completion

Test your recall of new vocabulary terms.

_____ 1. the study of ocean water and its movement

_____ 2. the study of the ocean floor

_____ 3. the study of life forms in the sea

_____ 4. maps that show the elevations and shape of the sea floor

_____ 5. a cross section of the ocean crust, generated by sonar, that shows the different layers that make it up

_____ 6. the area where the land meets the sea

_____ 7. variations in the depth of ocean waters

_____ 8. large regions of the ocean that have the same water depth

_____ 9. the shallow part of the ocean, no deeper than 500 km, that borders continental margins, whose floor slopes seaward at about 0.3°

_____ 10. the part of the ocean that begins at the seaward edge of the continental shelf and where the ocean floor slopes seaward, at an angle of about 2°, descending to depths of about 4 km

_____ 11. the part of the ocean floor that slopes gently (less than 2°) from the seaward edge of the continental slope to the almost horizontal plain that's found at a depth of 4.5 km.

_____ 12. margins of continents that are not plate boundaries

_____ 13. margins of continents that coincide with plate boundaries

_____ 14. the crest of the submarine mountain chain that winds through the oceans of the world

_____ 15. a deep, elongate trough that marks the edge of a subducting ocean plate and borders a volcanic arc

_____ 16. narrow, linear zones of rough topography that run parallel to each other, extend outward from the mid-ocean ridge, and make up transform faults on the ocean floor

_____ 17. broad regions of ocean floor that are underlain by old, cool, flat-surfaced lithosphere

_____ 18. a thick layer, composed of microscopic plankton shells and flakes of brown clay, that blankets the abyssal plain

_____ 19. narrow deep valleys that dissect continental shelves and slopes

_____ 20. flows of sediment mixed with water that cause extensive erosion as they race down submarine slopes

_____ 21. deposits of graded beds produced by turbidity currents

_____ 22. a triangular-shaped structure, composed of turbidites, that builds up at the base of a continental slope

_____ 23. an extinct volcanic island that has sunk below sea level

_____ 24. a seamount that has become capped with a layer of limestone, created by coral reefs growing on its summit

_____ 25. the concentration of salt in water

_____ 26. the boundary between surface-water salinities and deep-water salinities

_____ 27. the ability of a substance to absorb or release heat energy yet undergo little change of temperature

_____ 28. the temperature boundary between relatively warm, sun-heated water and much colder, deeper water

_____ 29. the depth at which density of sea water changes rapidly

_____ 30. fairly well defined flow patterns (streams) within a larger body of water

_____ 31. currents that affect the upper 100 m of ocean water

_____ 32. currents that may be found anywhere from 100 m deep all the way down to the ocean floor

_____ 33. isolated circular patterns (swirls) of water that form along the edges of currents

_____ 34. the deflection of a fluid (wind or water) flowing across Earth's rotating surface, so that north-flowing currents in the Northern Hemisphere deflect to the east and south-flowing currents deflect to the west (Deflections are just the opposite in the Southern Hemisphere.)

_____ 35. a spiral-like pattern of flow, progressing downward through a water column, caused by wind shear on the surface water and successive shearing motion between deeper and deeper water layers

_____ 36. the movement of a mass of water at an average angle of 90° to the wind direction

_____ 37. large circular flow patterns of surface currents, clockwise in northern seas and counterclockwise in southern seas, that are the result of the interaction of ocean basin geometry and patterns of wind directions

_____ 38. a tropical seaweed

_____ 39. places where near-surface water sinks

_____ 40. places where deep water rises

_____ 41. a complex phenomenon involving winds, surface currents, and upwelling currents, that ultimately affects global atmospheric circulation and rainfall

_____ 42. upwelling and downwelling that is driven by density differences, which were created by temperature and salinity differences

_____ 43. a vertical subdivision of the Atlantic Ocean, the waters of which sink along the coast of Antarctica

_____ 44. a vertical subdivision of the Atlantic Ocean, the waters of which sink in the north polar region

_____ 45. the daily rise and fall of sea level

_____ 46. the difference in sea level between high tide and low tide

_____ 47. the yearly average between high and low tide, which is what is really meant when we talk about sea level

_____ 48. the rising tide

_____ 49. the falling tide

_____ 50. the boundary between water and land

_____ 51. a coastline area, broad and level, on which the shoreline may migrate as much as a few kilometers as the tide rises and falls

_____ 52. a flood tide that arrives as a visible wall of water

_____ 53. the coastal area across which the tide rises and falls

_____ 54. the producer of tides (a combination of gravitational attraction of the Sun and Moon and centrifugal force produced by Earth's spin)

_____ 55. extra low tides that occur when the Sun's gravitational attraction counteracts the Moon's

_____ 56. extra high tides that occur when the Sun's gravitational attraction adds to the Moon's

_____ 57. water driven landward by high winds

_____ 58. the maximum depth to which wave motion can occur (roughly equal to half the wavelength)

_____ 59. how fast air is moving

_____ 60. the distance across which a wind blows

_____ 61. pointed waves created when the wind begins to blow

_____ 62. the height of a wave, measured from crest to trough

_____ 63. the horizontal distance between two wave troughs

_____ 64. large waves, with amplitudes of 2–10 m and wavelengths of 40–500 m

_____ 65. the overlap of swells that may lead to giant waves

_____ 66. a wave in which the top of the wave has curved over the base

_____ 67. the area close to shore in which you find breakers

_____ 68. the upward surge of water onto the beach that happens when a breaker crashes to shore

_____ 69. the retreat of a swash

_____ 70. the bending of a wave as it approaches shore, due to frictional slowdown of the first water to touch bottom, so that its angle of approach is diminished

_____ 71. a wave that flows parallel to the beach

_____ 72. places where higher land protrudes into the sea

_____ 73. places along the shoreline that are set back from the sea

_____ 74. strong, localized currents that flow seaward perpendicular to the beach

_____ 75. a gently sloping band of sediment along the shore

_____ 76. a cross section of a beach drawn perpendicular to the shore

_____ 77. the steeper, concave part of the foreshore (intertidal) zone where the swash actively scours the sand

_____ 78. the area of a beach that extends landward, starting at a small step (escarpment) cut by high-tide swash and reaching inshore to the front of the dunes or cliffs that mark the inward edge of beach

_____ 79. horizontal to landward-sloping terraces that receive sediment only during a storm

_____ 80. the layer of sand on the sea floor that is moved by wave action daily

_____ 81. the layer of sand, buried below the active layer, that may possibly be moved during severe storms

_____ 82. the gradual migration of sand along a beach

_____ 83. extension of a beach out into open water that occurs where the coastline indents but beach drift doesn't

_____ 84. sand spits that have grown across the opening of a bay

_____ 85. narrow ridges of sand, parallel to the shoreline, that are created by the scouring action of waves

_____ 86. extra-high offshore bars that have risen above the mean high-water level

_____ 87. the area of quiet water between a barrier island and the mainland

_____ 88. a stirring-up of sediments due to the movement of burrowing organisms (like clams and worms) through it

_____ 89. a comparison of the amount of sand supplied to the amount of sand removed in an area

_____ 90. the process that is the result of the combined shattering, wedging, and abrading by waves

_____ 91. the undercut on a cliff face created by wave erosion

_____ 92. the gradual migration of a shoreline cliff inland due to undercutting, breaking off, and removal of rock

_____ 93. the level layer, exposed at low tide, that resulted from cliff retreat

_____ 94. a weathering pattern in shoreline rocks that is a grid of small hollows separated by narrow ridges, resulting from salt wedging

_____ 95. a headland that has been eaten through to become an arch and is still connected to the mainland

_____ 96. a former sea arch in which the bridged, or arch, section collapsed and left an isolated offshore column

_____ 97. a narrow ridge of sand that links a sea stack to the mainland

_____ 98. a flat-lying stretch of coast that floods with shallow water but experiences no wave action

_____ 99. wetlands that are dominated by trees

_____ 100. wetlands that are dominated by grasses

_____ 101. wetlands that are dominated by moss and shrubs

_____ 102. a bay in which seawater and river water mix

_____ 103. the process in which clay particles clump together and form pieces that are large enough to settle out

_____ 104. deep, steep-sided, flooded glacial valleys along an ocean shoreline

_____ 105. small, colonial invertebrate animals related to jellyfish

_____ 106. an individual coral animal

_____ 107. a shallow, warm submarine environment consisting of mounds of solid limestone, living corals, and associated organisms

_____ 108. a coral reef that forms directly along a coast

_____ 109. a coral reef that develops offshore and is separated from the mainland by a lagoon

_____ 110. a circular ring of coral reef that surrounds a lagoon

222 | Chapter 18

_____ 111. a broad, flat land area that merges with the continental shelf, such as the Gulf Coast and the southeastern Atlantic Coast

_____ 112. the slow rise of land, formerly covered by a glacier, after the ice has melted

_____ 113. a rise or fall of the ocean surface over the entire globe

_____ 114. coastal areas where the land is rising relative to the sea level, or the sea level is falling relative to the land

_____ 115. coastal areas where the sea washes sediment away faster than it can be supplied, so the coastline migrates landward

_____ 116. coastal areas where more sediment is received than is eroded away, so the coastline migrates seaward and broad beaches develop

_____ 117. concrete or stone walls built perpendicular to the shore to prevent beach drift

_____ 118. pairs of walls built to protect harbor entrances

_____ 119. an offshore wall, built parallel or at an angle to a beach, to protect it from wave action

_____ 120. stone or concrete walls built on the landward side of the backshore zone to protect buildings built alongside the beach area

_____ 121. large stone or concrete blocks used to build seawalls

_____ 122. the procedure of increasing sediment supply artificially by trucking or shipping in huge amounts of sand to replenish a beach

_____ 123. coasts in which living organisms control landforms

_____ 124. the process in which coral reefs lose their color and die, possibly due to the death of the organism zooxanthella

Matching Questions

COASTS AND EARTHQUAKES

Label M if the location hosts Many earthquakes; F if the location hosts Few quakes.

_____ 1. east coast of South America
_____ 2. west coast of South America
_____ 3. east coast of North America
_____ 4. west coast of North America
_____ 5. west coast of Central America
_____ 6. west coast of Europe
_____ 7. west coast of Africa
_____ 8. north coast of Africa, along the Mediterranean Sea
_____ 9. any passive continental margin
_____ 10. any active continental margin

OCEAN SALINITY

Ocean salinity ranges between 1 and 4.1% because it's influenced by various factors. Label H if the factor named tends to cause a High salinity or L if the factor contributes to a Low salinity.

_____ 1. location near a river mouth
_____ 2. within a restricted sea
_____ 3. location at a high latitude with heavy rainfall
_____ 4. location in a polar area with much frozen water
_____ 5. location near the equator

Short-Answer Questions

Answer with a few words, numbers, or short sentences, or choose from the words in parentheses.

1. What major feature exists because 70% of Earth's lithosphere is denser and thinner and lies deeper than the other 30%? _____

2. a. What is *Alvin*? _____
   b. When did it start exploring the oceans? _____

3. a. What is the HMS *Challenger*? _____
   _____
   b. When did it start exploring the oceans? _____

4. What type of crust is formed on the bottom of:
   a. the Mediterranean and the Black Seas? _____
   _____
   b. the North Sea and the northern Red Sea? _____
   _____
   c. the Sea of Japan and the Philippine Sea? _____
   _____

5. a. What is the age of the oldest oceanic crust? _____
   _____
   b. Why isn't there older oceanic crust? _____
   _____

6. Present day ocean basins cover about 70% of Earth's surface, but they haven't always existed. Over the last 2 billion years, about what percent of Earth's surface has been covered by ocean basins? Explain your answer. _____
   _____
   _____

7. What is the mean depth of the ocean floor beneath sea level, in both km and miles (1 km = 0.62 mi)?

8. Briefly explain, in terms of a plate tectonics phenomenon, why active continental margins host numerous earthquakes and passive continental margins don't.

9. Describe the material that sits on top of a continental shelf along a convergent plate boundary. _____

10. What is true about the lithosphere along mid-ocean ridges that makes them sit higher than adjoining ocean floor? _____

11. What happens to the density and elevation of lithosphere as it moves away from the ridge axis?

12. What is the average depth of ocean trenches? _____

13. What is the name, location, and depth of the deepest ocean trench? _____

14. Give two reasons oceanic lithosphere older than 80 million years forms such a flat ocean-bottom surface.

15. a. Why is the layer of pelagic sediment thicker where it's far from the mid-ocean ridge axis? _____

  b. Why is it thicker in equatorial regions than it is in high-latitude waters? _____

16. Why are the largest submarine canyons located offshore of major river channels? _____

17. What is the average salinity of:
  a. freshwater? _____
  b. ocean water? _____
  c. What is the maximum salinity allowable for drinking water? _____

18. Dissolved salt ions fit between water molecules without changing the volume of the water.
  a. What does this do to the density of saltwater as compared to freshwater? _____

  b. How does this affect your ability to float in it?

19. a. Where do the cations (positive ions, like sodium, potassium, calcium, and magnesium) of sea salts come from? _____
  b. Where do the anions (negative ions, like chlorine and sulfate) of sea salts come from? _____

20. Rivers constantly bring salts to the ocean. Why doesn't the salinity of ocean water continue to rise without limit? _____

21. a. If all ocean water were suddenly to evaporate, how thick a salt layer would be left on the ocean floor? _____

  b. What would the components of this layer be? ___

22. About how many different kinds of salts are dissolved in sea water? _____

23. Name two common gases that are dissolved in sea water. _____

24. What is true about the evaporation rates and rainfall amounts in places where the sea is saltiest? _____

25. There are differences in seawater salinity down to about what depth? _____

26. What physical trait of water makes the ocean an excellent moderator of temperatures along coastal regions? _____

27. Supply the temperatures requested:
  a. average *global* sea-surface temperature _____

  b. average sea-surface temperature in polar regions _____

  c. average sea-surface temperature in restricted tropical seas _____
  d. seasonal temperature range of the sea-surface waters in the Tropics _____
  e. seasonal temperature range of the sea-surface waters in temperate latitudes _____
  f. seasonal temperature range of the sea-surface waters near the poles _____
  g. seasonal temperature range on land in central Illinois _____
  h. approximate temperature on the bottom of the sea _____

28. a. What is the average density of seawater at 4°C? _____

  b. What is the average density of freshwater at 4°C? _____

29. a. At what temperature is water densest? _____

b. Why does water get less dense when it gets colder than this? _____

30. a. Which factor has the greater influence on water density, temperature, or salinity? _____
    b. Because of the above influence, where on Earth's surface is seawater the densest? _____

31. In addition to the direction of the prevailing winds, what factor did skippers of sailing ships have to consider when planning their routes from Europe to North America, and why was this necessary? _____

32. Surface currents are produced by the interaction between what two things? _____

33. What Earth motion creates the Coriolis effect? _____

34. What type of flow is created by the merger, at the equator, of gyres of the Northern and Southern Hemispheres? _____

35. What is the Sargasso Sea? _____

36. How long does it take all of the water of the ocean to transfer between the ocean's surface and its floor? _____

37. Despite the fact the Sun is much larger than the Moon, it contributes only 46% as much tide-driving force as does the Moon. Why? _____

38. What two bulges in the global ocean are created by the tide-generating force and where are they located? _____

39. List five factors that influence the timing and reach of tides. _____

40. Waves form because of the interaction between what two groups of molecules? _____

41. What effect does oil have on wave formation, and why does it have this effect? _____

42. What is the basic motion of a particle of water within a wave, and where is this motion at its maximum? _____

43. What two properties of the wind influence the character of the waves produced? _____

44. What is the amplitude of:
    a. a very large hurricane swell? _____
    b. the largest documented swell in the open ocean? _____

45. What is the velocity range of ocean swells? _____

46. What effect do waves have on the ocean floor:
    a. when the floor lies below the wave base? _____
    b. when the wave base just touches the floor? _____

47. If you're swimming and get caught in a rip current, what should you do? _____

48. _____ (Erosion, Deposition) happens at headlands, forming a cliff; _____ (erosion, deposition) happens in embayments, forming beaches.

49. Name a logical source for the following types of sands:
    a. quartz-dominated beach sands of the Atlantic coast of North America _____
    b. carbonate sands _____
    c. black sands _____

50. Despite their small area, wetlands account for 10–30% of marine organic productivity. Why? _____

51. What does an estuary look like on a map? _____

52. How can dredging in an estuary harbor seriously harm the environment? _____

53. Name several geographical locations where you can find fjords. _____

54. Why do you find coral reefs only along clean coasts at latitudes of less than 30°? _____

55. What general area of a coral reef is the only part that is alive? _____

56. What happens to sea level if the number or size of mid-ocean ridges increases, and why is there this effect? _____

57. Why can channeling or damming rivers result in landward migration of the shoreline? _____

58. Name some coastal cities and areas of the United States that are in danger of being flooded if the current rise of sea-level and extraction of groundwater continue. _____

Making Use of Diagrams

Fill in the blanks on the diagrams that follow.

OCEAN BATHYMETRY ADJACENT TO A PASSIVE CONTINENTAL MARGIN

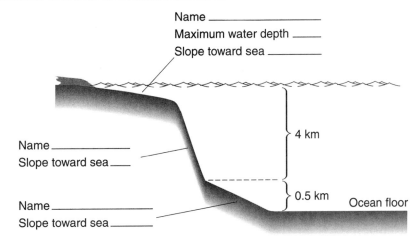

OCEAN BATHYMETRY ADJACENT TO AN ACTIVE CONTINENTAL MARGIN

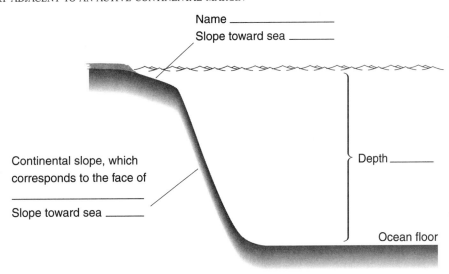

Name for the trough formed along the subducting plate _____

Short-Essay Questions

1. Discuss the origin of a continental shelf along a passive continental margin.
2. Write a paragraph about turbidity currents that includes definitions and a discussion of current velocity, graded beds, submarine fans, and continental rises.
3. Discuss how a subaerial hot-spot volcano becomes a submarine seamount or guyot.
4. Why does high tide arrive 50 min later each day?
5. Explain how upwelling and downwelling zones are created.
6. It is an accepted fact that Earth's rotation is slowing down. What effect does this have on the length of days and years and on the distance between the Moon and the Earth?
7. Most beaches consist of sand. Why don't they also contain silt, mud, boulders, and pebbles?
8. How can something as mild as salt spray and as weak as small organisms in intertidal zones break up rock and knock down cliffs?
9. Discuss the permanency of a sandy beach site.
10. Discuss the similarities and differences between flow patterns and ecologies of quiet estuaries and of turbulent estuaries.
11. Discuss measures people take to improve and protect their beachfront properties.

12. Discuss the details of how human activity has been hard on coastal environments, both wetlands and reefs.

Practice Test

1. Pick out the FALSE statement. Ocean crust
   a. is intermediate to silicic in chemical composition.
   b. is thinner but denser than continental crust.
   c. is composed of gabbro and basalt overlain by sediment.
   d. gets older the farther you go from the ridge axis.
   e. can be no older than 200 million years.

2. Pick out the FALSE statement. Seamounts
   a. develop as hot-spot volcanoes.
   b. that are capped with limestone coral reefs are called guyots.
   c. occur in chains that are progressively older farther from the place of origin.
   d. have steep sides composed of submarine pyroclastic flows.
   e. are on plate material that sinks as it gets older and colder.

3. Pick out the FALSE statement. The abyssal plains
   a. are underlain by cool oceanic lithosphere older than 80 million years.
   b. of the western Atlantic Ocean stretch from the base of the continental rise to the base of the mid-ocean ridge.
   c. are covered by pelagic sediment.
   d. are covered by microscopic plankton shells and fine flakes of brown clay.
   e. are dissected by submarine canyons.

4. All seas
   a. have salinity ranging between 1 and 4.1%.
   b. are underlain by ocean crust.
   c. have similar floors consisting of wide, gently sloping continental shelves, steeper continental slopes, gentle rises, and flat plains.
   d. have yearly temperature fluctuations of about 15°.
   e. have calm bottom waters with no apparent circulation.

5. Trenches
   a. result from sea-floor spreading.
   b. are found at divergent plate boundaries.
   c. border continental and island volcanic arcs.
   d. may be over 100 km (62 mi) deep.
   e. All of the above are true.

6. Pick out the FALSE statement. Continental crust
   a. is more buoyant than ocean crust.
   b. sits higher than ocean crust.
   c. has an average thickness of 35–40 km.
   d. consists of igneous, sedimentary, and metamorphic rocks.
   e. can be no older than 200 million years.

7. Pick out the FALSE statement. Turbidity currents
   a. created the continental slopes by depositing turbidites.
   b. and river erosion create submarine canyons.
   c. travel at a maximum speed of 60 km/h.
   d. leave deposits of graded beds called turbidites.
   e. may deposit overlapping turbidites that form a submarine fan.

8. Pick out the FALSE statement. Ocean water
   a. has an average salinity of 3.5%.
   b. receives its salt from groundwater and river water.
   c. has increased in salinity by about 0.5% each century.
   d. contains positive ions like sodium, potassium, calcium, and magnesium derived from the chemical weathering of rocks.
   e. contains negative ions like chlorine and sulfate derived from volcanic gases.

9. Which location is associated with high salinity of sea water?
   a. near the mouth of a river
   b. in restricted seas which do not mix freely with the main ocean
   c. areas of low evaporation, high rainfall
   d. near the equator
   e. None of the above is an area of high salinity.

10. Pick out the FALSE statement.
    a. Deeper sea water is fairly homogenous worldwide.
    b. Saltier water generally floats above fresher water.
    c. Water from rivers or rain tends to stay in a layer above saltier water for quite a while before it mixes.
    d. The boundary between surface-water salinities and deep-water salinities is called the halocline.
    e. Drinking water should not contain more than 0.05% salt.

11. Surface currents
    a. affect only the upper 30 ft of water.
    b. flow smoothly, with little or no turbulence.
    c. are caused by different salinities at different locations.
    d. are influenced by Earth's rotation.
    e. are found only in low- and mid-latitude areas.

12. Pick out the FALSE statement.
    a. The thermocline is the level below which water temperature drops sharply.
    b. The thermocline occurs at a depth of 300 m in the Tropics but only 100 m in the polar regions.
    c. The depth at which density changes rapidly is called the pycnocline.
    d. Freshwater has a density of 1.000 g/cm^3; saltwater is denser.
    e. Water is at its densest at 4°C.

13. Pick out the FALSE statement. Sea-surface temperature
    a. ranges between freezing and 35°C.
    b. averages 17°C.

c. has greater seasonal variation in the Tropics than in temperate latitudes.
d. varies within a narrow range because of the high heat capacity of water.
e. varies much less than seasonal temperature changes on land.

14. Pick out the FALSE statement. The Coriolis effect
    a. is a phenomenon created by the rotation of Earth.
    b. is a deflection of wind or water flowing over Earth's surface.
    c. causes opposite directions of deflection in the Northern and Southern Hemispheres.
    d. causes north-flowing currents in the Northern Hemisphere to curve to the west.
    e. must be taken into account when aiming artillery shells over long distances.

15. Pick out the FALSE statement. Water that is part of an Ekman spiral
    a. moves at average 90° angle to wind direction.
    b. at a depth of 100 m moves at about a 180° angle to the surface water.
    c. is said to be undergoing Ekman transport.
    d. moves either away from or toward a coast, even if the winds there blow parallel to the coast.
    e. at a depth of 100 m moves about twice as fast as surface flow.

16. Pick out the FALSE statement. Upwelling zones
    a. are areas where water flows in a vertical direction.
    b. result when water deflected in toward the coast creates an oversupply there.
    c. bring nutrients up from depth.
    d. are normal along the coast of South America.
    e. can be driven by density differences caused by temperature and salinity differences.

17. Pick out the TRUE statement.
    a. "Tidal reach" means the farthest inland the water goes at high tide.
    b. A flood tide that arrives as a visible wall of water is called a tidal bore.
    c. The tide-generating force is a combination of only two forces, the gravitational attractions of the Sun and Moon.
    d. Since the Sun is much larger than the Moon, its gravity is the biggest contributing factor to the tide-generating force.
    e. Ebb tide causes the shoreline to migrate inland.

18. What force causes the tidal bulge on the side of the Earth opposite the Moon?
    a. the centrifugal force due to Earth's rotation
    b. the centripetal force due to Earth's rotation
    c. the Sun's gravitational attraction
    d. the Moon's gravitational attraction
    e. the Sun's and the Moon's gravitational attractions combined

19. In the accompanying diagram, which letter(s) designate(s) the Moon's position(s) for neap tides?

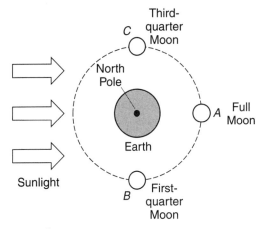

    a. *A* only
    b. *B* only
    c. *A* and *C*
    d. *B* and *C*
    e. All of them could be neap tide situations, depending on the month.

20. Pick out the FALSE statement.
    a. A particle of water in a wave is going in a circular motion as viewed in cross section.
    b. Submarines traveling deeper than about half the prevailing wavelength in an area would be experiencing smooth water.
    c. Interference of comparable-sized swells tends to cancel them out and produce smooth water.
    d. Waves form because of shear between molecules of air in the wind and molecules of surface water.
    e. The character of waves in open seas depends on the strength and fetch of the wind.

21. Pick out the appropriate pair of terms to match this statement: A surfer loves these; a swimmer hates these.
    a. swash, longshore currents
    b. rip currents, backwash
    c. breakers, rip currents
    d. embayments, tombolos
    e. eustatic changes, swash

22. Which of the following are NOT living organisms?
    a. mangroves
    b. sargassum
    c. cnidarians
    d. polyps
    e. groins

23. Choose the proper order to name the following types of coastal wetlands: those that are dominated by grasses, by moss and shrubs, and by trees.
    a. swamps, bogs, marshes
    b. marshes, bogs, swamps
    c. bogs, marshes, swamps
    d. swamps, marshes, bogs

e. The classification is outdated; any name will do for any type of dominating vegetation.

24. Pick out the FALSE statement. Estuaries
    a. are valleys flooded by seawater.
    b. have brackish water (brine) in them.
    c. originated as glacial valleys.
    d. are recognized on maps by their jagged coastlines.
    e. are found in the Chesapeake Bay; Hudson River, New York; and Columbia River, Oregon.

25. Pick out the FALSE statement.
    a. Coral reefs are complex environments consisting of colonies of cnidarians, zooxanthellae algae, shell debris, and associated organisms.
    b. An individual coral animal is a polyp.
    c. Corals grow only in clear, well-lit, warm water with normal oceanic salinity.
    d. Coral animals are tiny invertebrates that secrete silicic shells.
    e. Coral reefs are classified on the basis of their shapes, fringing, barrier, or atoll.

26. What term has nothing to do with protecting beaches from erosion?
    a. "tombolos"
    b. "groins"
    c. "breakwaters"
    d. "seawalls"
    e. "jetties"

27. *Alvin* and the HMS *Challenger* are deep-sea submersibles used to explore the ocean bottom.
    a. True
    b. False

28. Pulses of sound reflected off subsurface layers create a seismic-reflection profile that shows ocean crust structure.
    a. True
    b. False

29. *Alvin* can dive to depths of more than 4 km (2+ mi).
    a. True
    b. False

30. Even small seas, like the North Sea, Red Sea, and the Sea of Japan are underlain by oceanic crust.
    a. True
    b. False

31. The mean depth of ocean water is 3.75 km.
    a. True
    b. False

32. No ocean crust is older than 200 million years because by that time it has been destroyed by subduction.
    a. True
    b. False

33. The locations of ocean basins have remained the same for the past 2 billion years, but basin sizes have changed, so the percentage of ocean crust on the Earth has varied.
    a. True
    b. False

34. The continental slope descends at a steeper angle than either the continental shelf or the continental rise, but even so its slope is only a few degrees.
    a. True
    b. False

35. Saltwater is denser than freshwater, so objects (including you) float higher in it.
    a. True
    b. False

36. If all ocean water were to evaporate, it would leave behind a 6-m-thick layer on the ocean floor.
    a. True
    b. False

37. The freezing or evaporating of ocean water produces salty ice or salty water vapor.
    a. True
    b. False

38. Wind shear across surface waters initiates an Ekman spiral; shear between successively deeper water layers keeps the spiral going.
    a. True
    b. False

39. The Sargasso Sea, located at the center of the North Atlantic gyre, is an area of rapidly rotating surface water and tangled sargassum seaweed.
    a. True
    b. False

40. Cold, salty water sinks; warm, less salty water rises. This phenomenon is called thermohaline circulation.
    a. True
    b. False

41. On average, it takes about 1 million years for all water in the ocean to transfer between the surface and the floor.
    a. True
    b. False

42. During an El Niño event there are increased north winds, increased northwest currents, and increased upwelling in the Peruvian coastal waters, all of which changes atmospheric circulation and alters global weather patterns.
    a. True
    b. False

43. A combination of surface currents and vertical thermohaline circulation acts like a conveyor belt and moves water and heat among ocean basins and from ocean surface to ocean floor.
    a. True
    b. False

44. Spring tides are extra high tides that occur when the Sun, the Moon, and Earth are aligned; they are at new- and full-moon times.
    a. True
    b. False

45. At any particular location, high tide arrives 50 min later each day because it takes Earth 24 h and 50 min to catch up with the Moon in its orbit and be back in high-tide position.
    a. True
    b. False

46. Beaches usually consist of sand grains because waves winnow out finer sediment and carry it to deeper water, and break larger pieces down to sand-grain size.
    a. True
    b. False

47. Bioturbation is an oceanic worm that stirs up sediment on the abyssal plains.
    a. True
    b. False

48. Embayments are river valleys flooded by seawater.
    a. True
    b. False

49. Glacial rebound causes glaciers to advance toward the ocean and produce eustatic sea-level changes.
    a. True
    b. False

50. Property owners may try beach nourishment if their land is on an accretionary coast.
    a. True
    b. False

ANSWERS

Completion

1. oceanography
2. marine geology
3. marine biology
4. bathymetric maps
5. seismic-reflection profile
6. coast
7. bathymetry
8. bathymetric provinces
9. continental shelf
10. continental slope
11. continental rise
12. passive continental margins
13. active continental margins
14. mid-ocean ridge
15. trench
16. fracture zones
17. abyssal plains
18. pelagic sediment
19. submarine canyons
20. turbidity currents
21. turbidites
22. submarine fan
23. seamount
24. guyot
25. salinity
26. halocline
27. heat capacity
28. thermocline
29. pycnocline
30. currents
31. surface currents
32. deep currents
33. eddies
34. Coriolis effect
35. Ekman spiral
36. Ekman transport
37. gyres
38. sargassum
39. downwelling zones
40. upwelling zones
41. El Niño
42. thermohaline circulation
43. Antarctic bottom-water mass
44. North Atlantic deep-water mass
45. tide
46. tidal reach
47. mean sea level
48. flood tide
49. ebb tide
50. shoreline
51. tidal flat
52. tidal bore
53. intertidal zone
54. tide-generating force
55. neap tides
56. spring tides
57. storm surge
58. wave base
59. strength of the wind
60. fetch of the wind
61. ripples
62. amplitude
63. wavelength
64. swells
65. interference
66. breaker
67. surf zone
68. swash
69. backwash
70. wave refraction
71. longshore current
72. headlands
73. embayments
74. rip currents
75. beach
76. beach profile
77. beach face
78. backshore zone
79. berms
80. active sand
81. inactive sand
82. beach drift
83. sand spit
84. baymouth bar
85. offshore bars
86. barrier islands
87. lagoon
88. bioturbation
89. sediment budget
90. wave erosion
91. wave-cut notch
92. cliff retreat
93. wave-cut bench (or wave-cut platform)
94. honeycomb-weathering pattern
95. sea arch
96. sea stack
97. tombolo
98. coastal wetland
99. swamps
100. marshes
101. bogs
102. estuary
103. flocculate
104. fjords
105. cnidarians
106. polyp
107. coral reef
108. fringing reef
109. barrier reef
110. atoll
111. coastal plain
112. glacial rebound
113. eustatic sea-level change
114. emergent coasts
115. erosional coasts
116. accretionary coasts
117. groins
118. jetties
119. breakwater
120. seawalls
121. riprap
122. beach nourishment
123. organic coasts
124. reef bleaching

Matching Questions

COASTS AND EARTHQUAKES

1. F
2. M
3. F
4. M
5. M
6. F
7. F
8. M
9. F
10. M

OCEAN SALINITY

1. L
2. H
3. L
4. H
5. L

Short-Answer Questions

1. ocean basins

2. a. a deep-sea research submersible   b. in the 1970s

3. a. a British Navy ship converted to do ocean research
   b. 1872

4. a. oceanic crust trapped between colliding continents
   b. rifted continental crust
   c. back-arc basins formed on the opposite side of a volcanic island arc from the trench

5. a. 200 million years   b. It's been subducted.

6. The percent of Earth's surface that's underlain by oceanic crust has remained about the same, 70%. The ocean basin boundaries have changed.

7. 3.75 km; about 2.3 mi

8. The convergence of plates with resulting oceanic plate subduction produces earthquakes; passive continental margins don't have subduction.

9. an apron of sediment spread out over the top of an accretionary prism (material scraped off the downgoing subducting plate)

10. It's hotter and less dense.

11. It gets denser and sinks.

12. 8 km

13. Mariana Trench, in the western Pacific, 11,035 m deep

14. It's cooling so slowly that its density increase and subsequent sinking is almost undetectable.
    It's blanketed by pelagic sediment, which covers any irregularities.

15. a. The farther ocean floor is from the ridge, the more time there's been for pelagic sediment to accumulate on it.
    b. Warm, well-lit equatorial waters produce more plankton to contribute shells to the sediment layer.

16. The rivers cut into the continental shelf at times when sea level was lower and the shelf was exposed.

17. a. 0.02% dissolved salt ions   b. 3.5% dissolved salt ions   c. 0.05% dissolved salt ions

18. a. Increases the density.   b. You (or any object) will float easier and higher.

19. a. the chemical weathering of rocks   b. volcanic gases

20. As concentration increases, salt precipitates out to form solid mineral crystals. The amount of salt added by rivers is balanced by the amount removed by mineral crystallization.

21. a. 60-m   b. 75% halite (sodium chloride), lesser amounts of gypsum (calcium sulfate, hydrated) and anhydrite (calcium sulfate without water), and traces of other salts

22. 70

23. oxygen and carbon dioxide

24. There are high evaporation rates and minimal rainfall.

25. 1 km

26. its high heat capacity

27. a. 17°C   b. freezing   c. 35°C   d. 2°C   e. 8°C
    f. 2°C   g. 70°C (from a high of 40°C to a low of −30°C)   h. freezing

28. a. 1.025 g/cm³   b. 1.000 g/cm³

29. a. 4°C   b. because ice crystals form and take up more space that the liquid water molecules did; therefore the volume increases while the amount of mass stays the same

30. a. temperature   b. at the poles

31. directions of surface currents, because sailing against a current slowed the voyage considerably

32. sea surface and the wind

33. Earth's rotation

34. an equatorial westward flow

35. the center of the North American gyre, where the noncirculating waters allow the accumulation of the tropical seaweed sargassum

36. about 1,000 years

37. because the Moon is much closer to Earth

38. the sublunar bulge, on the side of Earth closer to the Moon; the secondary bulge, on the opposite side of Earth

39. Earth's tilt; the track of the Moon's orbit over Earth; the Sun's gravitational pull on Earth, which varies with the relative positions of the Sun and Moon; the shape of the ocean basins; and the weather (high- and low-pressure masses)

40. molecules of air in the wind and molecules of water in the sea's surface

41. Oil inhibits wave formation because it reduces the frictional shear between air and water.

42. circular; at the ocean's surface

43. strength and fetch of the wind

44. a. over 25 m   b. over 35 m

45. 20–90 km/h

46. a. none   b. Waves cause a slight back-and-forth motion that builds small ripples.

47. Swim parallel to the shore until you're out of the current; then let the waves carry you back in to shore.

48. erosion; deposition

49. a. glacial deposits eroded from metamorphic and igneous rocks of Canada and northern New England
    b. limestone or recent corals and shell beds
    c. basalt

50. because many marine species spawn in wetlands

51. The shoreline is jagged, like a river pattern.

52. It may stir up organic chemicals like polychlorinated diphenyls (PCBs), which are dangerous pollutants.

53. on the coasts of Norway, British Columbia, New Zealand, Maine, and southeastern Canada

54. Corals require clear, well-lit, warm (18–30°C) water with normal oceanic salinity.

55. only the top surface, which lies just below the low-tide level

56. The sea level rises; the increased volume of the ridge displaces the water.

57. The amount of sediment that's brought to the shore by a dammed or channeled river is decreased, so sediment may erode away faster than it is being supplied.

58. Washington, D.C.; New York; Philadelphia; and the Texas and Louisiana coasts

Making Use of Diagrams

OCEAN BATHYMETRY ADJACENT TO A PASSIVE CONTINENTAL MARGIN

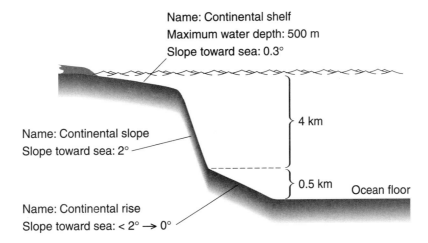

OCEAN BATHYMETRY ADJACENT TO AN ACTIVE CONTINENTAL MARGIN

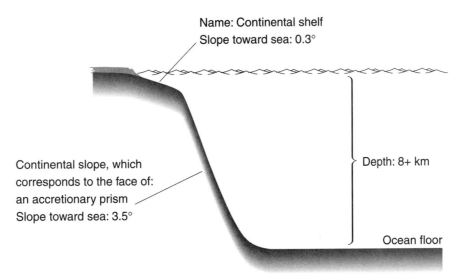

Name for the trough formed along the subducting plate: Trench

Short-Essay Questions

1. Rifting splits a continent apart, then ceases. The stretched crust at the boundary between ocean and continent gradually cools, sinks below sea level, and accumulates sediment washed off the continent and shells of marine life that die and sink in the ocean waters. This layer of sediment, which can get up to 18 km deep, is the surface of the continental shelf.

2. Turbidity currents are avalanches of sediment and water that race down the slopes of submarine canyons and cause erosion as deep as 1,000 m. The currents can move at speeds up to 60 km/h. At the base of the continental slope they deposit their sediment, coarsest first (graded beds) which over time build up into triangular aprons called submarine fans. Overlapping fans create the continental rise of passive continental margins.

3. A hot-spot volcano forms as an island above a mantle plume. When plate movement carries it off the plume, volcanic activity stops and erosion takes over. The plate carrying the volcano and the volcano itself cool and sink with age, until the volcano becomes a submerged peak (a seamount) or, if coral reefs grow on top, a flat, limestone-capped guyot.

4. Earth spins on its axis once every 24 h, and while it's doing this, the Moon advances a days' worth (roughly 1/28 of a circle) along its orbital path around Earth. Because of the Moon's advance, when Earth's day is done, the Moon is no longer in the same position relative to Earth as it was 24 h earlier, and it takes Earth 50 min to catch up and achieve that position. Since tides depend mainly on the Moon's gravitational pull, and that is tempered by Moon-Earth positions relative to each other, high-tide position (or any other tidal phase) is also 50 min later each passing day.

5. Wind blowing parallel to a coastline drags surface water along. If this water is deflected toward the coast by the Coriolis effect, the result is an oversupply of water along the shore, and excess water sinks (downwelling). If the Coriolis effect happens to deflect surface water away from the coast, the resulting deficit of coastal water is compensated for by water that rises to fill the gap (upwelling).

6. Daily collisions between tidal bulges and continental edges, friction between moving waters in the ocean, and the slight pull of the Moon on the side of tidal bulges all act as a brake on rotating Earth and slow it down 0.002 s/century. These factors have no braking effect on the speed of Earth's revolution around the Sun, so the length of a year stays the same. Over millions of years the tiny increase each century has added up significantly. In mid-Devonian times, about 375 million years ago, days were only 21.9 h long, so there was time for 400 complete rotations (days) per year. Just as a spinning ice skater slows down as he stretches his arms outward, Earth's slowing has been accompanied by the Moon's gradual moving outward (away) from Earth at the rate of 4 cm/year. Over time this small amount has added up significantly. Back in mid-Devonian times, when Earth was spinning so much faster, the Moon was about 1,500 km closer to Earth than it is today, and long before that, 3.9 billion years ago, it may have been over 15,000 km closer.

7. Waves winnow out finer sediment, like silt and mud, and carry it farther out to sea, to quiet waters where it settles. Boulder and pebble beaches form only where cliffs are nearby and can keep supplying these larger particles. A constant supply is necessary because storm waves smash boulders and pebbles together and break them down to smaller (sand) size.

8. The evaporation of ocean spray water leaves behind salt, which deposits in the pore spaces in the rock. The salt crystals grow, push apart the rock grains, and thus weaken the rock, all of which is a process called salt wedging. Plants and animals also weaken rock by boring into it. By weakening rocks, both salt wedging and plant and animal activity set them up for future extensive weathering and erosion.

9. Sandy beach sites are definitely not permanent features; instead, they're in a state of constant change. Local rivers, wind, offshore waves, and the process of beach drift may supply new sand, or offshore waves and beach drift may remove sand and redeposit it locally or far away in submarine canyons. Local events, such as the exhaustion of glacial moraine sand, or seasonal changes, such as alternating strong winter storms and milder summer ones, affect the amount of sand available. The relationship between the amount of incoming sand and the amount of outgoing sand is termed the sediment budget of a beach. As the budget changes, beaches may grow wider or narrower or disappear completely.

10. Both quiet and turbulent estuaries have jagged coastlines and a mix of seawater and freshwater. In quiet estuaries, such as on the Hudson River of New York or the Columbia River of Oregon, the water becomes stratified, and the denser oceanic saltwater flows as a wedge underneath the less dense fluvial (river) freshwater. Fresh river water always flows downstream; saltwater may flow either upstream or downstream, depending on whether the tide is rising or falling. In turbulent estuaries, like the Chesapeake Bay, the salt- and freshwater combine to create a nutrient-rich brine (brackish water). All estuaries support complex ecosystems that include unique species of shrimp, clams, oysters, worms, and fish that can tolerate large changes in salinity. All estuaries tend to encourage the settling

out of clay that had remained suspended in fresh river water but clumps (flocculates) and thus becomes big and heavy enough to sink in brackish water and create mudflats and salt marshes.

11. Property owners along beachfronts construct artificial barriers to protect against wave action and beach drift. The barriers may be groins (concrete or stone walls perpendicular to the shoreline), jetties (paired walls bracketing harbor entrances), breakwaters (offshore walls), or seawalls (walls on the landward side of the backshore zone). There are always results and sand does get redistributed, but often the results are unexpected and what aids one property owner may hurt his neighbor. Therefore some people, rather than trying to stop erosion, have resorted to importing (by truck or by ship) huge amounts of sand to replenish their beaches. This is called beach nourishment. All methods are controversial, and whichever human method is employed, every beach eventually changes in response to natural events.

12. Human activity has destroyed between 20 and 70% of coastal wetlands worldwide. The exact causes of damage are numerous. Sewage, chemical pollutants, and agricultural runoff have contaminated the habitat of burrowing marine life such as oysters, and damaging effects have moved up the food chain. The dredging of navigation channels has stirred up contaminants and dispersed them over large areas. Fertilizers and sewage have increased the nutrient content of water, which in turn supports runaway algal growth (algal blooms) that robs water of oxygen needed for indigenous organisms. Wetlands have been drained to create dry land for both farming and cities or to supply irrigation water. Some wetlands have been made into garbage dumps. Human activity has been just as tough on reefs, where delicate coral polyps are easily destroyed by even slight environmental changes. Pollutants, hydrocarbons, organic sewage, suspended sediment, agricultural runoff, and wastewater have all taken their toll by direct poisoning and suffocating of coral organisms or by temperature and salinity changes that indirectly kill the polyps. Direct damage has been done by touching reefs, by dragging anchors across them, and by quarrying them for use as construction material.

Practice Test

1. a Ocean crust is mafic (rich in magnesium and iron) chemical composition.

2. d They have gently sloping (< 4°) slopes of basalt.

3. e Submarine canyons dissect the continental shelves and slopes.

4. a Ocean floor can be continental crust in rifted areas or back-arc basins. Sea-floor bathymetries vary greatly. Yearly temperature fluctuations vary between 2 and 8°C. There are bottom currents.

5. c Trenches are found at convergent ocean-continent boundaries, and the deepest is about 11 km (6+ mi) deep.

6. e It is ocean crust that can be no older than 200 million years, because by then it has subducted.

7. a Turbidites are deposited by turbidity currents that have raced down the continental slopes; they do not form the slopes.

8. c Ocean water salinity stays fairly constant because the amount of salt continuously added balances the amount of salt that precipitates out.

9. b Restricted seas would experience high evaporation rates and little influx of new waters. Evaporation leaves salt behind.

10. b Freshwater is less dense so it floats on saltwater.

11. d They affect the upper 100 m, commonly have eddies, are caused by the interaction of wind and surface water, and are found worldwide.

12. b There is no pronounced thermocline in polar seas; the water there is all very cold.

13. c The opposite is true; there is a 2°C seasonal change in the Tropics and an 8°C change in temperate regions.

14. d They curve to the east.

15. e The water's rate slows with increasing depth and at 100 m is less than 5% of the rate of surface flow.

16. b Upwelling zones result when water is deflected away from the coast, and deep water rises to fill the deficit created.

17. b Tidal reach is the difference between high and low tides. Centrifugal force caused by Earth's rotation contributes to tides. The Sun is so distant it accounts for only 46% of the force. Flood tide causes the shoreline to migrate inland.

18. a The centrifugal force due to Earth's rotation causes the tidal bulge.

19. d At quarter-moon times, the gravitational forces of the Sun and Moon are acting on Earth at right angles, so each partially offsets the influence of the other.

20. c Interference would create larger waves.

21. c Breakers are the waves that curve over the base which surfers ride; rip currents are strong, localized seaward flows that trap and drown swimmers.

22. e Groins are concrete or stone walls that protrude perpendicularly to the shore and are built to prevent

beach drift. The other choices, in order, are trees, seaweed, tiny invertebrates that form coral reefs, and the individual coral animal.

23. b Only choice b presents the names in the proper order. The classification is not outdated.

24. c Flooded glacial valleys are called fjords.

25. d Coral animals secrete calcite shells, so the coral mounds are limestone.

26. a Tombolos are narrow ridges of sand that link sea stacks to the mainland.

27. False *Alvin* is a submersible, but the HMS *Challenger* was a British Navy ship that did ocean research as early as the 1870s.

28. True Reflections of sound waves off subsurface layers do create profiles that show ocean crust structure.

29. True The deep-sea submersible *Alvin* can dive deeper than 4 km (2+ mi).

30. False All of these are underlain by continental crust of rift valleys or back-arc basins.

31. True The measurement 3.75 km is the mean depth of ocean water.

32. True All ocean crust subducts and therefore is destroyed by the time it is 200 million years old.

33. False Just the opposite is true; basins have changed, but the percentage has remained almost constant.

34. True The continental slope is steeper than the continental shelf and the continental rise, but its slope is still just a few degrees.

35. True Any object floats higher in saltwater than it does in freshwater, because saltwater is denser than freshwater.

36. False The layer would be 60 m thick.

37. False Both processes leave the salt behind and produce freshwater ice or vapor.

38. True The Ekman spiral is initiated by wind shear and perpetuated by shear between water layers.

39. False The location is properly stated, but the waters are noncirculating and have therefore allowed the seaweed to accumulate.

40. True Thermohaline circulation is the flow pattern in which cold, salty water sinks and warm, freshwater rises.

41. False The average time required is only about 1,000 years.

42. False All of these flows decrease in strength, and thus cause the changes we call El Niño.

43. True Water and heat move horizontally and vertically through the ocean because of surface currents and thermohaline circulation.

44. True Extra-high spring tides occur when the Sun, the Moon, and Earth are aligned.

45. True High tide is 50 min later each day because it takes Earth 24 h plus 50 min to reach high-tide position.

46. True Waves move fine sediment from beaches to deep water and break large pieces down, so beaches usually consist of only sand-sized particles.

47. False Bioturbation is the process in which organisms ("bio") like clams and worms stir up tidal flat sediments.

48. False The definition applies to estuaries.

49. False Glacial rebound is the rise of land after the ice has melted and its weight no longer depresses the surface. Eustatic sea-level changes simply are worldwide sea-level changes.

50. False Beach nourishment (trucking or shipping in huge amounts of sand) may be tried if the properties are on an erosional coast, where the sea washes sediment away faster than it can be naturally supplied.

CHAPTER 19 | A Hidden Reserve: Groundwater

GUIDE TO READING

Chapter 19 continues the study of the compound that makes Earth unique in our solar system, liquid water. As explained in the previous two chapters, the greatest amount of Earth's water is salty and is found in oceans (Chapter 18), and the obvious freshwater supply flows on the surface in streams and lakes (Chapter 17). A much larger quantity of freshwater, hidden below the surface and rarely seen, is the subject of this chapter.

Your author begins by reminding you water moves between various reservoirs on, above, and under Earth's surface as part of a phenomenon known as the hydrologic cycle. Most groundwater begins as moisture that falls on land and infiltrates into the ground. Some remains close to the surface as soil moisture; more percolates downward through narrow, crooked channels in the bedrock. The reason water can exist and travel through "solid rock" involves a discussion of pore spaces, porosity (primary and secondary), permeability, and mobile and immobile water. A rock layer is classified as an aquifer (confined or unconfined) or as an aquiclude (or aquitard) on the basis of its porosity and permeability (that is, its ability to hold and conduct water).

Freshwater is a vital natural resource. To successfully and efficiently tap a groundwater supply, hydrologists study geochemistry and subsurface structure in the potential resource area. They are concerned with the ion content of the water (saturated, unsaturated, and oversaturated solutions; precipitates; and hard water). They try to discern where the unsaturated zone (or zone of aeration), capillary fringe, water table, and saturated zone lie. An ordinary well must be drilled down past the first three layers and into the fourth if is to become a dependable, long-term water producer and not just a seasonal well or a dry well. They identify special features like perched water tables and artesian wells (flowing and nonflowing) and special regions like recharge and discharge areas, so they can be used advantageously and not create problems. They consider the head of the water table, the hydraulic gradient and hydraulic conductivity of a region, and the effects of local drawdown and cones of depression when deciding how much groundwater an area will yield. They use the formula called Darcy's law to calculate the discharge figure. Sometimes hydrologists don't have to search for groundwater; it reaches the surface on its own. You read about several types of geologic settings that yield springs and about the special setting needed to produce an artesian spring.

Geothermal areas offer unique groundwater-caused surface features including hot springs, mud pots, and geysers that serve sometimes as tourist or recreational attractions and sometimes as energy sources.

There are problems associated with groundwater usage, like water depletion. Are we carelessly exhausting a nonrenewable resource and ruining land areas in the process? Your author uses several examples in discussing this complex issue. In many locations it is true that:

- Mining of groundwater is occurring (withdrawal faster than natural recharge), so the water table is being lowered and the flow direction of groundwater is being reversed.
- Saline intrusion is causing wells to yield useless salty water.
- Irreversible pore collapse and land subsidence are occurring.

Groundwater contamination is another problem, one that's especially serious because it's so hard to clean up once it's occurred. Society is growing increasingly concerned about putting any wastes in the ground that might pose a threat to local groundwater. You read about specific issues (like putting nuclear wastes into welded tuff at Yucca Mountain,

Nevada) and about general matters like categories of contaminants, terms for measurement (parts per million, PPM), injection wells, contaminant plumes, and bioremediation.

The chapter concludes on a lighter note, with a discussion of unusual karst landscapes (with sinkholes, natural bridges, disappearing streams, and tower karst), and the fascinating world spelunkers explore, caves. Since groundwater is naturally slightly acidic, it dissolves rock (chiefly limestone) to produce cave networks of passages and chambers that sometimes, due to the precipitation of limestone out of solution, get decorated with dripstone speleothems (soda straws, stalactites, stalagmites, and limestone columns) and flowstone and populated by unusual life forms that have adapted to life away from light.

In review, Earth's subsurface region is, like its surface region, a world of water, but in a subtle way. There are few streams that disappear into the ground and flow there, but there is a great quantity of water moving slowly underground through connecting microscopic pores, transporting materials in solution, indirectly sculpting the land, and providing fresh water for human needs when we're wise enough to figure out where and how to tap into the supply.

Completion

Test your recall of new vocabulary terms.

_____ 1. the water that resides under the surface of Earth in cracks and between grains of sediment or rock

_____ 2. a circular depression on Earth's surface caused by collapse into underground caverns

_____ 3. the movement of water between various reservoirs (the ocean, atmosphere, rivers, lakes, cracks and pores in soil and rock, glaciers, and living organisms) on Earth's surface and above and below it

_____ 4. the process of precipitation sinking into the ground and occupying cracks and pore spaces there

_____ 5. the groundwater that remains just below Earth's surface and occupies the space between soil grains and organic material there

_____ 6. any open space in sediment or rock

_____ 7. the total volume of empty space in a material, expressed as a percentage

_____ 8. the percentage of space that remains between solid grains or crystals immediately after sedimentation or rock formation

_____ 9. new pore spaces formed in existing rocks as they are affected by faulting and groundwater flow

_____ 10. the water that doesn't move through pore spaces because it's held in place by electrostatic attraction to mineral surfaces

_____ 11. the groundwater that flows freely through rock and soil

_____ 12. the quality of a material that allows water to flow easily through it

_____ 13. the quality of a material that allows water to flow slowly or not at all through it

_____ 14. the process in which a liquid meanders through tiny, crooked channels

_____ 15. geologists who study groundwater

_____ 16. layers of permeable rock or sediment through which water easily flows

_____ 17. layers of rock or sediment that are not very permeable and therefore restrict the motion of water through them

_____ 18. the rock or sediment unit that serves as the primary source of groundwater in an area

_____ 19. aquifers that intersect Earth's surface and thus can directly receive water from it or release water to it

_____ 20. aquifers that are separated from Earth's surface by aquitards (aquicludes) and thus can't exchange water with the surface

_____ 21. the boundary between the zone of aeration and the zone of saturation

_____ 22. the region of the subsurface in which pore spaces contain mostly air

_____ 23. the region of the subsurface in which pore spaces contain only water

_____ 24. the thin layer between saturated and unsaturated zones in which water seeps upward from the water table due to electrostatic attraction between water molecules and mineral surfaces

_____ 25. the elevation of the surface of the water table above a reference horizon

_____ 26. a quantity of groundwater that lies above the regional water table because it's underlain by a lens of impermeable material

_____ 27. a location where water enters the ground and flows downward

_____ 28. a location where groundwater flows back up to the surface

_____ 29. the slope of the water table

_____ 30. a number used as a coefficient in an equation that calculates discharge and based on the permeability of the ground and the viscosity of the fluid

_____ 31. an equation that calculates the volume of water that passes through a specified area in a given time (the discharge)

_____ 32. an expression that describes groundwater that is carrying as many ions as possible under the existing environmental conditions

_____ 33. an expression that describes groundwater that has the capacity to dissolve more ions

_____ 34. an expression that describes groundwater that temporarily holds more ions than environmental conditions dictate it should

_____ 35. the process in which ions in solution combine to form solids, which come out of solution

_____ 36. the groundwater that, due to its passage through limestone or dolomite, contains large amounts of calcium and magnesium ions

_____ 37. holes people dig or drill to obtain water

_____ 38. natural outlets from which groundwater flows

_____ 39. a well drilled below the water table into an aquifer, the water surface of which is the water table

_____ 40. a well that cannot supply water because it's drilled into an aquitard or into rock that lies above the water table

_____ 41. ordinary wells that yield water only during rainy seasons, when the water table rises above the level of the base of the well

_____ 42. the process in which the water table sinks down around a well as water is pumped out faster than it can be replaced

_____ 43. the downward-pointing, cone-shaped surface of the water table around a well where water is being pumped out faster than it can be replaced

_____ 44. a well in which water rises, without pumping, to a level above the level of the local aquifer, because water pressure has built up in a confined aquifer that lies beneath a sloping aquitard

_____ 45. the level to which water has the potential to rise, due to the pressure of its own weight, in an artesian situation

_____ 46. a place where the ground surface intersects a joint that taps a confined aquifer, and the water pressure is great enough to push groundwater through the joint up to the surface

_____ 47. springs that yield water with temperatures between 30 and 104°C

_____ 48. areas where, because of current or recent volcanism, magma or very hot rock lies close to the surface

_____ 49. the geologic feature formed where hot water rises through soil that is rich in volcanic ash and clay and produces a bubbling, viscous slurry

_____ 50. a fountain of steam and hot water that erupts periodically from a vent in the ground

_____ 51. wells in which liquid is pumped down with enough pressure to cause it to leave the well and occupy pore spaces in the surrounding soil or rock

_____ 52. the mass of contaminated groundwater that moves away from the source of contamination

_____ 53. the technique of injecting oxygen and nutrients into a contaminated aquifer to encourage bacterial growth that can break down the molecules of the contaminant

_____ 54. the rock formed by precipitation of calcium carbonate (limestone) out of water in the upper reaches of caves

_____ 55. variously shaped formations that grow in caves due to the accumulation of dripstone

_____ 56. people who explore caves

_____ 57. a dripstone feature that looks like a hanging icicle

_____ 58. a delicate tubular structure that is the initial step in stalactite formation

_____ 59. an upward-pointing cone that forms where drips hit the cave floor

_____ 60. the feature that results when a stalagmite merges with an overhanging stalactite

_____ 61. broad sheets of limestone that form on cave walls due to precipitation from groundwater

_____ 62. surface water that disappears down a crack or hole, flows through a cavern as an underground stream, and emerges from the cavern entrance downstream

_____ 63. a region of numerous round sinkholes separated by round hills or steep spires, formed because the bedrock limestone is underlain by a network of caverns

_____ 64. a version of karst landscape that forms where vertical joints control the patterns of surface collapse

Matching Questions

HOT SPRINGS

Match each hot springs location with the geologic reason for its existence.

a. Hot Springs, Arkansas
b. Iceland
c. Rotorua, New Zealand
d. Salton Sea area, California
e. Yellowstone

_____ 1. the water that reaches the surface through cracks in a caldera, located above the magma chamber of a continental hotspot

_____ 2. the merging point of the San Andreas Fault with a mid-ocean ridge

_____ 3. the deep groundwater, not associated with a volcanic field, which is hot simply because of its depth

_____ 4. an island that's an oceanic hot spot on the Mid-Atlantic Ridge

_____ 5. an active volcanic field above a subduction zone

Short-Answer Questions

Answer with a word, number, phrase, or short sentence.

1. What geologic feature was responsible for the sudden disappearance of trees, houses, cars, and even a swimming pool in Winter Park, Florida, in 1981? _____

2. Although groundwater is usually hidden, there are some places where we can see it. List three. _____

3. Name several things that can happen to precipitation that doesn't infiltrate the ground it falls on. _____

4. State two factors that influence how much water infiltrates the ground in any specific location. _____

5. State two ways soil moisture returns to the atmosphere. _____

6. How long does groundwater remain in the ground before returning to the surface? _____

7. What fraction of the world's freshwater supply is groundwater? _____

8. a. What is the origin of most groundwater? _____

   b. How can seawater account for some of the groundwater that underlies land? _____

9. Refer to Table 19.1 in the text to answer the following:
   a. What geologic material has the lowest porosity, and what is it? _____
   b. What is the percentage difference possible between unfractured marble and highly fractured marble? _____
   c. Which can be more porous, uncompacted mud or unconsolidated sand? _____

10. Why is overall porosity of poorly sorted sediment less than that of well-sorted sediment? _____

11. Why is the primary porosity of igneous and metamorphic rock usually small? _____

12. Name three ways secondary porosity can develop. _____

13. What characteristic of pore spaces in rock would assure that all groundwater in that rock would be immobile? _____

14. Name three factors that affect the permeability of a material. _____

15. In what type of location is the water table
    a. right at the surface? _____
    b. within a few meters of ground surface? _____
    c. tens of meters below ground surface? _____

16. Why is the bottom of the water table never deeper than about 10–15 km below the surface? _____

17. Groundwater in the zone of aeration moves in what direction, and why? _____

18. What causes pressure on groundwater that's in the zone of saturation of an unconfined aquifer? _____

19. a. How fast does groundwater move? _____
    b. Why does it move so slowly? _____

20. What are the two major factors that influence the rate of flow of groundwater? _____

21. What mineral, dissolved in groundwater, precipitates out in the cells of wood to create petrified wood? _____

22. Name two problems people have to deal with when they use hard water. _____

23. A city water supply in which gravity distributes water stored in a high tank is imitating what natural water-producing situation? _____

24. Describe two geologic settings where hot springs are found. _____

25. Why can hot springs serve as "mineral baths," but you can't take an ordinary spring, heat its water, and have a "mineral bath"? _____

26. What causes the brilliant green, blue, and orange colors of hot pools? _____

27. What is the structural difference between the "plumbing system" of geysers and the "plumbing system" of hot springs that causes only geysers to erupt? _____

28. What is meant by the expression "mining groundwater"? _____

29. State two things people do that lower the water table of an area. _____

30. What do the Leaning Tower of Pisa, Italy; the San Joaquin Valley, California; and the city of Venice, Italy, have in common? _____

31. Rock and sediment act as natural filters to remove what type of groundwater contamination? _____

32. Name groundwater pollutants that are
    a. soluble and toxic _____
    b. soluble, not toxic, but undesirable because they smell bad, stain what they touch, make water "hard" and in general are harmful to humans _____
    c. insoluble, less dense than water, and harmful to life _____

33. a. If a groundwater contaminant is present in a concentration of 5 parts per million (ppm), for every 5 molecules of contaminant how many molecules of water are present? _____
    b. What percentage of the groundwater is this? _____

34. Name several contaminants introduced into groundwater by human activities. _____

35. Describe three ways contaminants can enter groundwater. _____

36. a. What type of wastes will be disposed of at Yucca Mountain, Nevada? _____
    b. What rock type will hold these wastes? _____

37. List some effects of rising water tables that are damaging to people or their property. _____

38. Why is groundwater acidic? _____

39. Why are there no caves in Arctic climates? _____

40. State four conditions necessary to form large caves. _____

41. List four steps in the establishment of karst landscape. _____

42. Name several people or groups well known in history that have used caves for shelter. _____

43. a. What chemical is commonly added to water supplies to make it pure enough to meet national drinking water standards? _____
    b. What chemical is added to strengthen tooth enamel and help prevent tooth decay? _____

44. List three factors that determine the mineral content of groundwater in an area. _____

Making Use of Diagrams

WELLS

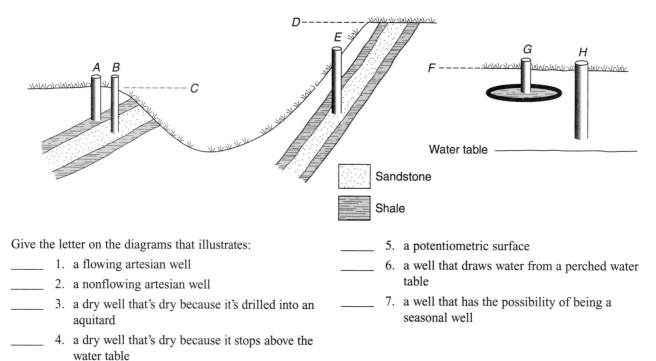

Give the letter on the diagrams that illustrates:

_____ 1. a flowing artesian well

_____ 2. a nonflowing artesian well

_____ 3. a dry well that's dry because it's drilled into an aquitard

_____ 4. a dry well that's dry because it stops above the water table

_____ 5. a potentiometric surface

_____ 6. a well that draws water from a perched water table

_____ 7. a well that has the possibility of being a seasonal well

SPRINGS

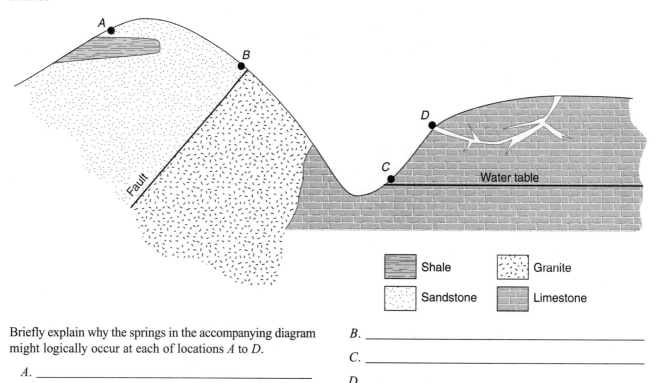

Briefly explain why the springs in the accompanying diagram might logically occur at each of locations A to D.

A. _____

B. _____

C. _____

D. _____

Short-Essay Questions

1. Discuss the origin of submarine groundwater and the role it plays in cooling Earth.
2. Two neighboring farmers are drilling for wells. One farmer has to drill to a depth of 100 ft to hit the water table, while his neighbor reaches the water table at only 30 ft. What's a possible explanation?
3. The water table is not a horizontal, planar surface. Explain what the shape of its surface is, and why this is so.
4. Describe two ways the rate of groundwater flow can be measured.
5. Explain why geysers erupt.
6. Discuss the cause and permanency of pore collapse.
7. Discuss why cave networks have different types of areas, including large open spaces, narrow passageways, and different levels.
8. Describe some life forms that live in caves.
9. There is significant water in the Sahara Desert region. Discuss where it's located, how it's recovered, and how the supply is recharged.

Practice Test

1. Which of the following statements is TRUE?
    a. A sinkhole developed at Winter Park, Colorado, and damaged a large section of a ski lift.
    b. The Leaning Tower of Pisa, Italy, is a classic example of a sinkhole collapse.
    c. Withdrawal of groundwater faster than natural recharge occurs has caused extreme land subsidence in the San Joaquin Valley, California.
    d. Pore collapse causes temporary, reversible loss of porosity and permeability in aquifers.
    e. Venice, Italy, has canals rather than streets because its impermeable ground surface causes almost constant flooding.

2. Which of the following statements is FALSE? A well in which the water rises on its own to a level above its aquifer
    a. consists of a confined aquifer beneath a sloping aquitard.
    b. is called an artesian well.
    c. always pushes water higher than the ground surface.
    d. must be drilled so it ends in the aquifer, not the aquitard.
    e. is analogous to a city water supply in which water flows from a high storage tank.

3. Which of the following statements is FALSE?
    a. Porosity is a measure of how much water rock can hold; permeability is a measure of how well water can travel through that rock.
    b. In order to produce water, a well must be drilled so it stops above the water table, within the zone of aeration.
    c. The world obtains roughly two-thirds of its freshwater supply from groundwater.
    d. The water table is defined as the top of the zone of saturation.
    e. An aquifer is a porous, permeable layer of rock that holds and conducts water.

4. Groundwater
    a. completely fills the zone of aeration above the water table.
    b. moves underground mainly in underground river channels.
    c. dissolves very porous rock like obsidian faster than it dissolves limestone.
    d. typically moves slowly within the ground through porous rock layers like sandstone.
    e. moves fastest through layers of clay called aquicludes (or aquitards).

5. Groundwater contaminants
    a. like sulfur, iron, calcium carbonate, and methane may come from the rock the water flows through.
    b. are all eventually removed by rock and sediment acting as natural filters.
    c. move so slowly they are usually detected and removed before they travel far.
    d. must be present in amounts greater than 10 parts per million (ppm) before they're considered harmful.
    e. All of the above are true statements.

6. Which of the following could NOT be a potential groundwater contaminant?
    a. pesticides, because they act only against specific insects
    b. animal sewage, because it's a natural product and will naturally and quickly dissipate
    c. petroleum products, because they're less dense than water, float on top, and therefore don't contaminate the water
    d. acid drainage from mines older than 100 years, because by then the waste minerals have been neutralized and can no longer cause acidic drainage
    e. All of the above are groundwater contaminants.

7. Which of the following statements is FALSE? In an area of rolling hills, groundwater moves
    a. from where the water table is at high elevation to where the water table is at low elevation.
    b. from regions under hills to regions under valleys.

c. along a curved path.
d. sometimes parallel to the slope of the water table and sometimes straight down, but never upward.
e. under the influence of both gravity and pressure differences.

8. Which of the following statements is FALSE? Primary porosity decreases
   a. with increasing depth of the rock layer.
   b. as sediment is changed to sedimentary rock.
   c. with increasing compaction of sediments or rock.
   d. with the cementing of sediments by mineral grains from groundwater.
   e. when rocks develop joints or faults.

9. Which of the following statements is FALSE?
   a. Water can be held stationary in pore spaces by electrostatic attraction to mineral surfaces.
   b. In order to be mobile, groundwater must be in spaces that are larger than 1 mm in diameter.
   c. The permeability of a rock depends on the number, size, and straightness of available conduits.
   d. Cork is very porous but very impermeable.
   e. Water moves upward in the capillary fringe due to electrostatic attraction between its molecules and mineral surfaces.

10. The water table
    a. is defined by the surface of a permanent stream, lake, or marsh.
    b. becomes a downward-pointing cone-shaped surface around the bottom of a well.
    c. lies within a few meters of the surface in humid areas.
    d. mimics the topography of the land it underlies.
    e. All of the above are true statements.

11. Which of the following statements is FALSE?
    a. A perched water table is one that sits high on a hill rather than low in a valley.
    b. The chief source of groundwater is precipitation.
    c. The water table of an area goes down as more wells are established there.
    d. The slope of the water table is called the hydraulic gradient.
    e. The water table fluctuates as the amount of precipitation changes.

12. Rock layer C in diagram 3 of the figure below cannot produce an artesian well at location X because
    a. it's all sandstone, which is aquifer material, so the water present can't be contained and build up pressure as is necessary for an artesian situation.
    b. it's composed of sandstone, and sandstone is an aquiclude.
    c. it doesn't have a city water supply available to recharge it.
    d. the recharge area is at a higher elevation than the discharge area.
    e. This question is misleading; rock layer C would make a very good artesian well.

13. Which of the following statements is TRUE?
    a. An artesian well never has to be pumped to draw water to the surface.
    b. In diagram 1 of the figure below, if there were severe drought conditions in Kansas but lots of rain in Colorado, the well in Kansas would be naturally recharged and would not run dry.

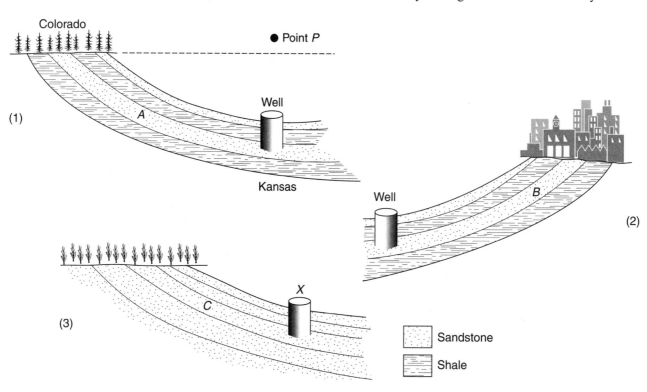

c. In diagram 1 of the accompanying figure, water from the well could rise to point *P*, above the potentiometric level.
d. Diagram 3 of the accompanying figure would be showing an artesian situation if layer *C* were shale because then you'd have the aquiclude present to satisfy the definition.
e. All of the above are true statements.

14. Which of the following statements is TRUE of the figure above?
    a. Since sandstone is an aquiclude, all of the wells shown would be dry.
    b. The farmlands in diagrams 1 and 3 are better recharge areas than the city in diagram 2 is because city pavements keep water from infiltrating.
    c. Layers *A*, *B*, and *C* are all too slanted to hold moisture; it would flow quickly through the area and leave all the wells dry.
    d. The wells in all three diagrams are seasonal.
    e. The potentiometric level in diagram 1 is wrong; it should be drawn at a higher elevation.

15. Groundwater in underground crooked passageways is heated past its normal boiling temperature by surrounding hot rocks. The superhot liquid expands in volume; this reduces its confining pressure, and it flashes into steam. This is an explanation of which geologic feature?
    a. a hot spring
    b. an artesian spring
    c. a mud pot
    d. a geyser
    e. karst

16. Features that are created when water dissolves surface and subsurface limestone, like sinkholes, troughs, caverns, natural bridges, and towers, are collectively called
    a. potentiometric surfaces.
    b. geothermal landscapes.
    c. karst landscape.
    d. artesian surfaces.
    e. undersaturated surfaces.

17. Stalagmites
    a. are a type of flowstone.
    b. are composed of limestone precipitated out of cave water.
    c. begin as delicate, hollow structures called soda straws.
    d. are icicle-shaped cones that hang from cave ceilings.
    e. All of the above are true statements.

18. Rate of groundwater flow
    a. typically varies between 4 and 500 mi/year.
    b. is determined theoretically; it can't be measured, since it happens underground.
    c. depends in part on the differences in elevation of the water table at various points along the flow path.
    d. is comparable to the rate of flow in an average surface stream.
    e. All of the above are true statements.

19. Which of the following statements is FALSE? Darcy's law
    a. is an equation used by hydrologists.
    b. is a method to determine the degree of saturation of water at any specified depth.
    c. states that discharge equals the hydraulic conductivity coefficient times the hydraulic gradient times the area involved.
    d. takes into account the permeability of the rock and the viscosity of the fluid.
    e. can be used to decide practical issues, such as whether sufficient groundwater exists to supply a city's needs.

20. Which of the following statements is FALSE? Dissolved ions in groundwater
    a. can create hard water, rich in calcium and magnesium.
    b. can precipitate out as carbonate minerals and clog plumbing pipes.
    c. include iron, which leaves an orange stain on materials it touches.
    d. are more abundant in young groundwater than in old groundwater, which has precipitated out most of its ions over time.
    e. can make water too salty to be used for irrigating crops.

21. Interconnected fractures that open onto a hillside and an impermeable rock layer intersecting a hillside are both likely locations of
    a. springs.
    b. artesian wells.
    c. potentiometric surfaces.
    d. cones of depression.
    e. natural bridges.

22. The hot springs of both Yellowstone and Iceland owe their existence to
    a. subducting plates.
    b. groundwater rising from great depth.
    c. motion along transform faults.
    d. hot-spot activity.
    e. rifting of plates.

23. Which of the following statements is TRUE?
    a. Mining groundwater simply means pumping it out of wells.
    b. The lowering of water tables is a growing problem in the U.S. desert southwest.
    c. In coastal areas saltwater can enter an aquifer, float on the freshwater there, and be drawn up in wells.
    d. The United States plans to dispose of petroleum wastes in salt domes at Yucca Mountain, Nevada.
    e. All of the above are true statements.

24. Water combines with carbon dioxide
    a. as it falls as rain, producing acidic water.
    b. as it filters through organic-rich soil on its way down to the water table.
    c. and thus forms carbonic acid.
    d. creating slightly acidic groundwater that dissolves limestone and forms caves.
    e. All of the above are true statements.

25. What is a spelunker NOT likely to encounter?
    a. fish without eyes
    b. speleothems
    c. limestone columns
    d. geysers
    e. snotites

26. Soil moisture is precipitation that sinks deep, fills the zone of saturation, and resides there for many decades.
    a. True
    b. False

27. Submarine groundwater expelled from great depth and from black smokers causes significant cooling of Earth's temperature.
    a. True
    b. False

28. Because sand grains are larger than mud grains, primary porosity is greater in unconsolidated sand than it is in uncompacted mud.
    a. True
    b. False

29. Igneous and metamorphic rocks generally have low porosity because their crystals grow and interlock, leaving minimal space.
    a. True
    b. False

30. Unfractured granite has low porosity and is impermeable; jointed granite can be permeable if the joints interconnect.
    a. True
    b. False

31. If there are many spaces between rock grains and the spaces connect with each other, the rock is said to be porous.
    a. True
    b. False

32. Groundwater flows upward in a recharge area, downward in a discharge area.
    a. True
    b. False

33. Hydraulic gradient is the difference in elevation of the water table between two points.
    a. True
    b. False

34. Saturated groundwater becomes oversaturated when it percolates down through hotter rocks, and must therefore precipitate out minerals as cement between sediment grains or as veins in rock cracks.
    a. True
    b. False

35. A dry well that's drilled into an aquitard is also a seasonal well.
    a. True
    b. False

36. In terms of human usage and availability, groundwater is a renewable resource.
    a. True
    b. False

37. Bioremediation is a technique that injects oxygen and nutrients into a contaminated aquifer in order to foster the growth of bacteria that can break down contaminant molecules.
    a. True
    b. False

ANSWERS

Completion

1. groundwater
2. sinkhole
3. hydrologic cycle
4. infiltrate
5. soil moisture
6. pore
7. porosity
8. primary porosity
9. secondary porosity
10. immobile water
11. mobile water
12. permeable
13. impermeable
14. percolation
15. hydrogeologists
16. aquifers
17. aquitards (or aquicludes)
18. principal aquifer
19. unconfined aquifers
20. confined aquifers
21. water table
22. unsaturated zone (or zone of aeration)
23. saturated zone (or zone of saturation)
24. capillary fringe
25. head of the water table
26. perched water table
27. recharge area
28. discharge area
29. hydraulic gradient
30. hydraulic conductivity
31. Darcy's law
32. saturated solution
33. undersaturated solution
34. oversaturated solution
35. precipitation
36. hard water
37. wells
38. springs
39. ordinary well
40. dry well
41. seasonal wells
42. drawdown
43. cone of depression
44. artesian well
45. potentiometric surface
46. artesian springs
47. hot springs
48. geothermal regions
49. mud pot
50. geyser
51. injection wells
52. contaminant plume
53. bioremediation
54. dripstone
55. speleothems
56. spelunkers

57. stalactite
58. soda straw
59. stalagmite
60. limestone column
61. flowstone
62. disappearing streams
63. karst landscape
64. tower karst

Matching Questions

HOT SPRINGS
1. E
2. D
3. A
4. B
5. C

Short-Answer Questions

1. a sinkhole

2. in springs; when it's pumped from wells; on the floor of caverns

3. Evaporates back into the atmosphere; becomes glacial ice; becomes runoff that enters streams, lakes, and eventually the ocean.

4. type of vegetation present; composition of the materials composing Earth's surface there

5. simple evaporation; absorbed by plant roots and transpired back into the atmosphere

6. anywhere from a few months to tens of thousands of years

7. two-thirds

8. a. precipitation
   b. Seawater can be trapped with sediment during the processes of deposition and lithification.

9. a. unfractured obsidian, less than 1%
   b. up to about 57% (60% porosity possible for highly fractured rock, minus 3% porosity of unfractured marble)
   c. uncompacted mud, which can be up to 70% porous, whereas unconsolidated sand has a maximum porosity of only 30%

10. In poorly sorted sediments, the smaller grains fill in the spaces between the larger grains.

11. Crystals formed due to molten rock's freezing or to the metamorphic processes interlock with each other, leaving minimal pore space.

12. fracturing of rocks to produce joints and faults; faulting to create breccia, whose angular jumble of pieces allows lots of pore spaces to exist; dissolving of mineral grains and cementing material by groundwater

13. If all pore spaces or cracks were smaller than 0.05 mm in diameter, all groundwater there would be immobile.

14. the number, size, and straightness of available conduits

15. a. permanent stream, lake, or marsh   b. humid regions   c. arid regions

16. Below 10–15 km the pressure of overlying rocks squeezes pores closed, so there's no place for water to be.

17. Percolates straight down because of gravity.

18. the weight of the overlying water

19. a. 0.01–1.4 m/day
    b. It flows through a complex network of tiny conduits, where friction between water and conduit walls slows the flow.

20. the permeability of the material it's flowing through and the slope of the water table (hydraulic gradient) in the area

21. quartz

22. Carbonate minerals precipitate out and clog pipes the water passes through; it's difficult to produce soap lather and get effective cleaning action with hard water.

23. an artesian well

24. a. where faults or fractures serve as conduits for very deep, hot groundwater to reach the surface
    b. where, due to current or recent volcanism, magma or very hot rock lies close to Earth's surface

25. Because hot water dissolves minerals from the rock it's in much more effectively than cold water does, so hot spring waters are mineral rich, whereas cold spring waters aren't.

26. bacteria that metabolize sulfur-containing minerals in the groundwater

27. Hot springs are fed by straight, unconstricted conduits; geysers have complex conduit systems with constrictions that impede the smooth flow of water to the surface.

28. pumping water out faster than nature can replace it

29. Pump groundwater from wells faster than it can be resupplied.
    Divert surface water away from the recharge area.

30. the problem of land subsidence

31. suspended solids, like mud and solid waste

32. a. arsenic, mercury, and lead   b. sulfur, iron, lime, and salt   c. gasoline

33. a. 999,995   b. 0.0005%

34. agricultural wastes (including pesticides, fertilizers, and animal sewage), industrial wastes, effluent from sanitary landfills and septic tanks, petroleum

246 | Chapter 19

products, radioactive wastes (from weapons manufacture, power plants, and hospitals), and acids leached from sulfide-rich minerals of coal and metal mines

35. Seep into the groundwater from subsurface tanks, infiltrate from the surface, and be intentionally forced through injection wells.

36. a. U.S. nuclear wastes   b. welded tuff (consolidated volcanic ash)

37. flooded basements; landslides and slumps; slumps into reservoirs that produce waves over dams

38. It has dissolved carbon dioxide from the atmosphere as it rained down and from organic-rich soil as it has percolated downward. Carbon dioxide plus water yields carbonic acid.

39. because cold groundwater, even when acidic, doesn't dissolve limestone nearly as effectively as warm groundwater does

40. a thick layer of limestone, significant rainfall, uplift of land surface above sea level to ensure the water table is underground, and temperate to tropical warmth

41. the establishment of a water table in a thick layer of limestone, the formation of a network of caves, a drop in the level of the water table, and roof collapse to create sinkholes and troughs and a surface topography of hills, ridges, and natural bridges

42. Jesse James, in caves in the Missouri Ozarks
    Vietminh troops during the Vietnam conflict
    Castro's followers during the Cuban revolution
    Yugoslavian resistance fighters in World War II

43. a. chlorine   b. fluoride

44. composition of the regional bedrock
    composition of the sediment the water has passed through
    length of time the water has been in the ground dissolving the surrounding rock

Making Use of Diagrams

WELLS

1. *E*   4. *H*   7. *H*
2. *B*   5. *D*
3. *A*   6. *G*

SPRINGS

A. Water percolating down is stopped by an aquiclude, creating a perched water table that intersects the surface at *A*.

B. Water percolating down reaches the steep impermeable layer of igneous rock, brought there by movement along a fault. The pressure of its own weight pushes the water up along the fault plane to emerge at the surface at *B*.

C. The ground surface intersects the water table at *C*, in a valley, which is a typical discharge area.

D. A network of interconnected fractures channels groundwater to the surface along the hillside, to emerge at *D*.

Short-Essay Questions

1. Seawater can become trapped with sediments as they're deposited and can infiltrate the sea floor the same way as precipitation can infiltrate the land. Where it has infiltrated the ocean bottom near magma chambers along mid-ocean ridges, it heats up, rises, and emerges as black smokers. Any submarine groundwater seeping up from below releases some heat energy, and this, plus the extreme release at black smokers, accounts for significant cooling of Earth.

2. The shallow well might have hit a perched water table, where a mound of groundwater has accumulated above a lens of impermeable rock that serves as an aquitard (aquiclude).

3. The water table mimics the shape of the overlying topography but in a subdued fashion; its ups and downs are not as extreme as those of the land above it. If groundwater flowed freely, it would assume a horizontal surface as soon as it reached its maximum depth, the same as water at the surface does. Instead it moves so slowly its tendency to level itself is constantly interrupted by the next event of water input, and it can never catch up and achieve and maintain horizontality.

4. The rate of flow of groundwater can be determined by injecting a tracer of dye or radioactive material into the water at one location, then noting the time of appearance of the tracer at a well drilled at some known distance. Plug the numbers for distance and time into the formula rate = distance/time and solve for rate. Second method involves determining the age of groundwater at some particular site by radiocarbon dating. Rainwater contains some carbon dioxide, some of which is the isotope carbon 14. Since groundwater originates as rainwater, new groundwater has the same carbon 14 content as rainwater. As the water flows through the ground and ages, its carbon 14 decays so its level goes down. Comparison of the carbon 14 content of rainwater with that of a sample of groundwater will give the age of that groundwater. If the origination

point of the groundwater is known, the scientist can determine the distance (origination point to sample site), the time which radiocarbon dating has supplied, and the same formula can be used to solve for the rate.

5. Any geyser is underlain by a network of convoluted fractures, interlaced with constrictions. This structure applies confining pressure to the groundwater present, which, because of the high pressure, remains as liquid water even when it gets heated to temperatures above the boiling point. Eventually some of the water gets hot enough to become steam which rises, reducing pressure on the remaining water. This deeper water now flashes into steam and shoots upward, and steam and water are explosively ejected from the system. Once the system has been flushed of its contents, it starts to refill with water, and the sequence of events recurs.

6. Because water cannot be compressed, water in pore spaces maintains the shape of the pore. When water is removed and replaced by air, which can be compressed, the grains of rock or regolith get packed more tightly, thus lowering the overall volume of pore space. The subsequent losses of porosity and permeability are permanent.

7. The shape of a cave network depends on variations in permeability and rock composition of the area. Large open spaces result where the rock is most soluble and the water flows the fastest. Alternating layers of permeable and impermeable rock produce separate levels within a cave system. Joint patterns determine the orientation of cave passages, many of which are grid-like because joint sets often sit at right angles to each other.

8. Caves with openings to the air house bats, insects, and spiders. Streams that pass through caves contribute fish and crustaceans. Species that spend their entire lives in caves often evolve unusual traits, such as loss of pigment or loss of eyes. Colonies of sulfur-metabolizing bacteria may form large mucus-like gobs that drip from cave ceilings and are appropriately (and rather disgustingly) known as snotites.

9. Huge amounts of water are found in the Sahara Desert region, not at the surface but in a porous sandstone aquifer underlying the area. It can be pumped to the surface out of deep wells, and it leaks to the surface in places known as oases, where folds have brought the aquifer up close to the surface or where joints or faults connect the aquifer to the surface. Recharge waters come from highland areas surrounding the desert, from occasional downpours, and from the Nile River, particularly from the portion of it, known as Lake Nassar, behind the Aswan High Dam.

Practice Test

1. c The classic sinkhole case was in Winter Park, Florida, where a sinkhole swallowed part of a neighborhood; the Leaning Tower and Venice have problems because of ground subsidence; and pore collapse is a permanent nonreversible phenomenon.

2. c If water rises higher than the ground surface, it's a flowing artesian well; when the level it reaches is below ground surface, it's a nonflowing artesian well.

3. b The well must reach below the water table into the zone of saturation in order to be a producing well.

4. d The zone of saturation is below the water table; groundwater moves mainly through interconnected pore spaces in rock and sediment, rarely in underground lakes or streams; aquicludes like clay are impermeable; and unfractured obsidian is neither soluble nor porous.

5. a Contaminants in groundwater do move slowly, but they often travel long distances, can't all be naturally filtered out, and can be harmful in concentrations as low as 0.1 ppm.

6. e All of the reasons offered to explain why the four choices are harmless are untrue.

7. d Pressure differences, caused by different thicknesses of groundwater, can push water up against gravity's influence.

8. e Joints and faults provide large openings for water and thus increase porosity.

9. b Spaces need only be larger than 0.05 mm in diameter to allow water to flow.

10. e Statements a to d are all true.

11. a A perched water table can occur anywhere that an impermeable lens of rock stops the downward percolation of groundwater.

12. a An artesian situation requires a confined aquifer layer lying under a slanted aquiclude layer; sandstone is an aquifer, not an aquiclude; and recharge could happen by natural precipitation.

13. b A nonflowing artesian well has to be pumped; artesian water can never rise higher than its potentiometric level; if C were shale, which is an aquiclude, you'd have a dry well; and any producing well must open into an aquifer.

14. b Sandstone is a good aquifer; groundwater doesn't flow fast enough to "escape" the area; there's no

data given by which to judge the seasonality of these wells; and water has the potential to rise only to the level at which it entered the system.

15. d   An artesian spring is not a geothermal feature; hot springs and mud pots don't have crooked subsurface passageways; and karst is a type of topography in humid limestone areas.

16. c   Potentiometric surfaces are the levels to which artesian waters have the potential to rise; geothermal landscapes would involve geysers, hot springs, and mud pots; the other terms are meaningless.

17. b   Flowstone precipitates as sheets of material across broad surfaces; soda straws develop into stalactites, which hang from cave ceilings.

18. c   Groundwater flow rate is 4–500 m, not miles, per year, which is much slower than flow rate of surface waters. The rate can be measured by using tracer dyes or radioactive elements and by plugging results into the formula rate = distance/time.

19. b   Darcy's law has nothing to do with saturation levels; it's a formula commonly used by hydrologists to determine the quantity of water passing through an area.

20. d   Old groundwater contains more ions because it's had more time to dissolve them from bedrock.

21. a   Choice a fits both location descriptions; no other choice is reasonable for either.

22. d   Yellowstone is located on a continental hot spot, Iceland on an oceanic hot spot. Iceland also sits on a rift area, but Yellowstone doesn't.

23. b   "Mining" implies pumping out more groundwater than can be naturally replaced; freshwater is less dense than saltwater, so saltwater invading a coastal aquifer moves in under the freshwater; and nuclear wastes will be disposed of in welded tuff at Yucca Mountain.

24. e   Statements a to d are all true.

25. d   Geysers are a surface feature in geothermal areas; the other terms denote structures or organisms found in caves, which spelunkers explore.

26. False   Soil moisture remains close to the surface and returns to the atmosphere by evaporation and by transpiration through plants.

27. True   Submarine groundwater carries large quantities of heat from great depths and releases it through black smokers.

28. False   Uncompacted mud can have up to 70% porosity, unconsolidated sand only 50%. Only after mud lithifies and becomes shale does its porosity get low (less than 10%).

29. True   The crystalline structure of igneous and metamorphic rocks limits their porosities.

30. True   Jointed granite can be permeable if its joints interconnect; unfractured granite is neither porous or permeable.

31. False   Such a rock, with interconnecting spaces, would be permeable.

32. False   Just the reverse is true.

33. True   The difference in elevation of the water table between two points is known as the hydraulic gradient.

34. False   The events described happen when groundwater flows through cooler, not hotter rocks.

35. False   A well that's dry because it ends above the water table could yield water during rainy seasons and thus be seasonal, but a dry well drilled into an aquitard will never have water flowing into it.

36. False   The hydrologic cycle will resupply groundwater in an area where it has been depleted, but this can take thousands of years. Therefore, on the human time scale, groundwater is a nonrenewable resource.

37. True   Some bacteria can break down contaminant molecules in an aquifer; encouraging their growth by injecting oxygen and nutrients is termed bioremediation.

CHAPTER 20 | An Envelope of Gas: Earth's Atmosphere and Climate

GUIDE TO READING

Chapter 20 directs your attention upward, into the envelope of gas that surrounds Earth's solid and liquid body. The text has consistently shown that geology is the study of the entire Earth environment and interactions between all of its components. This interaction is especially well illustrated as this chapter opens with a discussion of Earth's early atmosphere and the major changes it's undergone over time. You learn that:

- Its first components (hydrogen and helium) were driven off as the solid planet heated up.
- Volcanoes supplied the ingredients of the next atmosphere (water vapor, carbon dioxide, sulfur dioxide, and nitrogen).
- The atmosphere supplied water to fill the ocean basins.
- The oceans absorbed carbon dioxide from the atmosphere and precipitated out large amounts as limestone.
- The chemical weathering of rocks withdrew carbon dioxide from the atmosphere.
- Photosynthetic organisms contributed oxygen to the atmosphere.
- The ozone layer, which forms from oxygen and filters out harmful ultraviolet radiation, had to develop before life could safely leave the protection of seawater.
- The atmosphere continues to change and in the future it could be very different from what it is today.

The chapter continues by presenting the composition and characteristics of the modern atmosphere. You read about:

- its gaseous components (mainly nitrogen and oxygen)
- its solid components (tiny particles called aerosols)
- its atmospheric pollutants (including sulfates and nitrates that produce acid rain, the greenhouse gas carbon dioxide which contributes to global warming, and chlorofluorocarbons [CFCs] that break down ozone and create the ozone hole)
- its air pressure at various elevations
- its atmospheric heat and temperature, and how these are changed adiabatically
- its water content (relative humidity and latent heat of condensation)
- its layers (including the homosphere, heterosphere, ionosphere, mesosphere, mesopause, stratosphere, stratopause, thermosphere, troposphere, and tropopause)
- the aurora borealis and the aurora australis

Next you look into how the atmosphere moves from one place to another (winds) and in doing so reacts with and changes the physical and biological Earth underneath. Even humans don't escape the influence of winds. For example, mankind's travels have been affected by winds from the days of sailing ships on the oceans to the days of flying planes in the jet stream. Your author examines many aspects of this topic, including both local and global circulation patterns. He begins by showing how incoming solar radiation (insolation) varies and is responsible for creating temperature differences and pressure gradients that set the stage for winds to blow. He continues with an explanation of the basic classification of both high altitude and surface winds and the geographic zones they flow through. There are divergence and convergence zones (including polar highs, polar fronts or subpolar lows, subtropical convergence zones or subtropical highs, and the intertropical convergence zone or equatorial low). Three large-scale, global-encircling convection cells exist in both the Northern and Southern Hemispheres (Hadley cells, Ferrel cells, and polar cells). Prevailing surface winds include the northeast and southeast tradewinds, doldrums, surface westerlies, horse latitudes, and polar easterlies. Even winds at high altitude (like the high-altitude westerlies and jet streams) have their effect on Earth below. Once again you read about the Coriolis effect, which deflects air flow in a manner analogous to its deflection of ocean waters (see Chapter 18).

The chapter continues with more topics that will remind you that the scope of geology is broad. You will read about atmospheric regions with recognizable physical characteristics (air masses), their boundaries (cold, warm, and occluded fronts) and their patterns of motion (cyclones and anticyclones). Weather phenomena (including various types of clouds and precipitation) are produced when lifting mechanisms cause air in the air masses to rise. Severe weather events, storms, can catastrophically change Earth's surface. Your author briefly discusses thunderstorms, tornadoes, nor'easters, hurricanes, typhoons, and cyclones.

The chapter concludes by discussing climates, which are the average weather conditions in a region over a year. Several physical characteristics of Earth serve as controlling factors of climate. These include latitude, altitude, proximity to water, proximity to ocean currents, proximity to orographic barriers, and proximity to high- or low-pressure zones.

The next two chapters return you to Earth's surface, and you take a look at deserts and glaciers. As you read about them, take a moment to realize that the existence of these special areas on Earth's solid surface are the result of atmospheric conditions you've read about in this chapter.

Completion

Test your recall of new vocabulary terms.

_____ 1. the layer of gases that surrounds Earth

_____ 2. the flow of air from one place to another across Earth's surface

_____ 3. tiny particles of liquid or solid material (such as water, acid, sea salt, volcanic ash, clay, soot, and pollen) that are light enough to stay suspended in the air

_____ 4. the precipitation that is acidic because of the introduction, naturally or by human activity, of carbon dioxide, sulfates, and nitrates

_____ 5. a rise in average atmospheric temperature due in part to increased amounts of carbon dioxide

_____ 6. an area of lower than normal concentration of ozone in the stratosphere

_____ 7. the push air exerts on its surroundings

_____ 8. changes in the temperature of an air mass caused not by the addition (or subtraction) of heat but by the compression (or expansion) of the air mass

_____ 9. the amount of moisture air is holding compared to the maximum amount of moisture it is capable of holding, expressed as a percentage

_____ 10. the temperature at which air becomes saturated and must start condensing out water on available surfaces

_____ 11. the heat released when water changes from the gaseous state to the liquid state

_____ 12. the boundary between the top of the troposphere and the bottom of the stratosphere; it occurs 9 km high at the poles and 12 km high at the equator

_____ 13. the layer of the atmosphere that extends from Earth's surface to a height of 9–12 km, in which convection currents and weather phenomena occur

_____ 14. the boundary between the top of the stratosphere and the bottom of the mesosphere, which occurs at a height of 47 km

_____ 15. the layer of the atmosphere which extends from a height of 9–12 km up to 47 km, which it is a stable, stratified layer, and in which convection doesn't occur

_____ 16. the boundary between the top of the mesosphere and the bottom of the thermosphere, which occurs at a height of 82 km

_____ 17. the layer of the atmosphere that extends from a height of 47–82 km, which doesn't absorb much solar energy and gets cooler with increasing distance from Earth's surface

_____ 18. the outermost layer of the atmosphere, extending up from 82 km, which contains less than 1% of the atmosphere's gas and gets hotter with increasing distance from Earth's surface

_____ 19. the troposphere, stratosphere, and mesosphere considered together, where the chaotic motion of the gas molecules creates a homogeneous mixture of gases

_____ 20. the thermosphere based on the fact its gases are separated into distinct layers

_____ 21. the atmospheric interval between 50 and 400 km, including most of the mesosphere and the lower part of the thermosphere, where nitrogen and oxygen atoms have been stripped of electrons and have become positive ions

_____ 22. ghostly curtains of varicolored light in the night sky at high northern latitudes, caused by the interaction of charged particles from the Sun and ions in the ionosphere

_____ 23. ghostly curtains of varicolored light in the night sky at high southern latitudes, caused by the interaction of charged particles from the Sun and ions in the ionosphere

_____ 24. the rate of air pressure change over a given horizontal distance

_____ 25. a line on a weather map that connects places of equal air pressure

_____ 26. all incoming solar energy

_____ 27. the apparent deflection of an object moving above the surface of the rotating Earth, where in the Northern Hemisphere, deflection is toward the right, as viewed from the point of origin, and in the Southern Hemisphere toward the left

_____ 28. a place where sinking air separates into two flows that move in opposite directions

_____ 29. a place where two surface air flows meet and rise together

_____ 30. the globe-encircling convection cells located between the equator and latitude 30° in both the Northern and Southern Hemispheres

_____ 31. the globe-encircling mid-latitude convection cells in both hemispheres

_____ 32. the globe-encircling high-latitude convection cells in both hemispheres

_____ 33. surface winds that generally flow in a consistent direction across each of the six major convective cells

_____ 34. a specific set of weather conditions (including temperature, air pressure, relative humidity, and wind speed) that affects a region for a period of time

_____ 35. a body of air, at least 1,500 km across, that has recognizable physical characteristics

_____ 36. the boundary between a cold air mass and the warm air mass it moves under

_____ 37. the boundary between a warm air mass and the cooler air mass it overrides

_____ 38. the geometric situation that exists when a cold front overtakes a slower-moving warm front and causes it to lift up and lose contact with the ground

_____ 39. the counterclockwise (in the Northern Hemisphere) flow of air around a low-pressure air mass, or a hurricane that occurs in the Indian Ocean

_____ 40. the clockwise (in the Northern Hemisphere) flow of air around a high-pressure air mass

_____ 41. a huge spiral mass of clouds, rotating counterclockwise around a large low-pressure air mass that's moving west to east in mid-latitudes

_____ 42. a region of the atmosphere composed of tiny droplets of water or crystals of ice

_____ 43. clouds that form at ground level

_____ 44. preexisting solid or liquid particles on which water condenses during cloud formation

_____ 45. conditions that cause air to rise

_____ 46. a snow-producing process in which supercooled water droplets evaporate and condense onto nearby ice crystals

_____ 47. puffy, cotton-ball- or cauliflower-shaped clouds

_____ 48. clouds that consist of relatively thin, stable layers and therefore look sheet-like

_____ 49. clouds with a wispy shape, tapering into delicate feathery curls

_____ 50. a huge rain-producing cloud that begins at a height of about 1 km, grows upward across altitude divisions, and spreads laterally at the top of the tropopause

_____ 51. broad, flat-topped clouds that form the top of a cumulonimbus cloud

_____ 52. the air that has no tendency to rise

_____ 53. the air that has a tendency to rise because it's warmer than its surroundings

_____ 54. shafts of upward-moving and downward-moving air within cumulous clouds

_____ 55. an almost vertical, funnel-shaped cloud in which air rotates violently around an extremely low pressure center

_____ 56. a large mid-latitude cyclone, with cold, counterclockwise winds, that comes out of the northeast and affects the Atlantic coast of North America

_____ 57. a low-pressure air mass that develops off the west coast of Africa at about 20°

latitude, rotates counterclockwise, and gains strength as it moves westward across the central Atlantic Ocean

_____ 58. a cyclonic wind system that is located over the central Atlantic, has with winds rotating a minimum of 61 km/h, and is the precursor to a hurricane

_____ 59. a cyclonic wind system that has extremely low pressure at its center and winds rotating a minimum of 119 km/h, develops in the east-central Atlantic and moves west, and may cause damage in the Caribbean and along the Gulf Coast and east coast of North America

_____ 60. hurricanes located in the western Pacific at latitudes 20°N and 20°S

_____ 61. a rotating vertical cylinder of clouds that surrounds the central zone of a hurricane

_____ 62. the excess of water that's carried landward by the strong winds of a hurricane

_____ 63. lines on a weather map that connect places with the same temperature

_____ 64. a landform, such as a mountain range, that diverts air and causes it to flow upward or laterally

_____ 65. elliptical zones of high or low pressure that circle the globe, roughly parallel to the equator, and exist all year but vary somewhat with changing seasons

_____ 66. a weather phenomenon, triggered by shifts in locations of high- and low-pressure cells in the equatorial Pacific, that produces temporary changes of wind patterns and ocean currents worldwide

Matching Questions

ATMOSPHERIC CIRCULATION BELTS AND PREVAILING WINDS

Match each term with its description.

a. doldrums
b. high-altitude westerlies
c. horse latitudes
d. intertropical convergence zone (or equatorial low)
e. jet streams
f. northeast tradewinds
g. polar easterlies
h. polar front
i. polar high
j. southeast tradewinds
k. subpolar low
l. subtropical divergence zone (or subtropical high)
m. surface westerlies

_____ 1. the convergence zone, located at 60° latitude,
_____   where two surface airflows meet, rise, and thus create a low-pressure zone (Two answers are required.)

_____ 2. the convergence zone of low pressure located at the equator

_____ 3. the divergence zone of high pressure at 30° latitude

_____ 4. the zone of high pressure and sinking air at 90° latitude

_____ 5. winds that blow out of the northeast between the equator and 30°N latitude and which carried trading ships westward from Europe to the Americas

_____ 6. winds that blow toward the northwest between the equator and 30°S latitude

_____ 7. the belt where the northeast tradewinds and the southeast tradewinds merge and flow, very slowly, west along the equator, so that sailing ships were often becalmed here

_____ 8. winds that blow out of the west or southwest in mid-latitudes across much of North America and Europe

_____ 9. the region at latitude 30° where air motion is chiefly down and winds are weak and variable, so that animal cargo on sailing ships often died here due to heat exhaustion

_____ 10. winds that begin blowing southward from the North Pole and are deflected to the west by the Coriolis effect

_____ 11. winds that blow northward from the top of the troposphere at the equator toward the top of the troposphere at the North Pole and are deflected eastward along their route

_____ 12. especially fast moving sections of the high-altitude westerlies, located over the polar front and the horse latitudes

Short-Answer Questions

Answer with a word, number, phrase, or short sentence.

1. Give two reasons oxygen is essential to life on Earth.
   _____
   _____

2. How is ozone related to the development of terrestrial life forms? _____
   _____

3. If life on Earth were to disappear and all volcanic activity were to stop, what would happen to Earth's atmosphere? _____
_____

4. What is the percentage composition of dry air?
_____

5. State two human activities that have added large quantities of pollutants to the air. _____
_____

6. What type of pollutants reacts with the Sun's ultraviolet rays and liberates chlorine, which breaks down ozone molecules? _____

7. What is the pressure of the atmosphere
   a. at sea level in bars? _____; in pounds per square inch? _____
   b. on top of Mt. Everest in bars? _____; in pounds per square inch? _____

8. a. More than 99% of Earth's atmosphere lies below what elevation? _____
   b. Where does interplanetary space start? _____
_____

9. What is the difference between the *heat* (thermal energy) of a gas and the *temperature* of that same gas?
_____

10. What is the numerical difference in moisture content between hot, dry desert air and rainforest air? _____
_____

11. List the four basic layers of the atmosphere, in order, beginning with the one closest to Earth's surface.
_____

12. In what layer of the atmosphere is the ozone layer?
_____

13. Which layer of the atmosphere is the weather layer?
_____

14. In which layer of the atmosphere do most meteors (shooting stars) begin to burn? _____

15. Which atmospheric "sphere" acts as a mirror and bounces radio waves back and forth between itself and Earth, thus enabling the waves to travel long distances?
_____

16. Why do the auroras occur primarily at high latitudes?
_____
_____

17. a. Which is less dense, cold air or warm air? _____
    b. Because of this, as you fly from an area of warm air into an area of cold air, maintaining a constant elevation, what happens to the air pressure you experience and why? _____
_____

18. a. What causes winds to form? _____
_____
    b. Air always flows from regions of _____ pressure to regions of _____ pressure.

19. Why can't isobars on a map ever touch? _____
_____

20. In respect to isobars, what direction does air begin to flow? _____

21. What deflects prevailing winds and causes them to blow almost parallel to isobars? _____

22. Give two reasons places at high latitudes receive less energy per square meter than places at low latitudes do.
_____
_____
_____

23. Why does the tilt of the Earth cause seasons? _____
_____

24. Name the three globe-encircling convection cells that occur in each hemisphere and give their latitude locations. _____
_____

25. A surface wind is what part of a convective cell? ___
_____

26. A wind is named according to the direction the air is _____ (blowing to, coming from).

27. Why do winds along the equator have very low velocities? _____
_____

28. a. Why is the top of the troposphere higher at the equator than it is at the poles? _____
_____
    b. How does this affect the direction of flow of high-altitude air in the Northern Hemisphere? _____
    c. As the high-altitude air leaves the equator, how does the Coriolis effect alter its flow? _____
_____

29. a. List the three basic types of weather fronts. _____
_____
    b. Which of these exists as a simple vertical plane, depicted on a map as a straight line? _____
    c. Which of these exists as a sloping surface? _____
_____
    d. Which has a steeper sloping surface, a cold front or a warm front? _____
    e. Which moves faster, a cold front or a warm front?
_____

30. Because there's a pressure gradient along the margin of an air mass, what happens here? _____

31. What direction does air move
    a. around the base of a low-pressure air mass? _____
    b. around a high-pressure air mass? _____

32. What happens to the volume, temperature, and moisture content of
    a. rising air? _____
    b. sinking air? _____

33. Which type air mass, high pressure or low pressure, is associated with
    a. stormy conditions (clouds and rain)? _____
    b. fair weather (clear and dry)? _____

34. Describe what a mid-latitude cyclone looks like from a satellite. _____

35. How and why do clouds affect Earth's temperature
    a. during the day? _____
    b. during the night? _____

36. State several ways air can become saturated and develop clouds. _____

37. a. How does rain develop in clouds? _____
    b. How does snow develop in clouds? _____

38. What do the suffix "-nimbus" and the prefix "nimbo-" mean in cloud names? _____

39. What is the elevation range for
    a. low-level clouds? _____
    b. midlevel clouds? _____
    c. high-level clouds? _____

40. When do storms form? _____

41. Give the following vital statistics of tornadoes.
    a. locations where most occur _____
    b. the maximum speed of the rotating winds _____
    c. the diameter of the base of the storm _____
    d. low or high air pressure _____
    e. the velocity of the storm across Earth's surface _____
    f. the tornado season in the United States _____
    g. the location of tornado alley _____

42. What is a tornado swarm? _____

43. Give two reasons tornadoes cause damage. _____

44. What is the Fujita scale? _____

45. Why are tornado funnels dark colored? _____

46. What is the difference between a tornado watch and a tornado warning? _____

47. a. What instrument do meteorologists use to detect both the presence of rain drops in an air mass and their speed of rotation? _____
    b. Describe the tornado "picture" this instrument produces. _____

48. Give the following vital statistics of hurricanes.
    a. the average diameter of a hurricane _____
    b. the calmest zone in a hurricane _____
    c. the place with the greatest rotary wind velocity _____
    d. the range of velocities of the storm along its track (storm-center velocity) _____

49. If a hurricane rotary wind velocity is 200 km/h and its storm-center velocity is 40 km/h, what is the effective wind speed on the side where
    a. the rotating winds and the entire storm are moving in the same direction? _____
    b. the rotating winds and the entire storm are moving in opposite directions? _____

50. List four ways hurricanes can cause damage. _____

51. Under what conditions do hurricanes die out? _____

52. What is the Saffir-Simpson scale used to describe? _____

53. What is the difference between weather and climate? _____

54. List six factors that control climate. _____

55. How do land and water differ in heat capacities? _____

Short-Essay Questions

1. Create a summary of the development of Earth's atmosphere by using the following questions as a guide. What gases made up Earth's first, short-lived atmosphere? What gases composed Earth's second

atmosphere, and what was their source? Generally speaking, why has the atmosphere changed over time? More specifically, why have the levels of water vapor, carbon dioxide, nitrogen, and oxygen changed?

2. Explain how clouds, dew, and frost form.

3. Why do people feel more uncomfortable in hot humid areas than they do in hot dry areas?

4. Briefly explain the process of atmospheric convection.

5. What happens to solar energy that reaches Earth?

6. Describe the motion of air masses across the Northern Hemisphere. Include the terms "Coriolis effect," "convection cells," "divergence and convergence zones," and "polar front."

7. Why do tropical rainforests form at the equator and major deserts form at 30°N and S latitudes?

8. Describe the jet streams and how they affect air travel.

9. List and describe four conditions that cause air to rise (lifting mechanisms).

10. Discuss thunderstorms. Include the reasons they develop, their stages of development, and why they can produce thunder, lightning, and hail.

11. What are the differences between a hurricane, a typhoon, and a cyclone?

Practice Test

1. Which of the following statements is FALSE? Earth's atmosphere
    a. at the time of Earth's beginning would not have sustained life as we know it.
    b. initially was water vapor and nitrogen.
    c. initially was composed of gases pulled from the surrounding solar nebulae but changed to one composed of volcanic gases.
    d. interacted with the solid and liquid Earth and thus changed its composition.
    e. released enough moisture as Earth cooled to fill the ocean basins.

2. Which of the following statements is FALSE? Atmospheric oxygen
    a. is necessary for the respiration of multicellular organisms.
    b. provides the raw material for ozone.
    c. was originally introduced by the photosynthesis of blue-green algae (cyanobacteria).
    d. is supplied mainly by chemical breakdown of carbonate rocks like limestone.
    e. reached its present concentration by about 0.4 billion years ago.

3. Earth's atmosphere
    a. is composed of 78% nitrogen, 21% oxygen, and 1% trace amounts of various gases.
    b. contains aerosols, which are tiny droplets of water and acid and particles of sea salt, volcanic ash, clay, soot, and pollen.
    c. is becoming more polluted due to the burning of fossil fuels.
    d. in some locations contains enough sulfates and nitrates to create acid rain.
    e. All of the above are true statements.

4. Air pressure
    a. is inversely related to elevation.
    b. on Mt. Everest's summit is only 1/10 as much as air pressure at sea level.
    c. at sea level is expressed as 14 bars.
    d. is nonexistent at elevations higher than 5 km.
    e. All of the above are true statements.

5. When air moves from a region of high pressure to a region of low pressure, with no addition or subtraction of heat,
    a. it contracts.
    b. its molecules speed up.
    c. it begins to move vertically upward.
    d. it undergoes adiabatic heating.
    e. All of the above are true statements.

6. Which of the following statements is FALSE?
    a. Saturated air has a relative humidity of 100%.
    b. If the dewpoint temperature is below freezing, frost forms instead of dew.
    c. Moist rising air cools adiabatically.
    d. Latent heat must be added to an air mass to cause water vapor in it to condense.
    e. Sweating will cool a person more efficiently when the relative humidity is 20% than when it is 50%.

7. If you are in the stratosphere, you
    a. can experience weather phenomena.
    b. are below the tropopause.
    c. will encounter the ozone layer.
    d. will be affected by convection currents.
    e. will see the beginning of the burning of meteors (shooting stars).

8. Which of the following statements is FALSE? The ionosphere
    a. is the atmospheric layer in which solar energy strips electrons from nitrogen and oxygen atoms.
    b. transmits radio waves by bouncing them back and forth between Earth and itself.
    c. hosts the aurora borealis and aurora australis.
    d. includes most of the stratosphere and the lower part of the mesosphere.
    e. gets its name because it's composed of electrically charged atomic particles called ions.

9. Which of the following statements is FALSE?
   a. Isobars are lines on a weather map that connect places of equal barometric pressure.
   b. Winds begin blowing in a perpendicular direction away from one isobar toward a lower-pressure isobar.
   c. Areas receiving oblique solar radiation experience more insolation than areas receiving direct solar radiation do.
   d. Light hitting the ground in polar regions has passed through more atmosphere than light hitting the ground close to the equator, so more has been backscattered into space.
   e. Because of the tilt of Earth's axis, the amount of solar radiation received by any point on its surface changes with the seasons.

10. Which of the following statements is FALSE? The Coriolis effect in the Northern Hemisphere
    a. deflects a northward-moving object to the left (west).
    b. causes exactly the opposite deflection as does the Coriolis effect in the Southern Hemisphere.
    c. affects the atmosphere, causing winds to curve.
    d. exists because an object moving across Earth's surface has two directions of travel: the direction it's aimed and the direction of Earth's rotation beneath it.
    e. deflects high-altitude winds that are moving north and causes them to move due east.

11. The air flow pattern at 60°N latitude
    a. is a divergence zone.
    b. is a zone of sinking air.
    c. is called the subpolar low.
    d. reverses itself with changing seasons.
    e. All of the above are true statements.

12. Which of the following statements is FALSE? Atmospheric circulation zones include
    a. an intertropical convergence zone that produces an equatorial low.
    b. a subtropical high at 30° latitude which has downward-moving, diverging air.
    c. a Hadley cell–Ferrel cell convergence at 60° latitude, which produces a zone of rising air called the subpolar high.
    d. an equatorial low that creates regions of heavy rainfall and tropical rainforests.
    e. a subpolar low, which has rising air and is located at 60°.

13. Which of the following terms is not a correct description of atmospheric conditions at the equator?
    a. "doldrums"
    b. "equatorial low"
    c. "intertropical convergence zone"
    d. "sinking air"
    e. "slow wind flow, almost due west"

14. Which of the following statements is FALSE? Prevailing winds
    a. are surface winds.
    b. are the base of convective cells.
    c. are named according to the direction they blow toward.
    d. include the northeast and southeast tradewinds, which blow westward between the equator and 30°N and S latitudes.
    e. blow approximately parallel to the high- and low-pressure belts surrounding Earth.

15. Jet streams in the Northern Hemisphere
    a. are zones of especially fast moving high-altitude easterlies.
    b. flow at speeds between 50 and 100 km/h.
    c. originate where pressure gradients are at their lowest.
    d. originate over the polar front (60° latitude) and over the horse latitudes (30° latitude).
    e. All of the above are true statements.

16. A maritime tropical air mass
    a. originates over warm seas.
    b. is a minimum of 1,500 km across.
    c. creates hot humid weather in the area it's affecting.
    d. expands when it rises, cools adiabatically, and condenses out clouds.
    e. All of the above are true statements.

17. A high-pressure air mass
    a. has rising air at its center.
    b. produces clear dry weather.
    c. in the Northern Hemisphere has counterclockwise circling winds.
    d. is associated with fog and clouds.
    e. is called a cyclone.

18. Which of the following statements is FALSE? Clouds
    a. at Earth's surface are called smog.
    b. form when water condenses on preexisting solid or liquid particles.
    c. form when the air is saturated.
    d. may form when the air cools at night.
    e. may form when the air rises and cools adiabatically.

19. Clouds that form over the plains of an ocean island do so because of
    a. orographic lifting.
    b. convergence lifting.
    c. convective lifting.
    d. frontal lifting.
    e. the Bergeron process.

20. Which of the following statements is FALSE? A cumulonimbus cloud
    a. is one of vertical development.
    b. produces rain.

c. may be topped by an anvil cloud.
   d. is filled with updrafts, not downdrafts of air.
   e. may extend to the top of the troposphere.

21. Which of the following statements is FALSE? Thunderstorms
    a. form where a cold front moves into a region of warm moist air.
    b. are common both in North American mid-latitudes and in tropical rainforests.
    c. last from one to a few hours.
    d. end when the updrafts become stronger than the downdrafts.
    e. with strong updrafts can produce ice balls of hail.

22. Which of the following statements is FALSE? Lightning
    a. occurs when a positively charged cloud base induces a negative charge in Earth's surface.
    b. involves a conductive path or leader of negative ions from cloud to ground.
    c. is composed of both upward and downward flows of ions.
    d. can stay within a cloud or travel back and forth between the cloud and the ground.
    e. may be accompanied by the sound of explosively expanding hot air called thunder.

23. If there's a 7 s time delay between observed lightning and the sound of thunder, how far away is the lightning?
    a. 7 mi
    b. 0.7 mi
    c. 2 mi
    d. 1.7 mi
    e. 1.4 mi

24. Which of the following statements is FALSE? Tornadoes
    a. have maximum wind speeds up to 500 km/h (300 mph).
    b. in North America generally travel from southwest to northeast at speeds up to 100 km/h (60 mph).
    c. can have pressure readings 100 millibars lower than normal air pressure.
    d. are rated on the Saffir-Simpson scale, which runs from 1 to 5.
    e. do damage because their winds batter the surroundings and their low pressures induce the outward explosion of buildings.

25. Which of the following statements is FALSE?
    a. Tornadoes occur most frequently during late summer and into late fall.
    b. Doppler radar detects rain and rotating winds in clouds and thus can "see" tornadoes.
    c. Tornadoes develop when opposing updrafts and downdrafts develop into spiraling winds.
    d. Tornado alley, an area of frequent tornadoes, runs from Texas to Indiana.
    e. Dozens of tornadoes may develop from a single thunderstorm system.

26. Which of the following statements is FALSE? Hurricanes
    a. begin as tropical disturbances off the east coast of South America.
    b. can be several hundreds of kilometers in diameter.
    c. are called typhoons in the western Pacific, cyclones in the Indian Ocean.
    d. track across the Atlantic and do damage in the Caribbean and along the Gulf Coast and eastern coast of North America.
    e. have their strongest winds in the eye wall, but their calmest region in the eye.

27. Which of the following statements is FALSE? Hurricanes
    a. by definition have minimum winds of 119 km/h.
    b. die out when they cross over cold waters or onto land.
    c. occur, on the average, several dozen times per year on the North American east coast.
    d. do damage with their intense winds and rainfall, huge waves, and storm surges.
    e. result when tropical depressions grow stronger.

28. A hurricane with a storm-center velocity of 50 km/h and rotary wind velocity of 180 km/h would have an effective wind velocity ranging from 50 to 180 km/h.
    a. True
    b. False

29. In order to maintain the composition of our current atmosphere, there must be volcanic activity and life activity.
    a. True
    b. False

30. Carbon dioxide is an atmospheric pollutant that reacts with and destroys ozone.
    a. True
    b. False

31. Air that is saturated at 60°F would become undersaturated if its temperature dropped to 50°F.
    a. True
    b. False

32. The aurora are curtains of light in both north and south polar regions caused by the interaction of charged particles (protons and electrons) from the Sun and ions in the ionosphere.
    a. True
    b. False

33. The three globe-encircling convections cells in each hemisphere, listed in order from the equator to the Pole, are Ferrel cell, Hadley cell, and polar cell.
    a. True
    b. False

34. The troposphere is thicker at the poles than it is at the equator, so high-altitude winds flow from the poles to the equator.
    a. True
    b. False

35. Occluded fronts form where faster-moving warm fronts overtake slower-moving cold fronts and slide under them.
    a. True
    b. False

36. The presence of clouds always results in cooler ground temperatures.
    a. True
    b. False

37. A cold front signals its approach long before it arrives with clouds possibly 800 km in front of its arrival; a warm front has a much steeper boundary plane and thus develops clouds only 50 km in advance of itself.
    a. True
    b. False

38. If your local radio station broadcasts a tornado warning, you should stay tuned to see if a tornado does develop. If it issues a tornado watch, take cover immediately, because a tornado has actually been seen.
    a. True
    b. False

31. Ferrel cells
32. polar cells
33. prevailing winds
34. weather system
35. air mass
36. cold front
37. warm front
38. occluded front
39. cyclone
40. anticyclone
41. mid-latitude (wave) cyclone
42. cloud
43. fog
44. condensation nuclei
45. lifting mechanisms
46. Bergeron process
47. cumulus cloud
48. stratus cloud
49. cirrus cloud
50. cumulonimbus cloud
51. anvil clouds
52. stable air
53. unstable air
54. updrafts and downdrafts
55. tornado
56. nor'easter
57. tropical disturbance
58. tropical depression
59. hurricane
60. typhoon
61. eye wall
62. storm surge
63. isotherms
64. orographic barrier
65. semipermanent pressure cells
66. El Niño

Matching Questions

ATMOSPHERIC CIRCULATION BELTS AND PREVAILING WINDS

1. h, k
2. d
3. l
4. i
5. f
6. j
7. a
8. m
9. c
10. g
11. b
12. e

Short-Answer Questions

1. It's the basic respiration gas of multicellular organisms. It's the raw material for ozone, which accumulates at an altitude of 30 km and absorbs harmful ultraviolet radiation.

2. Life couldn't leave the protection of seawater until the ozone layer was established.

3. Within a few million years much of the gas would leak into space and not be replaced.

4. 78% nitrogen, 21% oxygen, and 1% several gases in trace amounts

5. burning fossil fuels and industrial operations

6. chlorofluorocarbons (CFCs)

7. a. 1 bar; about 14 lb/in^2 b. 1/3 bar; about $4^{2/3}$ lb/in^2

8. a. 100 km (about 62 mi)   b. more than 1,000 km high (more than 620 mi)

9. Heat is the *total* kinetic energy of all the molecules of the gas; temperature is their *average* kinetic energy.

10. 0.3% in desert air; 4% in rainforest air

ANSWERS

Completion

1. atmosphere
2. wind
3. aerosols
4. acid rain
5. global warming
6. ozone hole
7. air pressure
8. adiabatic heating (adiabatic cooling)
9. relative humidity
10. dewpoint temperature
11. latent heat of condensation
12. tropopause
13. troposphere
14. stratopause
15. stratosphere
16. mesopause
17. mesosphere
18. thermosphere
19. homosphere
20. heterosphere
21. ionosphere
22. aurora borealis
23. aurora australis
24. pressure gradient
25. isobar
26. insolation
27. Coriolis effect
28. divergence zone
29. convergence zone
30. Hadley cells

11. troposphere, stratosphere, mesosphere, and thermosphere

12. stratosphere

13. troposphere

14. mesosphere

15. ionosphere

16. because Earth's magnetic field traps the solar particles (protons and electrons) that instigate the auroras and carries them to the poles

17. a. warm air
    b. It decreases, because in the warmer area the air had expanded upward, so there was more air above you pushing down than there is at the same elevation in the colder area.

18. a. differences in pressure at different locations (a pressure gradient)   b. high; low

19. They represent different air pressures, so they could never occupy the same point.

20. perpendicularly away from a high-pressure isobar toward a low-pressure isobar

21. Coriolis effect

22. Locations at high latitudes are hit by more oblique rays, which deliver less-concentrated energy than direct rays do. Incoming solar radiation at high latitudes, because it strikes at a more oblique angle than it would at low latitudes, has to traverse more atmosphere than it does at low latitudes, and atmosphere scatters radiation back into space.

23. Because of the tilt, the *angle of incidence* of solar radiation, and therefore the *amount* of solar radiation any particular point receives changes during the year. This results in seasons.

24. Hadley cells, from the equator to 30° latitude; Ferrel cells, from 30 to 60° latitude; and polar cells, from 60 to 90° latitude

25. the base of a convective cell

26. coming from

27. because air motion here is largely vertical, and winds are basically horizontally flowing air

28. a. Warm air at the equator expands and rises, thereby making its top sit higher than the top of colder air masses over polar regions (i.e., the troposphere is thicker over the equator than it is over the poles).
    b. The pressure gradient created by the thicker troposphere at the equator causes all high-altitude air in the Northern Hemisphere to flow due northward from the equator.
    c. The Coriolis effect deflects the northward flowing air to the east, resulting in high-altitude westerlies.

29. a. cold front, warm front, and occluded front
    b. Despite the fact they're all depicted as straight lines on a map, none are vertical planar boundaries.
    c. All are sloping.   d. cold front   e. cold front

30. Winds begin to blow.

31. a. in toward the center, then up   b. down and then out from the center

32. a. expands, cools, and condenses out moisture
    b. compresses, warms, and absorbs moisture

33. a. low pressure   b. high pressure

34. a huge spiral mass of clouds rotating counterclockwise, centered on the low-pressure mass

35. a. Ground temperature is lowered because clouds reflect and scatter incoming sunlight.
    b. Clouds prevent infrared (heat) radiation from escaping and thus keep the ground warmer.

36. when evaporation provides additional water; when dust storms, volcanic eruptions, smokestack effluent, or jet exhausts provide condensation nuclei; and when air cools and its water-holding capacity decreases

37. a. Tiny water droplets in warm clouds collide, coalesce, and eventually become heavy enough to fall.
    b. Supercooled (below normal freezing temperature) water droplets in cold clouds evaporate, then condense on preexisting ice crystals, and eventually become heavy enough to fall.

38. The cloud will produce rain.

39. a. below 2 km   b. between 2 and 7 km
    c. above 7 km

40. when contrasts of temperature and pressure between converging air masses are great, and prevailing conditions allow a continuous supply of hot moist air

41. a. in the mid-latitudes of the northern hemisphere
    b. about 500 km/h (300 mph)
    c. range between 5 and 1,500 m across
    d. low
    e. 0 to 100 km/h
    f. March through September
    g. in the midwestern states, from Texas to Indiana, and in Florida

42. dozens of tornadoes in sequence, generated by massive thunderstorm fronts

43. Strong winds lift and tumble anything in the way. Low air pressure can cause buildings to explode outward.

44. a scale to rate tornadoes on the basis of wind speed, path dimensions, and possible damage

45. The low end of the funnel cloud touches ground and sucks up dirt and debris.

46. "Watch" means conditions exist that could produce tornadoes; "warning" means a tornado has been spotted and people at risk should take cover

47. a. Doppler radar
    b. a hook shape along the backside (usually western edge) of a storm that's drifting to the east

48. a. 600 km b. in the eye c. in the eye wall
    d. ranges from 0 to 60 km/h, and rarely up to 100 km/h

49. a. 240 km/h b. 160 km/h

50. strong winds, huge waves, flooding due to storm surges, and intense rainfall

51. when they lose their source of warm moist air by crossing onto land or across cold waters at high latitudes

52. hurricanes

53. "Weather" means the physical conditions (temperature, pressure, moisture content and wind velocity and direction) of the atmosphere at a given time and location. "Climate" means the average weather conditions and range of conditions of a region over a year.

54. latitude (probably the single most significant factor), altitude, proximity to water, proximity to ocean currents, proximity to orographic barriers (like mountains), and proximity to high- or low-pressure zones

55. Land absorbs or loses heat quickly. Water can absorb and hold more heat than land can, but it does so more slowly.

Short-Essay Questions

1. Earth's first atmosphere was simply hydrogen and helium gases, some of which came from the surrounding nebulae, the rest from Earth's interior. This atmosphere lasted only about 0.1 billion years, because by 4.5 billion years ago the newly formed core heated the planet enough to cause these light-weight gases to expand, become even lighter, and escape into space. They were replaced by volcanic gaseous emissions consisting chiefly of water vapor, carbon dioxide, and sulfur dioxide, with minor amounts of nitrogen. Due to interactions with the solid Earth and with the hydrosphere, the atmosphere continued to change. By 3.8 billion years ago, when Earth had cooled enough for liquid water to exist on its surface, downpours filled the oceans and depleted the atmospheric water content. Large amounts of carbon dioxide were then lost to the atmosphere, as it dissolved in the newly formed ocean waters and eventually precipitated out to create limestone. The increased surface moisture increased the amount of chemical weathering, which in turn caused rocks to absorb carbon dioxide from the atmosphere. Atmospheric nitrogen is chemically inactive, so its concentration remained constant as the amounts of carbon dioxide and water vapor decreased. Therefore, relatively speaking, the percentage composition of nitrogen in the atmosphere grew. The photosynthetic process (food making, using light energy) releases oxygen to the atmosphere. Cyanobacteria (blue-green algae) were the first organisms to practice photosynthesis and thus contribute oxygen to the air. As life developed and proliferated, many other photosynthetic organisms (basically, green plants) released oxygen to the atmosphere and achieved our present concentration of 21% by 0.4 billion years ago.

2. Moist air cools adiabatically as it rises, causing some of its moisture to condense and form tiny water droplets or ice crystals. Clouds are accumulations of these droplets or crystals. At night, undersaturated warm air cools enough to become saturated, causing condensation to occur on available surfaces. It the temperature dictates condensation as liquid water, the result is called dew; if the condensation product is ice, the result is called frost.

3. The human body regulates its temperature by sweating. Drops of sweat evaporate from the skin and in the process draw heat energy from the body to accomplish the change of state, liquid to gas. In hot humid areas, the air already holds lots of moisture and it doesn't easily take up more. Hot dry air does readily absorb more moisture, so sweat evaporates rapidly, draws heat from the body, and makes the person feel cooler and more comfortable.

4. Solar radiation heats air. Heated air expands, becomes less dense, and rises. It's replaced by cooler, denser air that sinks and moves into the vacated space. A current of hot, rising fluid and cold, sinking fluid is called a convection current, whether it's happening in a pan of water on the stove or in Earth's atmosphere.

5. Roughly 30% of incoming solar radiation is reflected back to space by clouds, water, and land. Air absorbs 19% and land and water absorb 51% of the remaining solar radiation. Different wavelengths are absorbed at different levels. Gases of the upper atmosphere absorb short wavelengths of energy, while greenhouse gases (like carbon dioxide) absorb longer wavelengths of energy. Earth's surface absorbs energy and reradiates some of it back to the atmosphere as infrared (heat) energy. This gets trapped by a variety of greenhouse gases and contributes to global warming.

6. Warm air rises at the equator and flows toward the poles. In the Northern Hemisphere, northward-moving high-altitude air deflects to the right (east) due to the Coriolis effect, so by the time it reaches 30°N (the subtropics), it's moving due east, not north, and since it has cooled, it's sinking. Overall it's moving as a global convection cell which, in cross section, is a loop. As it sinks, the airflow divides into two parts, which makes this a divergence zone. One part moves back toward the equator and the other again moves north toward the Pole. At about 60°N the northward-moving surface air collides with cool surface air flowing south from the North Pole to create a convergence zone called the polar front. It has no place to go but up. When it reaches the top of the troposphere, the combined flow separates, some heading north as a polar convection cell, the rest returning, at high altitude, to the equator, thus completing a mid-latitude convection cell. The polar front shifts slightly with changing seasons. The Southern Hemisphere experiences a mirror-image version of all of the above.

7. Warm moist air rises at the equator, creating reduced air pressure and the zone known as the equatorial low. The rising air cools adiabatically and condenses out moisture, which forms clouds that produce rain. At 30°N and S latitudes, cooled dry air sinks, creating high-pressure belts. The sinking air heats adiabatically, absorbs moisture, and thus creates desert conditions.

8. Jet streams are zones within the high-altitude westerlies, over the polar front and the horse latitudes, where the air moves particularly fast, with speeds of 200–400 km/h. The more northerly zone, known as the polar jet stream, is faster and has greater influence on human activity than does the more southerly subtropical jet stream. The position of the polar front jet stream undulates in wavelike fashion and shifts from northern Canada to more southern latitudes of North America with changing seasons. The jet stream's strong, steady winds affect jet planes that fly at the same height, acting as helpful tailwinds on flights headed east and as slowing headwinds on flights headed west.

9. The following four conditions cause air to rise and thus act as lifting mechanisms:
   a. Convective lifting: Ground surface which is warmer than its surroundings, such as an ocean island, heats the air above it, which expands, becomes buoyant, and rises.
   b. Frontal lifting: Along fronts between air masses, the warmer air rises along the boundary surface, cools, and forms clouds. Cold fronts are steep, so clouds form only about 50 km in advance of the major portion of the cold air mass. Warm fronts are more gentle, so clouds form up to 800 km in advance of the main air mass. At occluded fronts clouds develop above the V intersection of the warm and cold fronts.
   c. Convergence lifting: Air masses push together and have nowhere to go but up. The rising air cools and condenses out moisture to form clouds.
   d. Orographic lifting: Humid air blown against a mountain range is forced to rise, causing moisture to condense, and this forms clouds above the mountains.

10. Thunderstorms are short lived (1 to a few hours) local episodes of intense rain, strong gusty winds, thunder, lightning, and possibly some hail. They develop in places like the mid-latitude of North America or over tropical rainforests, where cold fronts move into regions of warm humid air. The storm begins as a steady flow of warm moist air that rises, condenses out moisture, releases latent heat of condensation, and forms a cumulonimbus cloud that billows upward. Strong updrafts are created, which may even blow moisture high enough that it condenses out as ice crystals, which can develop into ice balls called hail. Eventually the cloud releases its moisture as falling rain, which pulls air down with it, creating strong downdrafts. The system is now in its mature stage with a "boiling" appearance, caused by concurrent strong updrafts and downdrafts, and a flattened spreading top region called an anvil cloud. Winds are gusty, and thunder and lightning are occurring. The lightning may flash back and forth within the cloud or may jump from the negatively charged bottom of the cloud, back and forth to the positively charged Earth surface beneath the cloud. The heat of a lightning flash, which can be as strong as a 30-million-volt pulse of electricity, causes air to expand explosively, which is the sound we call thunder. The thunderstorm eventually dissipates, as its downdrafts become strong enough to smother the updrafts that had been supplying the warm moist air.

11. Hurricane, typhoon, and cyclone are different names for huge rotating storms that develop over warm water. "Hurricane" applies to storms born in the east-central Atlantic, south of 20°N latitude, that drift first westward, then northward, and generally track across the Caribbean, Gulf Coast, and east coast of North America. "Typhoon" is the term applied to storms that form at latitude 20°N and 20°S latitude in the western Pacific. "Cyclone" is the name for storms that form at 20°N latitude in the Indian Ocean.

Practice Test

1. b The initial atmosphere was hydrogen and helium.
2. d It's supplied by the activity of photosynthetic organisms (green plants).

3. e Choices a to d are all true.

4. a The relation is an inverse one: the higher the elevation, the lower the air pressure: Mt. Everest's summit has 1/3 the air pressure found at sea level; sea-level pressure is about 14 lb/in^2 but this is equivalent to only 1 of the units called bars; and at 5 km the air pressure is about 1/2 sea-level pressure.

5. c The lower pressure would allow it to expand, slow its molecular motion (cool), rise, and experience adiabatic cooling.

6. d Latent (hidden) heat is released during the change of state, vapor to liquid.

7. c Weather phenomena and convection currents occur only in the troposphere, which is below the stratosphere; the tropopause is the boundary between troposphere and stratosphere; and meteors begin to burn in the thermosphere, which is above the stratosphere.

8. d It's at a higher altitude than choice d states, and includes most of the mesosphere and the lower part of the thermosphere.

9. c The more direct the radiation, the greater the insolation (incoming solar radiation).

10. a Deflection in the Northern Hemisphere is always to the right as viewed from the point of origin, so a northward-moving object would be deflected toward the east.

11. c The area at 60°N latitude is a convergence zone where surface polar air moving south collides with surface mid-latitude air moving north and rises; the exact position of the convergence shifts with seasonal changes but does not reverse.

12. c Hadley and Ferrel cells do converge, but at 30° latitude, to create a subtropical high with sinking air; at 60° a polar cell and a Ferrel cell converge, resulting in rising air of the subpolar low.

13. d Air at the equator is rising.

14. c Winds are named according to the direction they blow from.

15. d Jet streams are westerlies that originate where pressure gradients are extremely steep and flow between 200 and 400 km/h.

16. e Choices a to d are all true.

17. b A high-pressure mass, called an anticyclone, rotates clockwise in the Northern Hemisphere and has sinking air at its center; low-pressure cyclones are associated with fog and clouds.

18. a Surface clouds are called fog.

19. c Convective lifting does occur when air warms more over the ground surface than over the adjacent water surface; orographic lifting refers to air pushed up over mountains; convergence lifting happens when two air masses push together; and the Bergeron process is the growth of ice crystals in a cloud at the expense of water droplets.

20. d A cumulonimbus cloud has both updrafts and downdrafts, and therefore is stirred so violently it appears to boil.

21. d During the mature stage of the storm, both updrafts and downdrafts exist, but eventually downdrafts dominate and cut off the supply of moist air from updrafts, ending the storm.

22. a A cloud has a negatively charged base that repels negative ions of the ground surface, thus creating a positively charged ground surface.

23. e Lightning travels at the speed of light, so a person sees it essentially at the moment it happens; thunder travels at the speed of sound, which is roughly 1 mi every 5 s, so 7 s / 5 s/mi = 1.4 mi away.

24. d They're measured on the Fujita scale, which runs from F0 to F5; the Saffir-Simpson scale, which does run from 1 to 5, rates hurricanes.

25. a Tornado season is from spring through summer (March through September).

26. a Hurricanes begin in the central Atlantic, off the west coast of Africa.

27. c The North American east coast averages 5 hurricanes per year; the yearly total of 11 in 1995 was exceptionally high.

28. False On the side where the direction of rotating winds is in the same direction as the entire storm is moving, the effective wind velocity would be 180 + 50 = 230 km/h; on the opposite side, where the direction of rotating winds is opposite to the storm's direction of travel, the effective wind velocity would be 180 − 50 = 130 km/h.

29. True Atmospheric gases do leak away, and if these activities were to cease, within a few million years our atmosphere would become incapable of sustaining life.

30. False Chlorofluorocarbons (CFCs) release chlorine which breaks down ozone; carbon dioxide is a greenhouse gas that contributes to global warming.

31. False Cooler air can hold less moisture than warm air, so the reverse would be true, and moisture would condense out as the temperature dropped.

32. True   The statement is a correct summary of the cause of both the northern and the southern auroras.

33. False   The names are correct, but the correct order is Hadley (equator to 30° latitude), Ferrel (30 to 60°), and polar (60° to Pole).

34. False   The troposphere is thicker over the equator, because the hotter air here has expanded, and therefore the pressure gradient initiates flow from the equator to the poles; the Coriolis effect deflects these flows, toward the east in the Northern Hemisphere, toward the west in the Southern Hemisphere.

35. False   Cold fronts move faster than warm fronts; when a cold front overtakes a warm front, it flows under the warm air and lifts it.

36. False   During the day the clouds scatter and reflect some sunlight back into space and this does lower ground temperature, but at night clouds prevent infrared radiation (heat) from escaping and thus keep the ground warmer.

37. False   Reverse the terms "cold" and "warm," and then the statement is true.

38. False   Reverse the terms "warning" and "watch," and then the statements are true.

CHAPTER 21 | Dry Regions: The Geology of Deserts

GUIDE TO READING

Water is such an important factor in Earth's surface processes that the last four chapters have focused on it. The author considered fresh surface water in Chapter 17, ocean water in Chapter 18, groundwater in Chapter 19, and atmospheric water in Chapter 20. The chapter after this will again deal with water, but in its frozen form, glaciers. Is there no end to water? In this chapter you find it, for the subject is deserts. However, in a sense water is still an issue, for the defining characteristic of all deserts is the *lack* of water.

The chapter begins by dispelling the popular notion that all deserts are hot and sandy places. It lists and describes the five classes of deserts and points out that their only common factor is their aridity (dryness). The five classes are:

- subtropical: in the hot dry latitudes between 20 and 30°, both north and south
- rain shadow: on the landward side of coastal mountain ranges
- coastal: along coasts bordering cold ocean currents
- continental interior: deep within continents, far from major water sources
- polar: in the cold dry polar regions, both north and south

Do note the geographic locations given to illustrate the desert types because obviously their geographic locations explain why the regions are deserts and matching desert locations with desert types are frequently asked test questions.

Early in your reading you are reminded that plate tectonics activities, as usual, play a part in explaining geologic phenomenon. In the case of deserts, plate tectonics movements have:

- Created the mountains that produce rain shadow deserts by convergence, rifting, or collision.
- Sutured together small continents into large ones with interiors far from oceans and therefore dry.
- Moved land masses on shifting plates into the subtropics, where climatic conditions have turned them into desert regions.

The chapter continues with a discussion of how weathering and erosion processes are modified by desert conditions. Chemical weathering is made minimal and slow, but it does happen and it creates caliche-rich soils and some special desert features like case-hardened rock and desert varnish. Rainfall is minimal, so streams are intermittent (or ephemeral). Dry water channels and basins (washes, arroyos, wadis, and playas) are common. When water is present, it does a vigorous job of eroding the land and is a more important agent of erosion than the wind is. Flash floods are not rare. They scour the land, produce dramatic steep-sided channels, and polish the canyon walls with their sand-laden waters. Wind, like water, is a fluid and, like water, can carry its load of sediment suspended or as bedload, which it may bounce along the ground in the process of saltation. Wind isn't as powerful as water and can't move grains larger than coarse sand. Therefore it creates lag deposits and desert pavement. Wind may abrade rock surfaces to produce smooth faces (facets) on pebble-sized particles called ventifacts or may carve mushroom-shaped columns (yardangs) by the differential erosion of rock strata. Wind erosion can lower the land surface over large areas, a process called deflation, and can produce special circular depressed areas known as blowouts.

Material removed by erosion must be deposited somewhere, so logically deposition in desert environments is the next chapter topic. Deposits may be of varying sizes and may have been transported by a variety of agents. Large angular rocks tumble downhill due to gravity and pile up in talus aprons. Dust-sized particles lifted and then dropped by wind are called loess deposits. Desert streams drop materials when their gradients lessen and produce triangular shaped structures called alluvial fans. Overlapping alluvial fans may extend for miles along mountain fronts as elongate structures called bajadas. Streams carry various salts into desert lake basins (playas) where they are left as thick salt deposits when the water evaporates. Occasionally these interior basins (lakes with channel inlets but no channel outlets) are very large, like the Great Salt Lake of Utah and the Dead Sea along the Israel-Jordan border.

Desert landscapes are varied, and of course there is special vocabulary to describe all of them. "Hamada," "reg," and "erg" are terms used to designate very different aspects of the Sahara Desert region. Rocky desert areas change over time as scarp retreat forms pediments, mesas, buttes, chimneys, hoodoos, cuestas (hogbacks), dip slopes, inselbergs, and bornhardts. Depending on the amount of sand present and the constancy and velocity of the wind, sandy deserts are filled with different-shaped sand dunes. You read about barchan, star-shaped, transverse, parabolic, and longitudinal (seif) dunes, as well as the anatomy of an individual dune (windward slope, lee slope, slip face, angle of repose, and ripples).

Modern scientific thought tends to emphasize the interrelationships between the sciences. Therefore while this is a geology text and naturally concentrates on inorganic aspects of Earth study, it does point out interactions between the obvious realm of geology, the lithosphere (rocky Earth), and Earth's other spheres, the atmosphere (Chapter 20), the hydrosphere (Chapters 17, 18, 19, and 22), and the biosphere (interwoven through many of the chapters). The last part of this chapter concentrates on the biosphere. It discusses many adaptations life forms have developed that help them survive harsh desert conditions, and many human interactions with desert environments. Since the times of early civilizations in Egypt and Mesopotamia, humans have lived in desert regions, affected them, and been affected by them. You read about ancient Egypt and Mesopotamia, Ayers Rock in Australia, the Great Plains of North America, the Sahel region of Africa, and the Aral Sea in Asia.

The chapter concludes with a discussion of the process called desertification, in which semi-arid regions are changed into true deserts, partially due to natural causes, partially due to human activities. It is a somber reminder that geologic happenings influence human society, human society influences geologic happenings, and these interactions may be harmful to some fragile Earth environments and life forms.

LEARNING ACTIVITIES

Completion

Test your recall of new vocabulary terms.

_____ 1. piles of drifting sand

_____ 2. a region so dry it has little vegetation and no permanent rivers flowing through it

_____ 3. dry

_____ 4. a false image of a shimmering lake produced when a layer of hot air at ground surface refracts sunlight

_____ 5. regions of organic-rich mud, crusted with salt, that formed when land areas were temporarily flooded with seawater

_____ 6. arid regions that form on the inland side of coastal mountain ranges because precipitation falls on the coastal side

_____ 7. regions that are arid simply because they lie at the interior of large continents and air reaching them has already lost much of its moisture as precipitation

_____ 8. the process in which groundwater leaches out cementing material in a coarse-grained rock, redeposits it on the rock's exterior, and thus produces a hard-surfaced rock with a rotten interior

_____ 9. the dark, rusty-brown coating of iron and magnesium oxides deposited on rock surfaces in desert climates

_____ 10. drawings that are light-colored figures against dark backgrounds, created by Indians of the U.S. Southwest by chipping away desert varnish

_____ 11. a concrete-like material that develops in subsurface desert soil when calcite in downward-percolating water precipitates out and binds soil grains together

_____ 12. stream channels that contain flowing water only in times of exceptionally high rainfall

_____ 13. dry stream beds (names used in U.S. West)

_____ 14. dry stream beds (name used in Middle East and North Africa)

_____ 15. the entire quantity of fine-grained sediment (dust and silt) that is carried in suspension by the wind

_____ 16. the way sand grains are moved along the ground by moderate to strong winds, a combination of rolling and bouncing along

_____ 17. the entire quantity of sand-sized grains of sediment that are moved along the ground by the wind

_____ 18. sediment that is coarser than sand that's left behind after wind has removed the finer grains

_____ 19. a natural mosaic-like stone surface formed as lag deposits settle and fit together closely over time

_____ 20. the grinding away of surfaces in the desert by wind-blown sand and dust

_____ 21. smooth faces on pebbles, cobbles, and boulders created by wind abrasion

_____ 22. rocks whose surfaces have been faceted by the wind

_____ 23. mushroom-shaped columns of rock created when wind abrasion cuts into softer rock underlying more resistant rock layers at the surface

_____ 24. the lowering of the land surface throughout an area due to prolonged wind erosion

_____ 25. a specialized case of deflation, in which land has channeled the wind into a turbulent vortex that scours a deep, bowl-like depression

_____ 26. the rock debris that tumbled down a hill simply because of gravity and sits at the base of the hill in a fan-shaped mound

_____ 27. a wedge-shaped pile of sediment dropped by a stream where it entered a plain at the foot of mountains

_____ 28. an elongate wedge of sediment formed along the front of a mountain range by overlapping alluvial fans

_____ 29. a dry lake bed in a desert

_____ 30. lakes that have no outlet to the sea, lose water only by evaporation, and thus get saltier over time

_____ 31. fine-grained sediment layers that have been transported and deposited by the wind

_____ 32. desert regions in the Sahara that are barren, rocky highlands with elevations exceeding 3 km

_____ 33. desert regions in the Sahara that are vast stony plains formed by river deposition during wetter Pleistocene times

_____ 34. desert regions in the Sahara that are seas of sand filled with large dunes

_____ 35. the gradual retreat of a cliff (escarpment) as rock splits away from its face along vertical joints

_____ 36. the broad, almost horizontal bedrock surface formed along a retreating cliff face

_____ 37. large, flat-topped hills with several square kilometers of top surface area

_____ 38. medium-sized flat-topped hills (smaller than mesas, larger than chimneys)

_____ 39. flat-topped columns, whose vertical heights far exceed their top surface areas

_____ 40. bright-colored, strangely eroded chimneys (name used in Bryce Canyon National Park, Utah, and other parts of the U.S. Southwest)

_____ 41. asymmetric ridges that form where bedding dips at an angle to a horizontal line, analogous to mesas and buttes in areas of flat-lying strata

_____ 42. any rock slope whose surface sits at the same angle as the dip angle of the beds composing it (example: the gradual slope of a cuesta or hogback)

_____ 43. remnants of a hill after scarp retreat on all sides, so it sits isolated in the surrounding pediment or alluvium

_____ 44. inselbergs that have a loaf shape (steep sides, rounded crest)

_____ 45. crescent-shaped dunes in which the points of the crescents point downwind

_____ 46. sand dunes that constantly change shape, as wind direction shifts and causes a series of crescents to overlap each other in changing patterns

_____ 47. sand dunes that look like a sea of waves whose crests lie perpendicular to the wind direction, formed where the ground surface is completely covered by sand and only moderate winds blow

_____ 48. dunes whose ends point in the upwind direction, formed when transverse dunes were broken by strong winds

_____ 49. dunes whose axes lie parallel to the wind direction, formed where there's abundant sand and strong, steady winds

_____ 50. the steeper, lee face (opposite of windward) of a sand dune, on which sand becomes temporarily unstable and readjusts by sliding down the slope to achieve an angle of repose

_____ 51. the process of transforming nondesert areas into deserts

_____ 52. a choking storm of rolling black clouds formed when wind strips away silt and clay-sized particles of soil, lifts them high, and carries them away

Matching Questions

TYPES OF DESERTS

In the figure below, classify each of the deserts identified by type.

1. coastal _____

2. continental interior _____

3. polar _____

4. rain shadow _____

5. subtropical _____

Match the choices below with the appropriate description.

a. coastal
b. continental interior
c. polar
d. rain shadow
e. subtropical

_____ 1. large deserts which lie in climatic belts with little rain or cloud cover, intense sunlight, and high temperatures and which are located between latitudes 20 and 30°N and S

_____ 2. areas on the inland side of coastal mountain ranges

_____ 3. strips of continental land adjacent to cold ocean currents

_____ 4. areas deep within continental masses, far from ocean moisture

_____ 5. areas in the far north and south, beyond the Arctic and Antarctic Circles, where the area is constantly cold and dry

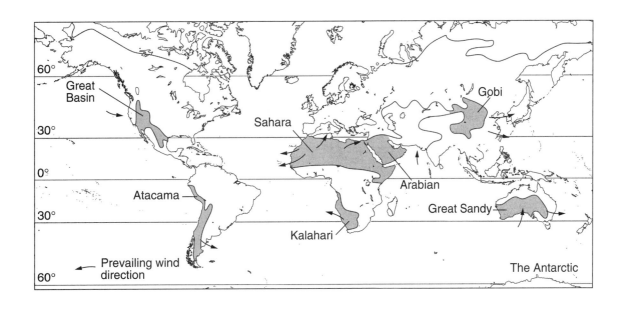

Short-Answer Questions

Fill in the blanks with a word, number, or brief phrase.

1. There are many varieties of deserts. What one characteristic do they have in common? _____

2. What is the typical (but not essential) rainfall criterion used to determine desert regions? _____

3. List several characteristics of cold deserts. _____
_____

4. List several characteristics of hot deserts. _____
_____

5. What causes a mirage? _____

6. What effect may the high evaporation rates in hot deserts have on falling rain? _____
_____

7. List three factors that contribute to the high amount of heat reradiation back into space at night? _____
_____

8. Daily temperature ranges in a desert may be extreme, possibly as great as _____

9. Deserts typically have _____ (thick, thin) soil, _____ (slow, fast) chemical reactions, _____ (very little, lots of) humus, and _____ (harsh and rugged, gentle and forested) landscapes.

10. If you took advantage of desert varnish, and carved a picture onto a desert rock face, you would end up with a _____ (light, dark) picture against a _____ (light, dark) background.

11. Which of the following causes the most erosion in deserts: wind, water, or temperature changes? _____

12. a. What might sand do to your car if your were driving through a desert sandstorm? _____
_____

    b. What might sand do to the second-story windows in a house in the desert during a sandstorm? _____
_____

13. Where would you be most likely to find salt deposits like halite, gypsum, or borax: in bajadas, ergs, playas, pediments, or inselbergs? _____

14. a. Which side of a sand dune is steeper, windward or leeward? _____

    b. Which side of a sand dune is the slip face, windward or leeward? _____

Making Use of Diagrams

SAND DUNES

Name the type of sand dune, and draw arrows marking the proper wind direction for each sketch.

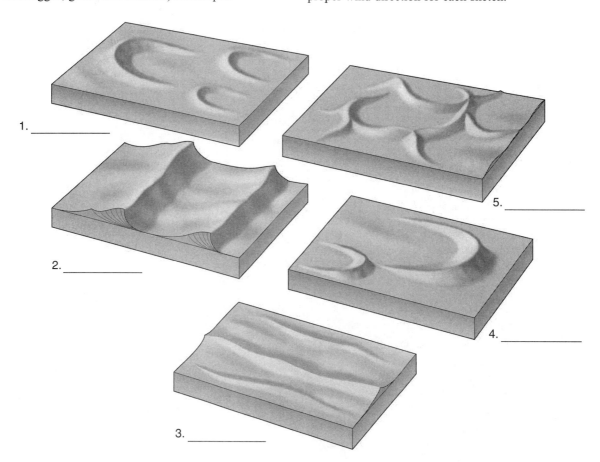

1. _____
2. _____
3. _____
4. _____
5. _____

Short-Essay Questions

1. Explain briefly, in terms of plate tectonics, why the following deserts exist: the Gobi, deserts in central Oregon and Washington States, the Basin and Range Province, and the Sahara.

2. Some of the storms of the Dust Bowl days in the 1930s carried material from the Midwest plains and deposited it on the Atlantic coast and even out to sea. Was some of this material sand? Explain your answer.

3. Discuss several physical characteristics of desert plants and animals that allow them to survive in such a harsh environment.

4. What lesson should we learn from the fact that the cradle of civilization 5,000 years ago was in the Middle East, and it's now barren desert land. Be sure to mention the term "desertification" in your answer.

5. Can desertification be reversed? Explain.

Practice Test

1. Deserts are
   a. hot.
   b. sandy.
   c. dry.
   d. located in the tropics.
   e. All of the above.

2. Which is not a term for a dry water channel or basin?
   a. "wadi"
   b. "wash"
   c. "arroyo"
   d. "mirage"
   e. "playa"

3. Choose the FALSE statement. Cold deserts occur
   a. along coasts bordered by warm ocean currents that cause much evaporation.
   b. where the temperature is usually below 20°C.
   c. at high latitudes.
   d. at high elevations.
   e. in places where sunlight strikes Earth obliquely.

4. Choose the FALSE statement. Deserts reradiate much of their heat back into space at night
   a. because dry desert air doesn't retain heat well.
   b. because they lack cloud cover, which would act as a blanket and reradiate heat back to Earth.
   c. but even then the temperature never goes below freezing.
   d. and thus may have daily surface temperature highs of as much as 80°C.
   e. even if they are the hottest of deserts.

5. Which of the following is a characteristic feature of deserts?
   a. thick soils
   b. flash floods
   c. abundant humus
   d. fast chemical reactions due to high temperatures.
   e. All of the above are true.

6. The subtropics
   a. lie between 5 and 15° latitude, both north and south of the equator.
   b. are climate belts where cloud cover blocks incoming solar radiation.
   c. are areas of hot, dry rising air currents.
   d. host the world's largest deserts, including the Sahara and Kalahari.
   e. All of the above are true.

7. Choose the FALSE statement. Because of typical convection cell movement in the atmosphere,
   a. hot moist air close to the equator rises.
   b. rising air expands and cools.
   c. cooled air spreads north and south and its moisture condenses out in latitude zones between 20 and 30°N and S.
   d. the cooled dry air sinks and warms.
   e. cooled sinking air absorbs moisture from the area below.

8. Desert varnish
   a. is a product of chemical weathering over time.
   b. is a dark, rust-brown surface coating of iron oxide and magnesium oxide.
   c. happens when precipitation dissolves ions out of the rock, then leaves them as evaporation residue on the rock surface.
   d. can be chipped away to create light colored petroglyphs.
   e. All of the above are true.

9. Which two compounds often cement desert regolith together to form a concrete-like material called caliche?
   a. iron oxide and magnesium oxide
   b. calcite and gypsum
   c. iron oxide and calcite
   d. halite and gypsum
   e. quartz and halite

10. Choose the FALSE statement. Moderate winds can
    a. move sand grains by saltation.
    b. raise dust and sand grains more than 30 m above Earth's surface.
    c. create vortices that lift fine sediment as dust devils.
    d. carry sediment as both suspended and bed load.
    e. round and frost sand grains by knocking them together.

11. Moderate wind
    a. can produce faceted ventifacts.
    b. deposits are poorly sorted.
    c. piles fine sediment into hills called lag deposits.
    d. can lift gravel about 2 m high.
    e. carves cliff bases into triangular alluvial fans.

12. Which feature might result where wind erosion acts on a resistant rock layer overlying a softer rock layer?
    a. desert pavement
    b. talus
    c. yardang
    d. bajada
    e. seifs

13. Which term has nothing to do with temporary desert lakes?
    a. "halite"
    b. "playas"
    c. "gypsum"
    d. "reg"
    e. "borax"

14. Which term applies to a region rich in sand?
    a. "erg"
    b. "hamada"
    c. "reg"
    d. "pediment"
    e. "desert pavement"

15. Flat-lying sedimentary strata or volcanic rock layers may erode to form isolated structures called
    a. hogbacks and cuestas.
    b. dip slopes.
    c. mesas, buttes, and chimneys.
    d. barchans.
    e. All of the above are true.

16. In the accompanying diagram,

    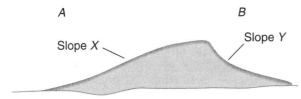

    a. the slip face is labeled "Slope Y."
    b. slope Y will adjust itself until it reaches the angle of repose.
    c. the wind is blowing from A to B.
    d. the entire dune will shift in the direction A to B.
    e. All of the above are true.

17. Which type of sand dunes would form where there's abundant sand totally covering the ground surface and only moderate winds?
    a. barchan
    b. star shaped
    c. transverse
    d. parabolic
    e. longitudinal (seif)

18. Which would be the steepest hike?
    a. up the slip face of a dune
    b. up the windward face of a dune
    c. across the pediment to the base of a cliff
    d. up the backside of a cuesta
    e. across the top of a butte

19. Sand dunes
    a. can migrate downwind more than 25 m in a year.
    b. may be stabilized if they are covered with vegetation.
    c. in cross-sectional view show cross beds of sand.
    d. have ripples on their surfaces.
    e. All of the above are true.

20. What characteristic is NOT an adaptation for desert survival?
    a. a pattern of active night life
    b. very small ears to expose little surface area to the sun
    c. minimal sweating and urination
    d. deep tap roots
    e. fleshy stems and leaves

21. What do the Sahel, the Aral Sea of today, and the U.S.-Canadian Great Plains of the 1930s have in common?
    a. sand dunes greater than 100 m high
    b. 3 in. or less annual rainfall
    c. being some of the sandiest places in the world
    d. experienced natural and human-induced desertification
    e. All of the above are true.

22. By definition, deserts are areas with 25 cm (10 in.) or less annual rainfall.
    a. True
    b. False

23. The hottest verified Earth temperature was 176°F in Death Valley, California.
    a. True
    b. False

24. Rain shadow deserts are located on the coastal side of mountain ranges because air has to rise here to cross the mountains and it absorbs moisture as it rises.
    a. True
    b. False

25. Continental-interior deserts exist because air masses have traveled so far in getting there they've precipitated most of their moisture along the way, even if they didn't cross mountains.
    a. True
    b. False

26. Alternate heating and cooling of desert rock causes it to expand and contract daily, and this is the chief cause of rock weathering in the desert.
    a. True
    b. False

27. Case hardening is the process in which calcite and quartz are leached out of coarse-grained rock and redeposited on the rock's surface, resulting in a hollow, very hard rock.
    a. True
    b. False

28. Colored rock layers of the Painted Desert, Arizona, result from oxidized iron.
    a. True
    b. False

29. "Loess" is the term for wind-blown sand.
    a. True
    b. False

30. It would be easier to climb to the top of a hogback than to the top of a chimney because the hogback has two sloping sides, one of which is gentle, but a chimney has two vertical sides.
    a. True
    b. False

31. Ayers Rock in Australia is a loaf-shaped inselberg called a bornhardt.
    a. True
    b. False

32. Sand dunes can migrate, but very slowly, possibly 25 m/century.
    a. True
    b. False

33. Thick-skinned seeds; shallow, broad root systems; deep taproot systems; and fleshy stems and leaves are all special adaptations of desert plants.
    a. True
    b. False

34. Match the desert location with the class of desert.
    a. coastal
    b. continental interior
    c. polar
    d. rain shadow
    e. subtropical

    _____ i. Atacama Desert, west coast of South America
    _____ ii. deserts of the states of Nevada and Washington
    _____ iii. Gobi Desert, Mongolia
    _____ iv. Great Australian Desert
    _____ v. land at 80° south latitude
    _____ vi. Sahara Desert, Africa

35. Match the plate tectonics reason for its development with the desert location.
    a. continental collision
    b. plate convergence and subduction
    c. climate changes due to plate movement
    d. continental rifting

    _____ i. in Chile and Argentina, east of the Andes
    _____ ii. in elongate valleys of the Basin and Range Province, western United States
    _____ iii. in the interior of Asia
    _____ iv. the Sahara, northern Africa

ANSWERS

Completion

1. sand dunes
2. desert
3. arid
4. mirage
5. sabkhas
6. rain shadow deserts
7. continental-interior deserts
8. case hardening
9. desert varnish
10. petroglyphs
11. caliche
12. intermittent (or ephemeral) streams
13. washes and arroyos
14. wadis
15. suspended load
16. saltation
17. bed load
18. lag deposit
19. desert pavement
20. wind abrasion
21. facets
22. ventifacts
23. yardangs
24. deflation
25. blowout
26. talus apron
27. alluvial fan
28. bajada
29. playa
30. interior basins
31. loess
32. hamada
33. reg
34. erg
35. scarp retreat
36. pediment
37. mesas
38. buttes
39. chimneys
40. hoodoos
41. cuestas or hogbacks
42. dip slope
43. inselberg
44. bornhardts
45. barchan dunes
46. star-shaped dunes
47. transverse dunes
48. parabolic dunes
49. longitudinal dunes (or seif dunes)
50. slip face
51. desertification
52. dust storm

Matching Questions

TYPES OF DESERTS

1. Atacama
2. Gobi
3. the Antarctic
4. Great Basin
5. Arabian, Great Sandy, Kalahari, Sahara

272 | Chapter 21

TYPES OF DESERTS

1. c
2. d
3. a
4. e
5. b, f, g, h

Short-Answer Questions

1. aridity (dryness)

2. less than 25 cm/year

3. temperatures below 20°C and locations at high latitudes with oblique sunlight, high elevations, or adjacent to cold oceans

4. temperatures above 35°C and locations at low latitudes with high-angle sunlight, low elevations, or far from cold oceans

5. refraction of sunlight by a very hot (up to 77°C or 170°F) layer of air right at the surface

6. May evaporate in midair, leaving the land surface below dry.

7. very dry air, lack of cloud cover, and lack of foliage

8. 80°C

9. thin; slow; very little; harsh and rugged

10. light; dark

11. water

12. a. Strip the paint and frost the windows.
    b. Nothing; the wind couldn't carry sand that high.

13. in playas, which are dry lake basins (ergs are sandy desert areas; bajadas are overlapping alluvial fans, pediments are bedrock surfaces along cliff faces, and inselbergs are rocky erosional remnants of cliffs)

14. a. leeward   b. leeward

Making Use of Diagrams

SAND DUNES

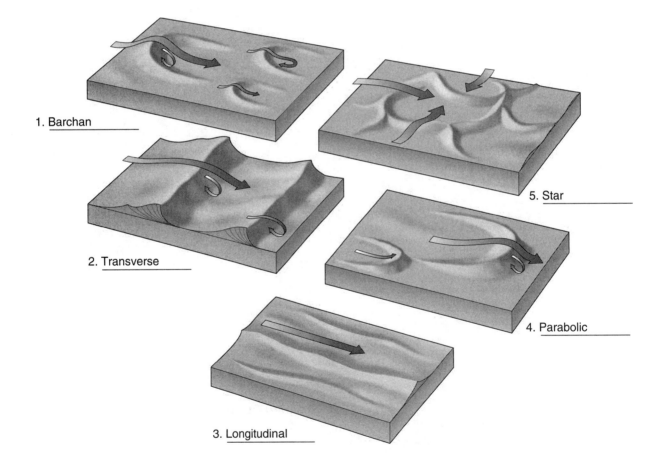

1. Barchan
2. Transverse
3. Longitudinal
4. Parabolic
5. Star

Short-Essay Questions

1. The Gobi Desert lies deep in the interior of Asia, far from oceans, due to the additions, by continental collision, of the European continent on Asia's western border and the subcontinent of India on its southern border. Central Oregon and Washington are in the rain shadow of the Cascade Mountains, which were formed by plate convergence, subduction, plate melting, and eventual surface volcanics. Deserts in the Basin and Range Province of the western United States are in the grabens (valleys) between horsts (elongate mountain blocks) created by continental rifting. The Sahara region, which used to be green and lush, was moved into its current subtropical location as the African plate moved northward.

2. The storms of the 1930s were true dust storms, not sand storms. Ordinary strong wind currents can't lift sand grains higher than about 2 m; it takes special severe storms like tornadoes and hurricanes to do this. Sediment carried hundreds of miles, from midcontinent to the East Coast, over all types of topography, including mountain ranges, was dust sized (silt and clay), not sand sized.

3. Desert plants and animals have numerous characteristics that enable them to survive the extreme temperatures and minimal water conditions of deserts.

   Plants may:
   Have thick-skinned seeds that can be dormant until sufficient rain falls to let them successfully germinate.
   Have deep-reaching taproot systems to stretch below even deep water tables into the zones of saturation.
   Have shallow broad-root systems that let a plant reach far from itself to procure any available water in the region.
   Be succulents, with thick, fleshy stems and leaves that store water for long periods of time and thorns and spines to ward off animals that would steal their water.

   Animals may:
   Be burrowers capable of dormancy; they can dig in and wait until it rains.
   Come out of cool hiding places only at night.
   Have large ears through which they radiate heat energy.
   Retain body water by not sweating and by producing concentrated urine.

   Until modern times, humans also had to make extreme adaptations to survive in desert conditions. They could live only in small groups, sometimes in carved rock (loess) caves underground, and had to use available food resources that don't sound very appetizing to most of us (like moisture-rich grubs).

4. Over time, regions naturally change from nondesert to desert areas, and vice versa. Over long geologic time periods these changes happen for plate tectonics reasons (as described in essay question 1). Short-term changes from nondesert to desert, desertification, can be naturally caused by drought, but changes are often speeded up considerably by human factors, including overpopulation, overgrazing, careless agriculture, and water diversions. Even in the modern world desertification can have tragic consequences, like mass starvation, because the people affected are not usually in industrialized countries equipped to deal with the changes or reverse them. The histories of the Sahel region on the southern margin of the Sahara; the Aral Sea in Kazakstan, Asia; and the Great Plains of the United States and Canada are full of examples of thoughtless human activities that adversely affect human civilization and Earth itself.

5. Desertification can be reversed but it takes time, effort, money, and water to do so. Basically, reversal is simple: plant and irrigate crops. This is not a simple accomplishment. In theory, necessary water may come from river diversion, groundwater pumping, desalinization of ocean water, or transporting icebergs from Antarctica, none of which are easy or problem free. A more reasonable answer to the problem of human-induced desertification is not to let it happen, and this will require intelligent, farsighted land use policies when dealing with semi-arid lands.

Practice Test

1. c There are many kinds of deserts; all they have in common is a dry condition.

2. d A mirage is a fake image of shimmering water, produced when a surface hot-air layer refracts sunlight.

3. a Cold ocean currents cool the air above, and cold air can't hold much moisture.

4. c Many desert areas, such as those at high elevations or high latitudes, have temperatures below freezing.

5. b Soils and humus are sparse to nonexistent; lack of water slows chemical reactions.

6. d The subtropics lie between 20 and 30° latitude, north and south; clouds and rain rarely occur; and the hot dry air is sinking.

7. c   The air between latitudes 20 and 30°N and S, already lost its moisture closer to the equator; it's now cool, dry, sinking, and absorbing moisture.

8. e   Choices a to d are all true statements.

9. b   Only choice b offers the two ingredients of caliche.

10. b   Fine-grained dust can be carried 30 m high by moderate winds; sand can be lifted only about 2 m.

11. a   Wind deposits are well sorted; lag deposits are materials left behind by wind erosion; wind can't lift gravel; and streams create alluvial fans.

12. c   Yardangs are mushroom-shaped columns formed under the conditions described.

13. d   "Reg" means stony desert plains formed by Pleistocene streams in north Africa; playas are the lake basins; and the other terms are possible evaporite deposits in the basins.

14. a   Hamada are barren rocky desert highlands; reg are stony desert plains; pediment is exposed bedrock at cliff bases; and desert pavement is closely packed coarse sediment on desert surfaces.

15. c   Hogbacks, cuestas, and dip slopes are tilted features; barchans are sand dunes.

16. e   Choices a to d are all true statements.

17. c   Barchans indicate scarce sand supply and steady winds; star shaped, scarce sand supply, and variable wind directions; parabolic, strong winds acting on transverse dunes; and longitudinal (seif), abundant sands and strong wind

18. a   The slip face is the steep, lee side of a dune; windward dune faces and cuesta slopes are gentle; and pediments and butte tops are horizontal surfaces.

19. e   Choices a to d are all true statements.

20. b   The desert adaptation concerning ears is large ears to radiate excess body heat.

21. d   The three locations do not have a common factor of large dunes, extremely low rainfall, or large amounts of sand.

22. False   Often that's true, but annual rainfall greater that this, but sporadic, or intensive evaporation can also produce sufficiently arid conditions to be classified as desert.

23. False   The hottest temperature in Death Valley was 133°F; the hottest in the world was 136°F in Libya, Africa.

24. False   Air rises and cools, and its moisture condenses and precipitates on the coastal side. Cool dry air sinks and absorbs moisture on the landward side, thus producing deserts.

25. True   The statement correctly explains why the interiors of large continents can be desert areas.

26. False   Alternate heating and cooling may contribute to physical weathering, but its importance is controversial.

27. False   The process is correctly described, but the resulting rock is a poorly cemented aggregate covered with a rust-brown hard coating.

28. True   Rock layers of the Painted Desert are colored by the oxidation of iron they contain.

29. False   "Loess" stands for wind-blown fine-grained sediment, that is, dust (silt- and clay-sized grains).

30. True   The sides of a chimney are vertical, and scaling it would require ropes; hogbacks have sloping sides, and it would be possible to walk up the gentler slope.

31. True   An inselberg is a remnant island of rock, surrounded by pediment or alluvium, left by scarp retreat. If it's loaf-shaped, it is called a bornhardt.

32. False   Sand dunes do migrate, but much faster than this, possibly more than 25 m/year.

33. True

34.
   i. a
   ii. d
   iii. b
   iv. e
   v. c
   vi. e

35.
   i. b
   ii. d
   iii. a
   iv. c

CHAPTER 22 | Amazing Ice: Glaciers and Ice Ages

GUIDE TO READING

You might think by now there would be nothing left to say about water. Not so. This chapter is all about water, but in its solid state, ice. More specifically, the chapter deals with glaciers, which are accumulations of snow that recrystallize into ice and begin to flow. They create an awesome but hostile environment. If they are widespread, they can dominate conditions over a large percentage of Earth for thousands of years. If they occupy smaller areas, they can rework the scenery, by erosion and by deposition, and create unique landscapes that were first correctly interpreted by Louis Agassiz in the mid-1800s. The chapter begins by recognizing his work and continues with a discussion of the nature and characteristics of ice, its albedo, crystal form, reactions to temperature and pressure changes, and similarities to metamorphic rocks.

A discussion of ice quite naturally leads to the many types of glaciers (mountain or alpine, cirque, valley, mountain ice cap, piedmont, and continental or ice sheet). The author makes the point that while pieces of glaciers may end up in the sea by the process of calving off icebergs, all true glaciers originate and move on land. They begin as accumulations of snow that change to firn and eventually to ice. (Masses of ice that originate as frozen seawater are called ice shelves.) Once formed, glaciers move. They may be wet-bottom or temperate glaciers and move by basal sliding, or they may be dry-bottom or polar glaciers and move by internal flow. They may break apart in their upper brittle zone as they move over uneven ground and create crevasses. Their average speeds may vary from 10 m to a few hundred meters per year, and they may show occasional periods of exceptionally fast movement called surging. The ultimate cause of glacial movement is the pull of gravity, which may create lateral movement called gravitational spreading or simple down-valley motion. To complicate the picture, even though glacial ice always moves in response to gravity, it doesn't always appear to keep advancing in the same direction. If it loses enough ice by melting, sublimation, or calving, its terminus (toe) may retreat, and its zones of ablation and accumulation may grow or diminish.

Glaciers are powerful agents of erosion. They plow through the landscape, incorporate and pluck (quarry) fragments from the bedrock, abrade the land surface, leave chatter marks, gouge out striations, and grind off rock flour to leave glacially polished surfaces. In their head (upper) regions, valley glaciers erode mountain peaks and valleys to create spectacular, jagged scenery composed of cirques, tarns, arêtes, horns, U-shaped valleys, hanging valleys, truncated spurs, roche moutonnées, and fjords.

Glaciers also cause major changes in the landscape by their depositional activities. First-time observers in glacial areas are often amazed at the amount of sediment and rock associated with glaciers, at times more obvious than the ice. Sometimes rocks even take over, and the glaciers become slowly moving jumbles of rock, impregnated with ice, called rock glaciers. There are always moraines—lateral, medial, end, terminal, ground, and recessional. Many other depositional features may occur, including kames, glacial drift (stratified and unstratified), till, lodgment till, erratics, drop stones, glacial marine sediment, outwash, glacial lake bed sediment, varves, drumlins, kettle holes, knob and kettle topography, and eskers.

Not surprisingly, glaciers greatly modify the climate. Average temperature can be up to 13°C cooler, and strong glacial winds (catabatic winds) pick up and transport and deposit fine-grained sediment called loess, creating immediate dusty conditions and future fertile farmland.

Glaciation leaves its mark on the landscape. Some previously glaciated areas are unique enough they have been made national parks (Glacier in Montana, Yosemite in California,

Voyageurs in Minnesota, Acadia in Maine, and Glacier Bay in Alaska). Other notable glacial scenery is found in the rugged mountains of the European Alps, the Rockies of western North America, the Andes of South America, and the fjord country of New Zealand, Alaska, and Scandinavia. The thousands of small lakes of northern Minnesota and the Great Lakes and the Finger Lakes in central New York owe their existence to the Pleistocene ice sheet.

Ice loading causes glacial subsidence, and the removal of ice causes glacial rebound. Sea level changes drastically when glaciers tie up great quantities of water, and this affects life in the area. Lowered sea levels in Pleistocene times exposed land bridges that allowed life (including humans) to migrate extensively. (For example, humans crossed the Bering Straits from Asia to Alaska and eventually spread throughout North America.) Sea level is higher now than it was in the Pleistocene, and it will get much higher if all current ice sheets melt. If that happens, coastal areas will be flooded, numerous new land lakes will form, and stream systems will be altered. You read of warming times toward the end of the Pleistocene that resulted in meltwater lakes, pluvial lakes, oversized valleys, and catastrophic floods like the Great Missoula Flood. Even areas around but not under the ice (periglacial areas) showed distinctive features like permafrost, patterned ground, and stone rings.

Roughly the last third of the chapter is devoted to ice ages, particularly the most recent, Pleistocene Ice Age, which began about 3 million years ago and ended (if it really did) about 11,000 years ago. You read about the Laurentide and Cordilleran ice sheets that covered northern North America, and how ice sheets covered roughly 30% of all land and greatly affected life on Earth. *Homo sapiens* was one of the life forms that had to cope with the harsh environment.

Why have ice ages happened? Milankovitch's ideas (which involve cyclical changes in Earth's orbit); plate tectonics phenomena such as shifting continents, mountain building, and continental rifting; and changes in the amounts of atmospheric carbon dioxide are all major factors to consider.

Are we still in an ice age, possibly an interglacial period of the Pleistocene? Will another great ice age come in the near future? How do scientists study these issues? In the 1800s Louis Agassiz could only interpret the local rock record. Scientists today can study tillites worldwide, apply radiometric dating to glacially killed trees, study biologic communities like ocean plankton, and interpret oxygen isotope ratios of marine shells. Data they have collected have changed some long-held ideas and produced some surprising theories.

Scientists now believe Earth has experienced four major ice ages, during the Mesoproterozoic, Late Proterozoic, Permian, and Pleistocene. They no longer think there were just four Pleistocene glaciations in North America (Nebraskan, Kansan, Illinoian, and Wisconsin), but instead they think there were at least thirty. The subject is very complex, and new research has produced more questions than answers about ice age issues. Equally qualified experts predict totally different glacial futures for planet Earth, and neither side can absolutely prove its viewpoint. Only time will tell.

Completion

Test your recall of new vocabulary terms.

_____ 1. boulders transported and dropped by glaciers far from their places of origin

_____ 2. sheets or streams of recrystallized ice that flow slowly and last many years

_____ 3. a time period during which a large percentage of Earth's land is covered with thick ice sheets

_____ 4. the ability of a substance to reflect radiation from its surface, measured as a percentage of the total radiation received

_____ 5. the category of glaciers that exist in or adjacent to mountainous regions, of which there are several subcategories

_____ 6. glaciers that fill the bowl-shaped depressions high on mountains in which glaciers originate

_____ 7. bowl-shaped depressions high in the mountains in which snow accumulates and eventually recrystallizes into glacial ice

_____ 8. rivers of ice that flow down valleys

_____ 9. sheets of ice that bury peaks and ridges along the crests of a mountain range

_____ 10. tongues of ice that extend out from valley glaciers onto the plains adjacent to a mountain range

_____ 11. enormous sheets of ice that spread over hundreds to thousands of square kilometers of continental crust

_____ 12. tongue-shaped extensions of ice along the front edge of a glacier that occur because not all of the glacial ice flows at the same speed

_____ 13. the ice version of a waterfall, which forms where glacial ice flows over a steep cliff

_____ 14. the process in which water changes from the solid state to the vapor state without going through the liquid state

_____ 15. the granular material, containing about 25% air, that is a stage in the transformation of snow to glacial ice

_____ 16. the melting of snow at the point where pressure is applied

17. a method by which a glacier moves on a layer of water under its base
18. the type of glacier for which basal sliding is the dominant style of movement
19. a style of glacial movement that results when glacial ice slowly changes its internal shape but does not break apart or completely melt
20. the type of glacier for which internal flow is the dominant style of movement
21. a large open crack in the brittle zone (upper 60 m) of a glacier
22. parts of a glacier in which the ice is flowing over water or wet sediment and therefore is moving faster than the ice flowing over adjacent dry-bottom parts
23. a pulse of rapid flow of a glacier, lasting no more than a few months, during which the glacier may move 10 to 100 times its usual speed
24. the spreading of a continental ice sheet from its thickest region outward in all directions due to its own weight
25. the loss of ice from a glacier due to sublimation, evaporation, and calving
26. the breaking off of large chunks of ice along the leading edge of a glacier
27. the upper area on a glacier where the amount of ice is growing due to added snowfall
28. the lower area on a glacier where it is growing smaller due to the loss of ice by sublimation, melting, and calving
29. the boundary between the zone of accumulation and the zone of ablation
30. the leading edge of a glacier
31. the obvious movement forward of a glacier when the rate of supply exceeds the rate of ablation
32. the *apparent* movement of a glacier back toward its source area when the rate of ablation exceeds the rate of supply
33. rocks carried out to sea in icebergs and dropped there when the ice melts
34. large floating sheets of ice created where the surface of the sea freezes
35. a type of glacial erosion in which flowing ice bulldozes rock loose and pushes it along
36. a method of glacial erosion in which ice surrounds loose debris and freezes it into itself
37. a method of erosion in which the glacier breaks off fragments of bedrock and freezes them into its bottom
38. triangular-shaped indentations on rock surfaces caused by glacial abrasion
39. the grinding away of bedrock by stone fragments embedded in the glacier
40. the fine-grained sediment produced by glacial abrasion (analogous to the dust produced by sanding wood)
41. shiny surfaces produced by glacial abrasion with embedded sand grains
42. elongate scratches or deep grooves gouged into the bedrock by large rock fragments frozen into the bottoms of glaciers
43. the highest area of a glacier, which is its starting edge
44. a lake that forms in a cirque after the glacier has melted
45. a knife-edged ridge of rock that separates two adjoining cirques or glacial valleys
46. a pyramidal pointed mountain peak created by ice in a minimum of three cirques eroding the peak
47. the typical shape of a youthful stream valley
48. the typical shape of a stream valley that has been reworked by glacial erosion
49. valleys high along the sides of a larger trunk valley, created by tributary glaciers that flowed into a large main glacier
50. masses of flat-faced rock protruding slightly into glacial valleys, created when the passing glacier sheared off greater protrusions
51. an asymmetric rounded bedrock hill shaped by a continental glacier flowing over it (French for "sheep rock")
52. deep, steep-sided valleys cut by coastal glaciers and later flooded by seawater
53. bands of sediment carried along on the ice at the sides of a glacier and, when the ice melts, dropped so they create elongate ridges along the sides of the glacial valley

278 | Chapter 22

_____ 54. a stratified sequence of sediment along a glacier's edge formed by flowing water's reworking the sediment of a lateral moraine

_____ 55. the stripe of debris created when two valley glaciers merge and their lateral moraines join to become an elongate ridge on the interior surface of the composite glacier

_____ 56. a mass consisting of rock debris and ice that moves down valley

_____ 57. any sediment deposited by glacial action

_____ 58. the sediment that has been carried on glacial ice and deposited, unsorted, along its bottom or margins

_____ 59. any glacial sediments that were first deposited, unsorted, by glacial ice, then sorted and redistributed by flowing water

_____ 60. the ocean-bottom sediment that is a mixture of drop stones and ordinary marine sediment

_____ 61. the till that was deposited by a glacier at its toe, then picked up by streams of glacial meltwater and transported and deposited as sorted sediment further from the glacier

_____ 62. a broad area of sorted sediment, fanning out from the toe of a glacier, that was deposited there by braided streams of glacial meltwater

_____ 63. lakes formed a short distance down valley from glacial fronts

_____ 64. a thick layer of fine-grained sediment that settles in meltwater lakes

_____ 65. a pair of thin layers of glacial lake bed sediment, one silt brought by spring floods, the other clay deposited in still winter waters

_____ 66. strong winds along glacial margins caused by the temperature differences between cold air over glaciers and warmer air over land

_____ 67. the fine-grained sediment that is transported and deposited by wind

_____ 68. any ridge of glacial till that forms at a glacier's toe when it stalls for a while

_____ 69. the end moraine that marks the position of the farthest advance of a glacier

_____ 70. the series of end moraines that mark the sporadic retreat of a glacier

_____ 71. the flattened layer of till created when a glacier advances over an end moraine

_____ 72. a thin, hummocky ground-cover that consists of lodgment till and till left behind during rapid glacial recession

_____ 73. a circular depression made by a block of ice calved off the toe of a glacier and buried in surrounding till

_____ 74. land surface made up of numerous kettle holes separated by round hills

_____ 75. twisting ridges of sorted sand and gravel that formed in tunnels of water flowing in the base of a glacier

_____ 76. the weighting down of the land by a large ice sheet

_____ 77. the sinking of the surface of a large area of land due to ice loading

_____ 78. the gradual uplift of land surface after its covering ice sheet melts

_____ 79. a lake formed along a glacial front where the ice has blocked the flow of a river

_____ 80. valleys which contain streams that obviously are too small to have created them and which were carved by outwash streams along former glaciers

_____ 81. the unusual landscape of eastern Washington: a semi-arid region of barren basalt bedrock with huge steep-sided valleys scoured into the flows

_____ 82. large vertical-sided valleys in basalt bedrock of Washington State

_____ 83. a monstrous flood that altered the landscape of the northwestern United States at the end of the Pleistocene when an ice dam gave way

_____ 84. lakes that were not directly associated with glaciers but existed in Pleistocene times simply because the climate was much wetter than it is now

_____ 85. permanently frozen ground

_____ 86. regions around the edges of glacial environments that are not covered by ice but do contain lots of permafrost

_____ 87. an unusual land surface, found in permafrost regions, that appears to consist of polygons laid down next to each other

_____ 88. unusual groupings of cobbles and pebbles that have been pushed to the surface by alternate freezing and thawing of ground above permafrost

_____ 89. the most recent ice age, responsible for the glacial landforms of North America and Eurasia

_____ 90. the current geologic time period, which began 11,000 years ago

_____ 91. the Pleistocene ice sheet that covered all of Canada east of the Rocky Mountains and extended as far south as southern Illinois

_____ 92. the Pleistocene ice sheet that covered the mountains of western Canada and the southern third of Alaska

_____ 93. treeless areas, underlain by permafrost, that support only low shrubs, moss, and lichen

_____ 94. the periods during ice ages when glaciers grow

_____ 95. the intervals of ice ages when glaciers retreat

_____ 96. deposits of glacial till that have hardened into rock

_____ 97. the gradual change of shape of Earth's orbit from more circular to more elliptical

_____ 98. the "wobble" of Earth's axis due to the slowing down of Earth's rotation

_____ 99. the amount of exposure to the Sun's rays

_____ 100. climate cycles predicted by an astronomer and geophysicist, that support a theory that explains the occurrence of ice ages

_____ 101. processes that are the result of some particular event, and perpetuate and enhance the event that caused them in the first place

_____ 102. the years between the 1300s and the mid-1800s, when average annual temperatures in the Northern Hemisphere fell enough to cause a significant advance of mountain glaciers

Matching Questions

GLACIAL LANDSCAPES

Match the geographical location with the description of its glacial landscape.

a. Acadia National Park, Maine
b. Glacier Bay National Park, Alaska
c. Glacier National Park, Montana
d. Illinois swamplands
e. Teton lakes, Wyoming
f. Voyageurs National Park, Minnesota
g. Yosemite National Park, California

_____ 1. an area of large roche moutonnées and small fjords created when continental glaciers reached the sea

_____ 2. the glacial scouring and deposition on the Canadian shield that produced glacial polish, striations, erratics, moraines, and outwash plains

_____ 3. moraine-dammed lakes

_____ 4. the spectacular, sharp, angular mountain scenery of cirques, U-shaped valleys, hanging valleys, terminal moraines, and many small glaciers

_____ 5. the U-shaped valley carved into the Sierra Nevada granite batholith, with its associated hanging valleys and waterfalls

_____ 6. valley glaciers calving into the sea and active inland glaciers

_____ 7. wetland depressions composed of glacial lake beds, ground moraines, and outwash plains separated by forested ridges of recessional and terminal moraines

MILANKOVITCH'S IDEAS

Fill in the blanks in the narrative below from the list of choices.

Degrees: a. 12–15°   b. 22.5–24.5°   c. 23.5–40.5°
Temperatures: a. 4°C   b. 5–7°C   c. 10–13°C
             d. 30–40°C
Terms: a. correct   b. eccentricity cycle
             c. incorrect   d. Polaris   e. precession
             f. Vega
Years: a. 25,000   b. 41,000   c. 100,000

Milankovitch discovered three types of cyclical changes in the geometry of Earth's orbit. First, the orbit changes from more circular to more elliptical every _____ years. This phenomenon is called the _____. Second, the tilt of Earth's axis changes from _____ every _____ years. Third, Earth wobbles on its axis, as does a top, because its rotation is slowing. This phenomenon is called _____. It causes the axis to trace a conical path in the sky and to complete a circle every _____ years. The direction the axis points changes over time, so our north star changes over time. Currently our north star is _____, but 12,000 years ago it was _____. On the basis of these orbital changes, Milankovitch made predictions of climate cycles, which turned out to be _____. His ideas offered a partial explanation for glacial advances and retreats within an ice age. They correctly predicted the timing of the events but not the severity of the temperature changes. Milankovitch's calculations predicted a decrease of about _____, but the actual temperature decreases during glaciation were _____ along coastal areas and _____ inland.

GLACIAL FORMATION VERSUS GLACIAL MELTING

Label each of the following situations F if it contributes to glacial Formation or M if it contributes to glacial Melting.

_____ 1. high rates of precipitation

_____ 2. ice loading causing glacial subsidence (thus lower elevations and warmer summers)

_____ 3. large expanses of frozen ocean surface

_____ 4. large areas of land with low temperatures and high elevations

_____ 5. widespread blanket of snow

_____ 6. warming temperatures and increasing precipitation

Short-Answer Questions

Answer with a word, number, phrase, or short sentence.

1. Name the Swiss paleontologist who in the mid-1800s convinced other scientists that great ice sheets had once covered the land. _____

2. a. What percentage of land is covered by glaciers today? _____
   b. What percentage of land was covered by glaciers in the last ice age? _____

3. What is the difference in appearance between pure ice and ice that is cracked or contains tiny air bubbles? _____

4. The solid form of water is _____ (more, less) dense that its liquid form. Therefore ice _____ (sinks in, floats on) water.

5. Where are continental glaciers (ice sheets) found today? _____

6. Why is there an ice sheet at the South Pole but not at the North Pole? _____

7. List three conditions that are necessary to allow glaciers to form. _____

8. What is the minimum elevation at which glaciers can form:
   a. in mid-latitudes (30–60°)? _____ km, which is about _____ ft
   b. in low latitudes (0–30°)? _____ km, which is about _____ ft
   c. in high latitudes (60–90°)? _____ km, which is about _____ ft
   (1 km = 0.62 mi; 1 mi = 5,280 ft)

9. a. Fresh snow is about what percentage air? _____
   b. Firn is about what percentage air? _____
   c. Glacial ice is about what percentage air? _____

10. Glacial ice tends to absorb red light; therefore the color of glacial ice is not truly white, but instead it's slightly _____.

11. How long does it take fresh snow to transform into glacier ice? _____

12. Name two factors that influence the speed of glacier flow. _____

13. Glacier X moves at 200 m/year; Glacier Y moves at 40 m/year. Which glacier is most logically the polar glacier? _____ the temperate glacier? _____

14. Choose the two words that describe where glacial ice moves the fastest: top, bottom, margins, center. _____

15. Name three ways glacial ablation can happen. _____

16. What zone of a glacier are you in if your elevation is higher than the snowline? _____

17. When during glacial retreat does glacial ice flow back uphill or back up to higher latitudes? Explain your answer. _____

18. Is the Antarctic ice sheet currently getting larger or smaller? Why? _____

19. Is seawater ice salty? Explain. _____

20. "Glacial drift" is the generic germ for any glacial deposit. Name six different categories of drift. _____

21. What is the origin of the ridge of sediment that makes up Long Island, New York, and Martha's Vineyard, Nantucket, and Cape Cod, Massachusetts? _____

22. Drumlins are asymmetric along their length. Which slope is the gentle one, the upstream slope (direction the glacier came from) or the downstream slope? _____

23. What layer of Earth's interior flows to allow the lithosphere to sink during glacial subsidence and rise during glacial rebound? _____

24. Arrange in order, from largest to smallest, the following three reservoirs of freshwater storage: surface water, groundwater, glaciers. _____

25. a. Sea level was how much lower during the Ice Age than it is now? _____
    b. How far did the coastline extended seaward onto the continental shelves? _____

26. Name two land bridges that existed because of lowered sea level during the Ice Age. _____

27. Name several U.S. cities that would flood if the continental glaciers of Antarctica and Greenland were to melt. _____

28. a. To what depth can permafrost extend? _____
    b. How old can it be? _____

29. Why is oil shipped warm through the trans-Alaska pipeline? _____

30. Why is the trans-Alaska oil pipeline built on a frame that holds it above ground? _____

31. What evidence exists in Central Park, New York, that shows it was buried in glacial ice during the Pleistocene epoch? _____

32. a. When did the Pleistocene epoch begin and end? _____
    b. When did the last Ice Age begin and end? _____

33. Of the seven continents, which four supported continental glaciers during the Pleistocene? _____

34. How do geologists know climatic belts shifted southward during the Pleistocene? _____

35. a. During the Pleistocene, where did rainfall increase? _____
    b. Where did rainfall decrease? _____

36. Why were Pleistocene times unusually windy? _____

37. List several animals that Pleistocene humans could have hunted (or, in some cases, been hunted by). _____

38. What do the following groups of names have to do with the Pleistocene?
    a. Wurm, Riss, Mindel, Gunz, and Donau _____
    b. Wisconsinan, Illinoian, Kansan, and Nebraskan _____

39. What material trapped in glacial deposits has been dated by radiometric methods and thus provided dates for glaciations? _____

40. Name two lines of evidence that have convinced geologists there were more than four or five major glaciations during the Pleistocene ice age. _____

41. Because of the above two lines of evidence, how many different glacial advances do geologists today believe occurred during the Pleistocene epoch? _____

42. What evidence do geologists search for in the stratigraphic record to identify ice ages? _____

43. List and date Earth's four ice ages. _____

44. List the three aspects of Earth's movement around the Sun that are involved in Milankovitch's explanation of glacial advances and retreats. _____

45. What fact suggests we live in an interglacial period that is about to end? _____

Short-Essay Questions

1. If water were like most other substances and contracted and got denser as it changed from the liquid to the solid state, there might eventually be no open bodies of water on Earth's surface. Explain.

2. Explain why the maximum depth of a crevasse can be only about 60 m.

3. Explain why meteorites buried in the Antarctic ice sheet keep coming to the surface.

4. What is the difference between an ice sheet and an ice shelf?

5. Give two reasons Pleistocene valley glaciers carved U-shaped valleys out into the sea and below sea level.
6. Is Hudson Bay, Canada, growing larger or smaller each year, and why is this so?
7. Discuss how glaciation affected the river drainage systems and the distribution of lakes in northeastern North America.
8. What is the origin of the Great Salt Lake?
9. How did Pleistocene conditions influence the development and geographical distribution of early *Homo sapiens*?
10. List and briefly discuss several factors (besides Milankovitch's ideas) that influenced the onset of ice ages and their glacial advances and retreats.
11. Will the future bring a return of a true ice age or will it bring global warming? What do scientists think? What do you think?

Practice Test

1. Ice
    a. is denser than water.
    b. made of frozen seawater incorporates salt into its crystalline structure.
    c. is transparent if pure but milky white if cracked or contains air bubbles.
    d. has a low albedo, which explains why it floats.
    e. All of the above are true.
2. Which of the following is NOT a type of mountain (alpine) glacier?
    a. mountain ice cap
    b. ice shelf
    c. valley glacier
    d. cirque glacier
    e. piedmont glacier
3. Choose the FALSE statement. Continental glaciers
    a. move because of the pull of gravity.
    b. always flow downslope.
    c. spread across the landscape laterally in response to their own weight.
    d. move by gravitational spreading.
    e. move because their basal ice can't support the weight of their overlying ice.
4. Which type of moraine can a glacier have only one of?
    a. recessional
    b. end
    c. medial
    d. terminal
    e. lateral
5. Which is NOT a feature resulting from glacial erosion?
    a. drumlin
    b. rock flour
    c. chatter marks
    d. truncated spur
    e. roche moutonnée
6. Choose the FALSE statement.
    a. During the Pleistocene the average temperature decrease was no more than 13°C.
    b. Settlers in the United States in the 1600s and 1700s endured exceptionally cold winters because those were little ice age times.
    c. If all of today's ice sheets melted, the global sea level would rise only about 20 ft and seawalls could protect any coastal cities threatened.
    d. A lake that occupies a former glacial cirque is called a tarn.
    e. Glaciers originate on land; therefore there are no glaciers at the North Pole.
7. Scientists can determine the direction of movement of continental glaciers by looking at glacial
    a. drift.
    b. cirques.
    c. horns.
    d. striations.
    e. firn.
8. Which of the following is NOT a name for a major glaciation of the Pleistocene?
    a. Minnesotan
    b. Nebraskan
    c. Wisconsinan
    d. Illinoian
    e. Kansan
9. According to Milankovitch, which of the following is a contributing factor to ice age cycles?
    a. sunspot cycles
    b. variations in ocean temperature
    c. shifting continental positions due to plate tectonics
    d. increased amounts of volcanic ash
    e. variations in the shape of the Earth's orbit
10. Choose the TRUE statement.
    a. The process of icebergs' breaking off coastal glaciers is called surging.
    b. Narrow, steep-sided, deep inlets of seawater in glacial valleys are called fjords.

c. Today glaciers cover 30% of the land; in the Ice Age they covered 70%.
d. Pluvial lakes are lakes that form on the surface of a glacier.
e. All of the above are true statements.

11. Continental ice sheets today are found only in
    a. Alaska and Iceland.
    b. Antarctica and Iceland.
    c. Antarctica and Greenland.
    d. Greenland and Iceland.
    e. Alaska and Antarctica.

12. Choose the TRUE statement.
    a. Snow that has lost much of its air and turned into packed, granular material is called firn.
    b. Terminal, recessional, medial, and lateral are all varieties of moraines.
    c. Large boulders carried long distances by glaciers and dumped where they obviously don't belong are called erratics.
    d. Glaciers grind down the land they move over, producing a fine material called rock flour.
    e. All of the above are true statements.

13. Choose the FALSE statement.
    a. Glaciers rework V-shaped stream valleys and change them into U-shaped valleys.
    b. Glaciers' average speeds are between a few tens of and a few hundred meters per year.
    c. A horn is a pyramid-like peak that results when several cirques eat away at a mountain peak.
    d. Tarn is rocky material that has been carried along and deposited by the ice of a glacier.
    e. Polar glaciers move mainly by internal flow, temperate glaciers mainly by basal sliding.

14. Which term has nothing to do with glaciers?
    a. "lag deposits"
    b. "hanging valleys"
    c. "Louis Agassiz"
    d. "zones of accumulation and ablation"
    e. "channeled scablands"

15. Choose the FALSE statement.
    a. Glaciers can move by basal sliding on water underneath them.
    b. A medial moraine forms when two glaciers join and their lateral moraines merge.
    c. The Antarctic ice sheet has been calving off huge icebergs over the last few decades.
    d. Regions covered by glacial ice are termed periglacial environments.
    e. The melting of ice sheets at the end of the Pleistocene caused glacial rebound in northern Canada.

16. Which of the following climate conditions would most likely allow glaciers to form?
    a. extremely cold winters
    b. high winds that produce low chill factors
    c. a heavy snowfall in winter coupled with relatively cool summers
    d. mild winters with lots of snow coupled with very hot summers
    e. temperatures below freezing for one month straight

17. Choose the TRUE statement.
    a. Glaciers can grow smaller by melting, sublimation, or calving.
    b. The zone of ablation is the area of the glacier where each year more snow falls than melts.
    c. Glaciers always retreat when they reach an elevation less than 5,000 ft.
    d. When a glacier retreats its ice contracts and flows back toward the glacier's point of origin.
    e. Loess is sand that was created by glacial abrasion and transported by glacial meltwater.

18. Choose the FALSE statement.
    a. Glacial abrasion can leave glacially polished rock surfaces.
    b. Glacial plowing and incorporation may remove regolith and expose the bedrock of an area.
    c. Some glaciers originate on land, some over the sea close to the shores of cold continents.
    d. Glacial abrasion can produce a fine sediment called rock flour.
    e. Ice crystals gradually move downward in the zone of accumulation and upward in the zone of ablation.

19. Choose the FALSE statement.
    a. When a tributary glacier melts it leaves a hanging valley above the main glacial valley.
    b. Continental glaciers create spectacular scenery filled with horns, arêtes, cirques, and truncated spurs.
    c. Modern streams flowing in oversized valleys are signs of past glaciation.
    d. The channeled scablands of eastern Washington were created by floodwaters from a huge glacially dammed lake.
    e. The Great Salt Lake is a remnant of the much larger pluvial Lake Bonneville.

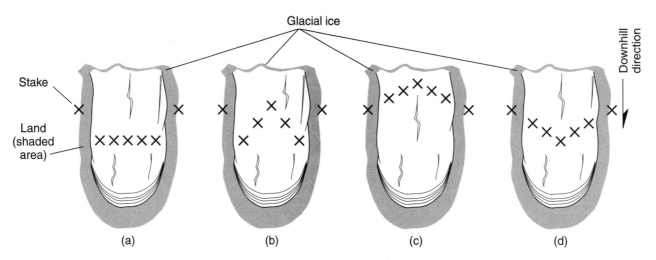

20. Which view in the diagram shows how a line of stakes planted straight across the toe area of a mountain glacier will look after a few months?
    a. A
    b. B
    c. C
    d. D
    e. None of the above, because glacial movement is erratic, with no predictable pattern.

21. Choose the FALSE statement. Drumlins
    a. are composed of glacial till.
    b. are elongate hills molded by glaciers flowing over them.
    c. are asymmetric along their length, with their gentler slope upstream.
    d. are created when a glacier overrides an end moraine and reshapes it.
    e. are depositional landforms of glacial environments.

22. Studies of assemblages of microscopic marine plankton and their oxygen isotope ratios suggest
    a. there were many more episodes of Pleistocene glacial advances and retreats than previously thought.
    b. the Pleistocene Ice Age began only 1 million years ago.
    c. there were thirty major ice ages in Earth's history.
    d. temperatures were 40C° colder during ice ages than now.
    e. All of the above are true statements.

23. Tillites
    a. are rocks composed of poorly sorted sediment.
    b. consist of large clasts in a matrix of sandstone and mudstone.
    c. contain clasts that have been glacially polished and striated.
    d. offer evidence of three ice ages on Earth before the Pleistocene Ice Age.
    e. All of the above are true statements.

24. Which is NOT a contributing factor to the onset of major ice ages?
    a. many continents' located at high latitudes
    b. most continents' sitting well above sea level
    c. a disruption of warm-water ocean currents flowing northward
    d. a high concentration of carbon dioxide in the atmosphere
    e. mountain ranges that promote cold temperatures and heavy precipitation in large areas

25. Milankovitch
    a. predicted climate cycles that have since been confirmed by ice core and marine sediment core studies.
    b. stated there are cyclical changes in Earth's orbit and axial tilt.
    c. stated there are changes in the amount and distribution of insolation received on Earth.
    d. offered ideas that explain the timing of ice age events but not the severity of temperature change associated with ice ages.
    e. All of the above are true statements.

26. The amounts of plankton and coral reefs in the ocean affect
    a. total insolation and its seasonal distribution.
    b. thermohaline circulation.
    c. the concentration of atmospheric carbon dioxide.
    d. Earth's albedo.
    e. the configuration of ocean currents.

27. Glacial striations running northeast-southwest indicate the glacier that produced them was moving from northwest to southeast.
    a. True
    b. False

28. Glaciers are analogous to metamorphic rocks because both involve the recrystallization of preexisting material in the solid state.
    a. True
    b. False

29. Ice melts under pressure, refreezes when pressure subsides, and thus can pluck rock fragments from the land it passes over.
    a. True
    b. False

30. Snow accumulates and forms glaciers where slopes are steep, windy, and cold.
    a. True
    b. False

31. On average, polar glaciers move faster than temperate glaciers and continental glaciers move faster than mountain glaciers.
    a. True
    b. False

32. Fast-moving portions of glaciers are called ice streams; episodes of fast movement of entire glaciers are called surges.
    a. True
    b. False

33. Varves are ridges of glacial drift deposited in tunnels of water flowing under a glacier's base.
    a. True
    b. False

34. Catabatic winds along glacial margins pick up small boulders and deposit them as drop stones on the outwash plain.
    a. True
    b. False

35. Kettle holes form when blocks of ice calve off the toe of a glacier, become buried in till, and then melt.
    a. True
    b. False

36. Most of Earth's freshwater exists as groundwater, and the next greatest quantity is stored in glacial ice.
    a. True
    b. False

37. Geologists have found ice as old as 235,000 years in the ice sheets of Greenland and Antarctica.
    a. True
    b. False

38. Scientists know we're entering a super interglacial period because most mountain glaciers are retreating and the Antarctic ice sheet is undergoing significant calving.
    a. True
    b. False

ANSWERS

Completion

1. erratics
2. glaciers
3. ice age
4. albedo
5. mountain (or alpine) glaciers
6. cirque glaciers
7. cirques
8. valley glaciers
9. mountain ice caps
10. piedmont glaciers
11. continental glaciers (ice sheets)
12. lobes
13. ice fall
14. sublimation
15. firn
16. pressure solution
17. basal sliding
18. wet-bottom (or temperate) glacier
19. internal flow
20. dry-bottom (or polar) glacier
21. crevasse
22. ice streams
23. surge
24. gravitational spreading
25. ablation
26. calving
27. zone of accumulation
28. zone of ablation
29. snow line
30. terminus (or toe)
31. glacial advance
32. glacial retreat
33. drop stones
34. ice shelves
35. glacial plowing
36. glacial incorporation
37. glacial plucking (or glacial quarrying)
38. chatter marks
39. glacial abrasion
40. rock flour
41. glacially polished surfaces
42. glacial striations
43. head
44. tarn
45. arête
46. horn
47. V-shaped valley profile
48. U-shaped valley profile
49. hanging valleys
50. truncated spurs
51. roche moutonnée
52. fjords
53. lateral moraines
54. kame
55. medial moraine
56. rock glacier
57. glacial drift
58. glacial till (or unstratified drift)
59. stratified drift
60. glacial marine sediment
61. glacial outwash
62. outwash plain
63. meltwater lakes
64. glacial lake bed sediment
65. varve
66. catabatic winds
67. loess
68. end moraine
69. terminal moraine
70. recessional moraines
71. lodgment till
72. ground moraine
73. kettle hole
74. knob-and-kettle topography
75. eskers
76. ice loading
77. glacial subsidence
78. glacial rebound
79. ice-margin lake
80. oversized valleys
81. channeled scablands
82. coulees
83. Great Missoula Flood
84. pluvial lakes
85. permafrost

86. periglacial environments
87. patterned ground
88. stone rings
89. Pleistocene Ice Age
90. Holocene
91. Laurentide ice sheet
92. Cordilleran ice sheet
93. tundra
94. glaciations
95. interglacials
96. tillites
97. eccentricity cycle
98. precession
99. insolation
100. Milankovitch cycles
101. positive-feedback mechanism
102. little ice age

Matching Questions

GLACIAL LANDSCAPES

1. a
2. f
3. e
4. c
5. g
6. b
7. d

MILANKOVITCH'S IDEAS

100,000; eccentricity cycle; 22.5–24.5°; 41,000; precession; 25,000; Polaris; Vega; correct; 4°C; 5–7°C; 10–13°C

GLACIAL FORMATION VERSUS GLACIAL MELTING

1. F   2. M   3. M   4. F   5. F   6. M

Short-Answer Questions

1. Louis Agassiz
2. a. 10%   b. up to 30%
3. Pure ice is transparent like glass; air bubbles or cracks diminish its transparency and turn it milky white.
4. less; floats on
5. Greenland and Antarctica
6. There's a continent at the South Pole but none at the North Pole.
7. cold climate so all winter snow doesn't melt each summer, lots of snowfall, and appropriate topography (gently sloping and wind protected)
8. a. 2.5 km; 8,200 ft   b. 5 km; 16,400 ft
   c. down to sea level
9. a. 90%   b. 25%   c. 20%
10. blue
11. varies greatly, from a few tens of years to thousands of years
12. steepness of the slope and whether or not water is present at the glacier's base
13. Glacier Y; Glacier X
14. top center
15. sublimation, melting, and calving
16. the zone of accumulation
17. Never, the position of the toe moves back toward the glacier's place of origin, but the ice itself never backs up.
18. smaller; possibly due to global warming
19. No, salt is not incorporated into the crystal lattice structure of ice.
20. glacial till, erratics, glacial marine (including drop stones), glacial outwash, glacial lake bed sediment (including varves), and loess
21. terminal moraine of the Laurentide ice sheet
22. downstream slope
23. asthenosphere
24. glaciers, groundwater, surface water
25. a. 100 m   b. more than 100 km
26. the Bering Strait, from northeast Asia to North America, and a land bridge between Australia and Indonesia
27. Miami, Houston, New York, and Philadelphia
28. a. 1,500 m below ground surface
    b. more than 1 million years
29. to prevent it from becoming too viscous to flow
30. so the warm oil doesn't melt the underlying permafrost
31. polished rock surfaces, striations gouged into bedrock and erratics
32. a. 1.8 million years ago; 11,000 years ago
    b. during the Pleiocence, between 3.0 and 2.5 million years ago; 11,000 years ago
33. North America, Europe, Asia, and Antarctica
34. by their examination of pollen preserved in the sediment of bogs
35. a. in North America   b. in equatorial regions
36. because of the greater than normal temperature contrasts between cold glaciated regions and warmer unglaciated regions
37. mammoths, mastodons, woolly rhinos, musk oxen, reindeer, giant ground sloths, bison, lions, saber-toothed cats, giant cave bears, hyenas
38. a. They're the five major Pleistocene glaciations in Europe.

b. They're the four major Pleistocene glaciations in Canada and the midwestern United States.

39. wood from trees that were destroyed by those glaciers

40. different assemblages of microscopic marine plankton and different isotope rations (^{18}O or ^{16}O), both of which indicate times of cold glacial seas and warm interglacial seas

41. at least thirty

42. presence of tillites

43. Pleistocene: began 3.0–2.5 million years ago, ended 11,000 years ago
    Permian: 280 million years ago
    Late Proterozoic: 600 million years ago
    Mesoproterozoic: 2.2 billion years ago

44. changes in the shape of Earth's orbit from more circular to more elliptical, changes in the tilt of Earth's axis, and precession (the wobble of Earth's axis because of slowing rotation)

45. Interglacial stages of the Pleistocene so far have lasted about 10,000 years; the current interglacial began about 11,000 years ago.

Short-Essay Questions

1. If solid water were more dense than liquid water, ice would sink, not float. Ice on lake bottoms would be insulated by liquid water from warm summer air and would not melt before the next winter freeze began. Over time it's possible that all natural water basins would get filled with ice, and there would be only minimal melting of surface ice during warm seasons.

2. The brittle portion of a glacier is roughly the upper 60 m. Below this the weight of all the surrounding ice keeps any ice from breaking apart, it simply deforms plastically. Since a crevasse is an open crack in ice, it simply can't exist below the brittle zone.

3. Ice crystals or any particles in the zone of accumulation of a glacier are continually moving downward as they are covered and depressed by new snowfall. As the glacial ice moves laterally (for a continental glacier) or downhill (for a mountain glacier), these crystals or particles are carried below the snow line into the zone of ablation. Here they begin a continual movement upward, as ice above is lost to melting or sublimation. This explains why meteorites that land on the ice and travel in the glacier eventually resurface.

4. "Ice sheet" is another way of saying "continental glacier." These may move toward the sea and calve into the sea, but they all form on land when snow accumulates and turns to ice. An ice shelf is a huge piece of ice floating in a polar region, created when the surface of the sea itself freezes. Ice shelves surround Antarctica and cover much of the Arctic Ocean.

5. A glacier's base remains in contact with the ground, and therefore can erode it, until the water depth exceeds about four-fifths of the glacier's thickness. When this happens the glacier floats. The sea level was lower during the Pleistocene, so the advances of coastal glaciers seaward were on land for greater distances than they would be now, and the glaciers then, as now, continued to erode the sea bottom until the water was deep enough to float them.

6. Hudson Bay is growing smaller. It formed because the area subsided due to ice loading during the Pleistocene, and since the ice melted, it has been experiencing glacial rebound. The floor of the bay slowly but continually rises and forces its waters back into the Arctic Ocean.

7. Before the Pleistocene several major rivers in northeast North America flowed north into the Arctic Ocean. The ice sheet buried these water channels, and the area had to establish drainage toward the south into the Mississippi-Missouri Rivers drainage network. After the ice melted, the new drainage network remained, and many new lakes and wetlands formed on the changed landscape. Knob-and-kettle topography in northern areas like Minnesota were filled with thousands of small lakes, and areas like Illinois were covered with swamplands. Some glacial valleys were filled with water and became elongate lakes like the Finger Lakes, New York. Glacially scoured depressions became lakes, some of them very large, like the Great Lakes.

8. The Great Salt Lake, Utah, is the remnant of a much larger pluvial lake, Lake Bonneville. The basin of that lake was created by normal faulting during continental rifting in the Basin and Range Province of Utah and Nevada. In the wetter Pleistocene climate, Lake Bonneville received abundant drainage from adjacent mountain ranges and eventually covered one-third of western Utah. Because it received much water but had no outlet to the sea, constant evaporation caused it to become a very salty lake.

9. Early human-like species lived at the beginning of the Pleistocene, and the modern human, *Homo sapiens*, lived on every continent but Antarctica by the end of the Pleistocene. Populations had to migrate in response to changing ice positions and associated harsh climates, and quite possibly human evolution was speeded along by the vital need to discover and develop comforts of life such as fire, shelter, and clothing. Pleistocene mammals were an abundant source of fresh meat, but humans had to develop hunting skills to procure the

meat and stay alive in the process. It is possible early humans hunted some Pleistocene mammal species to extinction. The lowered sea level exposed several land bridges which allowed human populations to migrate great distances. Humans no doubt crossed the Bering Straits from Siberia to Alaska and explored the Indonesian Islands, Papua New Guinea, and Australia because land bridges were exposed and ocean stretches were much reduced.

10. There were many contributing factors to the onset of ice ages and alternating glacial advances and retreats within them. Plate tectonics set the stage for climate change when continents moved to high latitudes or mountains were created that gave land greater elevation and increased snowfall. Shifting pieces of land altered ocean currents, sometimes cutting off heat from northern areas, but sometimes supplying these areas with warm waters that evaporated and caused increased snowfall. Extensive snow cover increased an area's albedo, which cooled the area and allowed even more snow (a positive-feedback mechanism). Decreased seawater salinity interrupted thermohaline circulation and decreased the amount of heat transport north. Several factors decreased carbon-dioxide concentration of the atmosphere, and this decreased the greenhouse effect and cooled Earth. Some examples of these factors were the increase in ocean plankton (which absorb carbon dioxide from seawater), the death of coral reefs (which cut the total amount of carbon dioxide contributed by growing corals), and the increased growth of large coal swamps (which withdrew carbon dioxide from the atmosphere). Several factors the increased carbon-dioxide concentration and thus increased the greenhouse effect and warmed Earth. Examples are the increased weathering of carbonate rocks and the increased volcanic activity. The issues are complex; there is as yet no clear-cut explanation as to why ice ages began or why they ended.

11. As the previous answer pointed out, the precise reasons for the onset of ice ages and their glacial advances and retreats are not known. The geologic record shows that for the last 2 million years interglacials have lasted about 10,000 years, and it has been 11,000 years since the last major glaciation of the Pleistocene. Maybe we are overdue. The world did experience a little ice age between about 1300 and 1850, but it was merely a short period of colder climate, not a major glaciation. Current events like increased calving along the edges of the Antarctic ice sheet suggest global warming, which may be enhanced by humankind's activities. Maybe we are on the brink of global warming; maybe we're about to begin a new glaciation. Currently no scientist can prove either point of view. Only time will tell.

Practice Test

1. c Ice is less dense, or it wouldn't float; salt doesn't fit in the crystal lattice; and ice's albedo (measure of light reflectivity) is high.

2. b An ice shelf is an area of frozen sea surface; the other glaciers are located along the mountain crest, in the valleys, in cirques, and along the adjacent plains.

3. b The weight of a glacier's own ice, pushing laterally, may force a glacier upslope.

4. d Only the end moraine extending farthest from the glacier's origin is also a terminal moraine; any others mark temporary halts of the glacier's advance.

5. a A drumlin is reworked till and therefore a depositional feature.

6. c The sea level would rise 70 m (230 ft), and it would be impossible to protect coastal cities or plains from flooding.

7. d Striations, gouged by material on a glacier's bottom, are parallel to its direction of motion.

8. a Choices b to e are names of major glaciations of the Pleistocene.

9. e All of Milankovitch's ideas involved aspects of Earth's movement around the Sun.

10. b Icebergs form by calving; today 10% of land is ice covered while in the Ice Age 30% was; and pluvial lakes were far from glaciers and existed because of wetter climate conditions.

11. c Antarctica and Greenland.

12. e All of the above.

13. d This is a definition of till; a tarn is a lake in a cirque.

14. a Lag deposits are desert deposits left behind when wind has blown away finer sediment.

15. d Periglacial environments are areas around the edges of glacial environments that have permafrost but no covering of ice.

16. c Glaciers need lots of snow that doesn't all slide or blow away when it falls or melt away in the summer.

17. a Ablation is wasting away; glaciers at high latitudes can reach to the sea; the position of the toe moves back, but the ice never does; and loess is fine-grained material transported by wind.

18. c All glaciers originate on land.

19. b Mountain (alpine) glaciers produce such spectacular features.

20. d The ice in the top center of a glacier moves the fastest; view C is wrong because glacial ice never backs up, even when the position of the toe retreats.

21. c The gentler slope of the drumlin is its downstream side.

22. a Studies of plankton and isotope ratios have allowed scientists to subdivide the major Pleistocene glaciations they first recognized.

23. e All of the above.

24. d Carbon dioxide is a greenhouse gas whose presence traps heat close to Earth's surface.

25. e All of the above.

26. c Plankton remove carbon dioxide from ocean waters; corals add it to ocean waters.

27. False Glacial movement is parallel to the direction of striations, not perpendicular.

28. True Glaciers and metamorphic rocks are analogous features because both result from the recrystallization of preexisting material in the solid state.

29. True Rocks are incorporated into glacial ice when it melts and then refreezes around the rocks.

30. False Slopes have to be gentle and wind protected so snow can accumulate, not avalanche or blow away.

31. False Just the opposite is true: Temperate glaciers move faster than polar glaciers, and mountain glaciers move faster than continental glaciers.

32. True Ice streams and glacial surges are correctly defined in the question.

33. False The description is of eskers; varves are seasonal pairs of thin layers of sediment, one silt and one clay, laid down in glacial lakes.

34. False They pick up fine clay sediment and glacial flour from the outwash plains and deposit them as loess far from the glacier's toe; drop stones are carried to sea by icebergs.

35. True Kettle holes do form in the manner described.

36. False Just the reverse is true: Most of Earth's freshwater is stored as glacial ice.

37. True Geologists have drilled deep enough in the ice sheets of Greenland and Antarctica to reach ice that is 235,000 years old.

38. False These things are happening, but it's a complex issue that nobody can prove. At the moment, the only honest answer is that time will tell.

CHAPTER 23 | Global Change in the Earth System

GUIDE TO READING

You've read 22 chapters about the physical makeup of Earth and the actions and interactions of its component parts. This last chapter presents some new ideas but also reviews material previously presented, to stress the idea that Earth has always been and will continue to be a dynamic planet—ever changing in its physical and biological features. Some new vocabulary is introduced to further develop this concept of change, and several topics from previous chapters are presented in the context of how they will shape Earth's future.

The chapter begins by reminding the reader that Earth is a unique planet for two basic reasons: it has a mobile asthenosphere that allows plate tectonics phenomena to occur, and its surface temperature straddles the freezing point of water, a fact that strongly influences surface processes. Numerous internal and external processes interact and create constant change on the planet. The term "Earth system" is commonly used to embrace this worldwide interconnecting web of physical and biological phenomena; some researchers use a more extreme term, the "Gaia concept," to infer the Earth system is analogous to a complex living entity. Change within the Earth system may be gradual or catastrophic, unidirectional or cyclical, biogeochemical or anthropogenic.

The unidirectional changes (transformations that progress in one direction and never repeat the same steps) you read about involve the evolution of the solid Earth, the atmosphere, the oceans, and life on Earth (the biosphere).

Cyclical changes involve the same steps, repeated over and over again, that may or may not produce similar results. Three *physical cycles* are discussed:

- The supercontinent cycle: You're reminded that geologists believe all continental material has been merged into one giant continent at least three different times in Earth's history.
- The sea-level cycle: The sea's transgressions and regressions have left a physical record of major sedimentary sequences and their minor subdivisions called cyclothems.
- The rock cycle: Earth's internal processes (including rifting, mantle plumes, subduction, sea-floor spreading, and convergence) and external processes (including weathering, erosion, and deposition) cycle the atoms of minerals through the three basic rock types (igneous, sedimentary, and metamorphic).

Two biogeochemical cycles are examined:

- The hydrologic cycle: Water may exist in any of its states (gas, liquid, or solid) as it cycles through the oceans, atmosphere, surface water, groundwater, glaciers, soil, and living organisms.
- The carbon cycle: Emphasis is placed on the role of greenhouse gases as they interact with rocks and influence climate.

The chapter continues with a discussion of the many facets of *global climate change*. There are long-term and short-term changes and global warming and global cooling to be considered when trying to predict climate change. Geologists look at past climates (paleoclimates) in order to predict future climates. They study paleoclimates by examining the stratigraphic record, paleontological evidence (including fossil pollen), oxygen-isotope ratios in ice and in plankton shells, air bubbles in ice, growth rings of trees (dendrochronology) and of corals and shells, and human history.

Long-term climate changes can result in greenhouse periods or ice-house periods (with or without ice ages). They're caused by changes in the positions of the continents, volcanic activity, uplift of land surfaces, and formation of coal, oil, and limestone. Short-term climate changes (such as the Younger Dryas, the Holocene climatic optimum, the medieval warm period, and the little ice age) may be explained by fluctuations in solar radiation, changes in Earth's orbit (Milankovitch's ideas), changes in the reflectivity (albedo) of Earth, and changes in the ocean currents.

Catastrophic climate changes seem to be linked to mass-extinction events, when large percentages of existing species disappeared and the biodiversity on Earth was greatly diminished. You read about two such mass-extinction events, one at the Permian-Triassic boundary and the other at the Cretaceous-Tertiary (K-T) boundary.

Before getting into a discussion about how humans affect Earth, the author presents some basic facts about human population. It is obvious that at first human population increased slowly, but now the population and the population growth rate are high enough to significantly impact Earth. Human activities have modified landscapes and ecosystems, and quite possibly they're starting to modify the global climate by contributing chemicals that enhance global warming.

There's general agreement some human-caused (anthropogenic) changes are undesirable. Slash-and-burn agricultural practices and the introduction of contaminants that cause pollution (smog, photochemical smog, water contamination, acid runoff, acid rain, radioactive materials, and ozone depletion) are detrimental to both Earth and humankind. Whether or not Earth is truly warming and whether or not human activities are playing a significant role in the process are both hotly debated issues. Individuals and governments worldwide are concerned because any changes will have political ramifications and will affect the welfare of the physical Earth and all life on it.

The chapter (and text) quite appropriately conclude by addressing the question of Earth's future. All through the text you've been reminded that if it's happened on Earth before, it will probably happen again, but what does this really mean? There's general agreement that human activity will play a significant role in the geologic near-term future and that over the long-term plate tectonics activity will slowly but inexorably change the looks of Earth's surface. Other long-term scenarios are less certain, and some of them range from unpleasant to disastrous. Severe inland flooding has happened and could happen again. Earth has been damaged in the past by bolide impacts and could be again. The creator of Superman probably had the right idea, that planets can be destroyed by large enough impacts, and that could be Earth's fate. Scientists do believe that in 5 billion years the Sun will run out of fuel and "die," and if Earth is still around, it will be engulfed and vaporized by the expanding Sun in its "death throes." Remember all of these predictions are not certainties, but they are the best scientific guesses possible today. Are they really going to happen? Only time will tell.

LEARNING ACTIVITIES

Completion

Test your recall of new vocabulary terms.

_____ 1. the worldwide interconnecting web of physical and biological phenomena that occur on Earth

_____ 2. modifications to the Earth system through time

_____ 3. the idea that the Earth system is comparable to a complex living entity (from the Greek "Earth goddess")

_____ 4. the global change that occurs on a time scale of millions to billions of years

_____ 5. the global change that takes place quickly (in seconds, or at the most, within centuries)

_____ 6. the global change that never repeats itself

_____ 7. the global change in which the same processes are repeated over and over again, but not necessarily with the same results

_____ 8. the global change that involves the exchange of chemicals among living and nonliving reservoirs

_____ 9. modifications of Earth's climate over time

_____ 10. the kind of global change that results from human activity

_____ 11. all regions of Earth, on, above, and below its surface, that are inhabited by living creatures

_____ 12. the entire sequence of small continents coming together, forming a giant continent, then breaking apart into small continents again

_____ 13. the blanket of sediment left behind by an episode of continent-wide advance and retreat of the sea, in which unconformities separate each blanket of sediment from its neighboring sediment blankets

_____ 14. subdivisions of a sedimentary sequence, each containing a specific succession of sedimentary beds, that represent short-term rise and fall of the sea level

_____ 15. a sea-level change that is experienced worldwide

_____ 16. the long-term cyclical process in which mineral materials are redistributed from one rock type to another due to plate tectonics processes and to weathering, erosion, and deposition

_____ 17. the situation in which the proportions of different chemicals in various natural reservoirs remain fairly constant, despite the fact there is a continual flow of chemicals between reservoirs

_____ 18. the situation in which the average atmospheric and sea-surface temperatures on Earth rise or fall

_____ 19. the results of computer programs that have taken factors of atmospheric composition, topography, ocean currents, and Earth's orbit and predicted how changes in them will affect Earth's climate

_____ 20. the climate of Earth in the past

_____ 21. dust-sized grains that are part of a plant's reproductive process

_____ 22. scientists who study pollen

_____ 23. scientists who study tree rings and use them to determine the age of the wood

_____ 24. periods of time during Earth's past when its atmosphere was significantly warmer than it is now

_____ 25. periods of time during Earth's past when its atmosphere was significantly colder than it is now

_____ 26. the feedback among components of Earth's system that slows or reverses an ongoing process

_____ 27. the feedback among components of Earth's system that enhances an ongoing process and possibly even accelerates its rate

_____ 28. a possible explanation of how positive feedback on Venus may have produced its current intensely hot, dense atmosphere

_____ 29. the cyclical increase and decrease of the number of black spots (possibly magnetic storms) on the Sun's surface

_____ 30. the degree to which sunlight is reflected back into space by Earth's atmosphere or by its surface

_____ 31. several episodes in Earth's past when large numbers of species abruptly disappeared

_____ 32. the number of different species that exist on Earth at any given time

_____ 33. any meteorite, comet, or asteroid that impacts Earth

_____ 34. a mantle plume that is much larger than is typical (The Hawaiian mantle plume is considered typical.)

_____ 35. the mass-extinction event that happened at the end of the Cretaceous period, 65 million years ago

_____ 36. both the physical environment and biological inhabitants of a particular region

_____ 37. the practice of cutting down forest and burning all vegetation present to clear the land for farming

_____ 38. an ozone-rich brown haze produced when the energy of sunlight acts on exhausts from vehicles

_____ 39. the precipitation produced when water falling through the atmosphere dissolves sulfur put there by power plant emissions

_____ 40. a condition of diminished ozone in the polar stratosphere

_____ 41. a huge star, much larger than our Sun, that scientists predict will be a future stage of our Sun during its waning years

Short-Answer Questions

Answer with a few words, numbers, or short sentences, or choose from the words in parentheses.

1. Give two reasons that Earth experiences extensive changes over time but no other planet or moon in our solar system experiences similar changes. _____
   _____

2. List four major unidirectional changes Earth has made over time. _____
   _____

3. How long did the proto-Earth remain a homogeneous body before it began to melt and differentiate into layers? _____

4. How did Earth's Moon form? _____
   _____

5. How did our current atmosphere
   a. come to be dominated by nitrogen? _____
      _____
   b. acquire its oxygen? _____
      _____

6. What was the first life form on Earth and when did it appear? _____

7. List four major physical cyclical changes Earth constantly undergoes. _____
_____

8. a. What is the minimum number of times in the last 2 billion years Earth's continental crust has been merged into a single supercontinent? _____
   b. Overall, has Earth's continental crust spent more time organized as a single supercontinent or as separate smaller continents? _____

9. What is the maximum amount, in meters, sea level has changed during the Phanerozoic eon? _____

10. State whether sea level would be *high* or *low* when the following conditions prevailed:
    a. Much sedimentary rock is being formed on continents. _____
    b. Unconformities are developing. _____
    c. There's extensive glaciation. _____

11. What is the sedimentary cycle chart (or Vail curve)? _____
_____

12. Name two factors besides ocean transgressions and regressions that affect the sedimentary record on land. _____
_____

13. List several chemicals that are involved in Earth's biogeochemical cycles. _____

14. a. How did the majority of Earth's surface carbon originate? _____
    b. List several ways carbon is removed from the atmosphere. _____
_____
    c. List ways carbon is returned directly to the atmosphere. _____
    d. List ways carbon is indirectly returned to the atmosphere. _____
_____
_____

15. a. Name three common greenhouse gases. _____
    b. The greater the atmospheric concentration of these gases, the _____ (hotter, cooler) the average temperature of the atmosphere.

16. What event caused the extremely cool conditions that made 1816 the "year without a summer"? _____
_____

17. Because evolution is a unidirectional change, what is true about new species that appear after a mass-extinction event? _____

18. List some possible effects of a bolide impact on Earth. _____
_____
_____

19. State two significant events that occurred on Earth at the Permian-Triassic boundary. _____
_____

20. What was the human population of the world:
    a. during the Stone Age? _____
    b. at the dawn of civilization, 4000 B.C.E.? _____
    c. in 1850? _____
    d. in 1930? _____
    e. in 1975? _____
    f. in 2000? _____

21. a. What is the current doubling time for the human population? _____
    b. In what year will the human population be 12 billion? _____

22. Give two reasons humans are affecting the Earth system (anthropogenic changes) more than they ever have. _____
_____

23. What have archaeological studies revealed as the earliest known example of human modification of an ecosystem, back in the Stone Age? _____
_____

24. List six different types of anthropogenic pollution. _____
_____
_____
_____
_____

25. What dangerous consequences result from a hole in the ozone? _____
_____

26. List four human activities that have disrupted ecosystems and led to a decrease of biodiversity.
_____
_____
_____

27. Why can't scientists agree on the future scenario in respect to global warming? _____
_____

28. What changes will happen to the map of our planet over the next 50 million years due to plate tectonics movements? _____
_____
_____

Short-Essay Questions

1. Discuss the meaning of the expression "Earth system," using several examples to illustrate the concept.

2. How did Earth's crust change from its original form (a homogenous thin skin of frozen mantle) to one of shifting plates of continental crust and ocean crust?

3. Discuss the orders of cycles of the sedimentary cycle chart (or Vail curve).

4. List and briefly explain six factors that are studied in order to analyze past climate conditions (paleoclimates).

5. List and briefly discuss four causes of long-term global climate change.

6. List and briefly discuss four factors that influence short-term climate changes.

7. What is the generally accepted scientific scenario for the end of Earth, assuming it avoids planet-destroying collisions?

8. Both the reality of human-induced global warming and the anticipated effects of global warming are issues of hot debate. Summarize both issues and offer your personal opinion as to what the current human population should do to address the problem.

Practice Test

1. Which of the following changes is NOT a unidirectional change?
   a. Life on Earth evolves.
   b. Earth's atmosphere lost huge amounts of water vapor and developed a high concentration of nitrogen.
   c. Small continents merged to become a supercontinent.
   d. Liquid iron alloy sank into Earth's interior.
   e. The frozen mantle crust changed to continental crust and oceanic crust.

2. Which of the following statements is FALSE?
   a. Earth began to differentiate into a layered planet about 0.5 billion years ago.
   b. The Moon formed from fragments of Earth's mantle that were gouged out when a Mars-sized body hit Earth.
   c. Carbon dioxide is absorbed from the atmosphere by the weathering of limestone and calcium-silicate rocks.
   d. Sea level is high during ice ages because floating icebergs displace lots of water.
   e. Cyclothems represent short-term rising and falling of sea level.

3. Which of the following elements is released into the atmosphere by all of these processes: volcanic outgassing, animal respiration and flatulence, burning of fossil fuels, and metamorphism of limestone?
   a. oxygen
   b. carbon
   c. nitrogen
   d. sulfur
   e. phosphorous

4. Earth's atmosphere
   a. began as nitrogen and argon only, left over from the solar nebula.
   b. acquires its carbon dioxide from photosynthesis by green plants.
   c. gets warmer as its carbon-dioxide concentration goes down.
   d. is being polluted by rain that falls through sulfur-containing aerosols from power plants.
   e. has a high albedo if it contains few aerosols and a low albedo if it contains many.

5. Sea level changes
   a. are reflected by blankets of sediment called sedimentary sequences.
   b. have been recorded on the sedimentary cycle chart (Vail curve).
   c. are termed eustatic if they are worldwide changes.
   d. have been as great as 300 m during the Phanerozoic eon.
   e. All of the above are true statements.

6. The hydrologic cycle
   a. is a biogeochemical cycle, involving both physical and biological phenomena.
   b. is the only biogeochemical cycle; all other element cycles are restricted to either the physical world or the biological world, never both.
   c. is an example of a unidirectional change.
   d. never results in global change because it maintains a steady state.
   e. All of the above are true statements.

7. Which of the following statements is FALSE?
   a. The oxygen-isotope ratios of ancient glacial ice indicate the atmospheric temperature of the snowfall which created the ice.

b. Bubbles of air trapped in old ice show the nitrogen concentration of the ancient atmosphere, and this is an indicator of its temperature.
c. The sunspot cycle is a cyclical rise and fall of the number of magnetic storms on the Sun and is associated with increased and decreased solar radiation received by Earth.
d. Ice-house periods are long periods of time when Earth's atmosphere was significantly cooler than it is now.
e. Ice ages were times during ice-house periods when Earth was cold enough to allow ice sheets to cover much land surface.

8. Which of the following statements is FALSE? Mass extinction
   a. events are times in Earth's history when large numbers of species abruptly vanished.
   b. of dinosaurs occurred at the Permian-Triassic boundary.
   c. events may reflect catastrophic changes in Earth's climate.
   d. may happen when a bolide impact starts a chain of events that blocks sunlight for weeks or even years.
   e. happened 65 million years ago and was probably caused by a bolide impact that produced a thin clay layer rich in iridium that has been found worldwide.

9. Which of the following statements is FALSE? Anthropogenic changes in the Earth system
   a. are increasing as population grows and the standard of living improves.
   b. include landscape modifications resulting from construction, mining, and farming.
   c. include the pollution of air and water.
   d. include global warming, which would not be happening if it were not for society's production of greenhouse gases.
   e. have affected the ecosystems of regions by deforestation, overgrazing, agriculture, and urbanization.

10. Which of the following statements is FALSE?
    a. Photochemical smog is produced by the action of sunlight on vehicle exhaust fumes.
    b. Acid runoff develops when precipitation flows through piles of sulfide-containing ores and coal.
    c. No radioactive pollution existed until humans produced nuclear weapons, nuclear power plants, and radioactive medical materials.
    d. The danger of the ozone hole is that it allows increased amounts of harmful ultraviolet radiation to reach Earth's surface.
    e. Stone Age hunters were the first humans to modify the ecosystem by hunting some large post–Ice Age mammals to extinction.

11. Human population
    a. has become a significant agent of global change.
    b. reached 1 billion in 1850.
    c. is currently a little over 6 billion.
    d. is doubling every 44 years.
    e. All of the above are true statements.

12. Which of the following statements is FALSE?
    a. Computer models prove global warming will continue and by 2050 the average annual temperature will have increased by 1.5 to 2°C.
    b. Carbon-dioxide concentration of the atmosphere has increased in the last century.
    c. In the year 2000, the ocean temperature at the North Pole was the warmest it's been in four centuries.
    d. Some scientists believe current temperature rises are part of natural cycles of short-term climate change.
    e. Global warming issues are emotional topics that are pitting corporations against environmentalists.

13. Which of the following statements is FALSE?
    a. The near-term future of the world depends heavily on human activities.
    b. Plate tectonics movements will alter the global map significantly over the coming millions of years.
    c. Shallow seas cannot ever again cover continental interiors because there has been enough sediment deposition to raise the average land surface high enough to prevent this.
    d. Five billion years from now scientists believe the Sun will begin to collapse, then swell to encompass the Earth.
    e. A collision with a planet-sized bolide could fragment the Earth and transform it into another asteroid belt.

14. Scientists believe cyanobacteria were the first life forms to appear on Earth about 0.5 billion years ago.
    a. True
    b. False

15. Many changes Earth undergoes don't happen to other planets because Earth is the only planet with a mobile asthenosphere and a surface temperature that straddles the freezing temperature of water.
    a. True
    b. False

16. The term "Gaia" is a synonym for "anthropogenic change."
    a. True
    b. False

17. If the concentration of a chemical is steadily increasing in one reservoir at the expense of the concentration of

that same chemical in another reservoir, it is termed a steady-state condition.
a. True
b. False

18. Studies of air bubbles and oxygen-isotope ratios in glacial ice, fossil pollen, and the stratigraphic record are all useful in determining paleoclimates.
a. True
b. False

19. Palynologists are scientists who study growth rings of trees, corals, and shells to decipher past climates.
a. True
b. False

20. Methane, water vapor, and carbon dioxide are all greenhouse gases that cause the atmosphere to cool by reflecting large amounts of solar energy back into space.
a. True
b. False

21. Large volcanic eruptions can put enough aerosols into the atmosphere and so cause global warming, as illustrated by the summer of 1816, which was the hottest summer on record.
a. True
b. False

ANSWERS

Completion

1. Earth system
2. global change
3. Gaia
4. gradual change
5. catastrophic change
6. unidirectional change
7. cyclic change
8. biogeochemical cycle
9. global climate change
10. anthropogenic
11. biosphere
12. supercontinent cycle
13. sedimentary sequence
14. cyclothems
15. eustatic sea-level change
16. rock cycle
17. steady-state condition
18. global warming (or global cooling)
19. climate-change models
20. paleoclimate
21. pollen
22. palynologists
23. dendrochronologists
24. greenhouse periods
25. ice-house periods
26. negative feedback
27. positive feedback
28. runaway greenhouse effect
29. sunspot cycle
30. albedo
31. mass-extinction events
32. biodiversity
33. bolide
34. superplume
35. K-T boundary event
36. ecosystem
37. slash-and-burn agriculture
38. photochemical smog
39. acid rain
40. ozone hole
41. red giant

Short-Answer Questions

1. Earth has a mobile asthenosphere and its surface temperature straddles the freezing point of water.

2. evolution of the solid Earth, its atmosphere, oceans, and life forms

3. 0.5 billion years

4. Collision with a Mars-sized body created a ring of mantle fragments and vapor that coalesced to form the Moon.

5. a. Atmospheric nitrogen did not readily react with any other substance and thus get removed from the atmosphere, as the other gases did, so its relative concentration increased over time.
   b. Oxygen was produced by green plants during photosynthesis.

6. cyanobacteria; 3.8 billion years ago

7. supercontinent, sea-level, rock, and landscape cycle

8. a. three   b. separate smaller continents

9. 300 m

10. a. high   b. low   c. low

11. a chart developed by an Exxon corporation team, lead by Vail, that lists all global transgressions and regressions that occurred during the Phanerozoic eon

12. a rise or fall of land surfaces due to plate interactions, and changes in sediment supply

13. water, carbon, oxygen, sulfur, ammonia, phosphorous, and nitrogen

14. a. as carbon dioxide released from volcanoes
    b. Dissolves in seawater, is absorbed by photosynthetic organisms (green plants) to use in making sugar, is absorbed during the chemical weathering of limestone, and is absorbed during the chemical weathering of calcium silicate.
    c. as carbon dioxide during the respiration of animals and as methane from the flatulence of animals; by the decay of dead organisms
    d. the carbon of dead plankton or plants stored in fossil fuels (oil and coal) is released when they're burned, and the carbon "stored" as limestone in seawater (extracted by sea organisms to form their calcite shells or deposited inorganically as limestone in seawater) is released when the limestone dissolves or is metamorphosed.

15. a. carbon dioxide, water vapor, and methane
    b. hotter

16. the eruption of Mt. Tambora in the western Pacific Ocean

17. They're different from the species that vanished during the extinction event.

18. debris in the atmosphere that blocks sunlight and produces night- or winter-like conditions for weeks or years, a disrupted food chain, aerosols that create acid rain, hot scattered debris that ignites forest fires, and debris in the oceans that chemically reacts and makes the water toxic or makes it so nutritious the resulting algal blooms deplete the water's oxygen supply

19. extinction of possibly 90% of all species, and extensive flood basalts in Siberia

20. a. less than 10 million
    b. a maximum of a few tens of millions
    c. 1 billion
    d. 2 billion
    e. 4 billion
    f. 6 billion

21. a. 44 years   b. 2044

22. The population is large and growing and the standard of living is improving, so per capita usage of resources is increasing.

23. hunting several species of post-Pleistocene-age large mammals (including mammoths, giant sloths, and giant bears) to extinction

24. smog and photochemical smog, water contamination (petroleum, gasoline, organic chemicals, radioactive wastes, acids, fertilizers, etc.), acid runoff (from sulfide-containing ores and coal), acid rain (largely from sulfur-containing aerosols from power plants), radioactive materials (from nuclear weapons, nuclear energy plants, and medical wastes), and ozone depletion (by chlorofluorocarbons released to the atmosphere through human activities)

25. Ozone shields Earth's surface from harmful ultraviolet radiation; less ozone results in increased ultraviolet radiation.

26. deforestation, overgrazing, agriculture, and urbanization

27. Predictions are dependent on complex computer models; researchers disagree on exactly what factors to consider and how to weight the importance of these factors in creating the models.

28. The Atlantic Ocean will get bigger, the Pacific Ocean will get smaller, the western part of California will migrate northward, and Australia will crash into the southern edge of Asia, squashing Indonesia in between.

Short-Essay Questions

1. Scientists in many disciplines have become more cognizant of the fact that geologic and biological phenomena of Earth interact and have named this idea of an interconnecting web of activity the "Earth system." One of many numerous examples involves photosynthetic organisms (green plants), which release oxygen into the atmosphere as a waste produce of food production. Atmospheric composition determines the particulars of the chemical weathering of rocks. Different weathering processes produce different soils, which support different vegetations. Types and quantities of vegetations influence mass wasting. The weathering of rocks may release greenhouse gases into the atmosphere, which contribute to climate warming.

2. Earth cooled so that its thin, frozen mantle skin couldn't completely melt on subduction. Instead, it melted preferentially, and the silicic minerals with lower melting temperatures began to dominate the composition of the melted rock. Rock from this silicic magma that solidified near the surface was buoyant, didn't subduct, and accumulated to form continental crust. At mid-ocean ridges partial melting produced basaltic magma which formed ocean crust. The top 100–150 km of the mantle beneath the crust cooled enough to become rigid. The crustal materials and the rigid upper-mantle material together became the material of the plates. The stage was now set for plate tectonics activities.

3. First-order cycles are the longest, lasting 200–400 million years. They correlate with the supercontinent cycle. When continental crust is joined in a supercontinent, there are fewer continental shelves, fewer mid-ocean ridges, larger ocean basins, and thus lower sea level. More land is exposed, and less sediment is deposited. The existence of several smaller continents produces just the opposite set of conditions. Second-order cycles are periods 10–100 million years long that correlate with changing rates of spreading at mid-ocean ridges. Fast rates produce much warm, light-weight, buoyant lithosphere, which displaces seawater and causes high sea level and associated increased deposition on land. Slow spreading rates result in lower sea level and less deposition of sediment on land. Third-order cycles, 1–10 million years long, may reflect ice ages. The shortest cycles, fourth-order cycles less than 500,000 years long, again correlate with glaciations and may relate directly to Milankovitch cycles of changes in Earth's orbit and tilt.

4. Six factors that are used to analyze paleoclimates and to predict future climates include:

The stratigraphic record: The very existence of sedimentary strata is a clue to the environmental conditions of their deposition. For instance, the presence of coal suggests past swamps, till means glaciers, cross-bedded sandstones form from sand dunes or river deltas, and iceberg drop stones tell of past tidewater glaciers.

Paleontological evidence: Because different assemblages of species inhabit different climatic environments, fossils present are a clue to past climates. Fossil pollen is particularly useful, because palynologists can determine what type plant produced it and therefore what the environment was like.

Oxygen-isotope ratios: The ratio of ^{16}O to ^{18}O in glacial ice and in calcium-carbonate plankton shells is a function of temperature. Analysis indicates the air temperatures when the ice formed or the seawater temperatures when the plankton lived.

Bubbles of atmosphere: Air bubbles in ancient glacial ice show the carbon-dioxide concentrations of past atmospheres. High concentration is associated with warm periods, low concentration with cold periods.

Growth rings: Dendrochronologists study the thickness of the growth rings of trees. Growth is faster and therefore rings are thicker during warm, wet years than in cool, dry years. Similar analysis can be done with the growth rings of corals and shells.

Human history: Written history and archaeological history (as recorded in paintings, stories, and records of crop yields) give direct reports of and indirect clues to past climates.

5. Four causes of long-term global climate change:

Positions of continents: Continent positions influence oceanic currents, and these distribute heat. Continents at high latitudes receive less solar energy and therefore are colder than continents at lower latitudes. Large continents are more likely to have extremely cold temperatures in their interior regions than small continental masses are.

Volcanic activity: Volcanic outgassing of carbon dioxide, an important greenhouse gas, contributes to global warming.

Uplift of land surfaces: Land thrusting upward to high elevations affects atmospheric circulation and rainfall amounts. Land at high elevation is very susceptible to weathering, and increased absorption of carbon dioxide due to chemical weathering may lower its atmospheric concentration and thus contribute to global cooling.

Formation of coal, oil, or limestone: The formation of all of these materials requires carbon, so carbon dioxide is removed from the atmosphere, with subsequent cooling.

6. Four factors that influence short-term climate change:

Fluctuation of solar radiation: No doubt solar radiation has fluctuated over time. One known example involves sunspot cycles. The number of sunspots (probably magnetic storms on the Sun) grows and diminishes on a cycle of 9–11.5 duration, therefore causing the amount of solar energy Earth receives to fluctuate with the same pattern.

Changes in Earth's orbit and tilt: These changes produce changes in the spatial relationship of Earth to the Sun and therefore alter the amount of solar energy received on Earth.

Changes in the reflectivity (albedo) of Earth: The increased concentration of aerosols produced by geologic activity (volcanic ash, soot, sea salt, oil, and sulfur) increases Earth's albedo and causes more energy to be reflected back into space, so less energy reaches Earth. Increased snow and ice cover also increase Earth's albedo, thus lowering the amount of solar energy Earth receives.

Changes in ocean currents: Since ocean circulation patterns are a major factor in heat distribution, changes in them significantly affect atmospheric temperatures worldwide.

7. Five billion years from now the Sun will begin to run out of nuclear fuel (hydrogen), the decreased thermal pressure from fusion reactions will allow gravitational pressure to dominate, and the Sun will collapse inward. The Sun isn't large enough to explode when this happens (a supernova event) but instead will heat up and expand to become a huge star, a red giant, whose radius will grow beyond Earth's orbit. Earth will be vaporized, and its atoms will become part of an expanding ring of gas that may eventually become part of a new solar system.

8. Statistics suggest that the global climate is warming. The 1990s were the warmest decade of the century, and the winter of 1999–2000 was the warmest winter on record in the United States. Ocean temperatures have been rising; temperatures in 2000 at the North Pole were the warmest in four centuries and produced open patches of water in the ice there. There are scientists who don't believe we have enough data to confirm this is true global change and not just normal temperature fluctuations. Even those who agree global climate is warming debate whether human activities (such as fossil fuel burning and extensive agriculture) are partially or largely responsible for the change. If global warming is occurring, the effects, too, are in doubt. The average annual temperature might rise by 1.5–2.0°C by 2050 and 5–11°C by 2150. This would make Earth as warm as it's been since the Eocene, 40 million years ago. If this happened, climate belts and associated vegetation would shift, storms might be stronger because there would be more evaporated moisture involved, sea level would rise and change

coastline features, and oceanic currents would change and redistribute heat. All of these changes would affect human lifestyles and consequently national and international matters. Already there is great concern at many levels, and the corporate world and environmental activists are at serious odds over the issues. The readers' personal opinions about what to do about the situation will, of course, vary. Thoughts range from believing humanity should continue its present course of action because Earth has always survived and is capable of taking care of itself, to serious concern that if humankind continues along its present course Earth will be irreparably damaged.

Practice Test

1. c Supercontinent formation is a cyclic change, meaning the same steps happen over and over again. Scientists believe a supercontinent has existed at least three times in Earth's past.

2. d Sea level is low during ice ages because so much water is tied up as glacial ice on land.

3. b All of these elements are involved in biogeochemical cycles, but it is only carbon that cycles between solid, liquid, and atmospheric Earth reservoirs via all of these processes.

4. d The early atmosphere was hydrogen and helium from the solar nebula; it acquired oxygen from the photosynthetic process; it warms when its carbon-dioxide concentration goes up; and it has a high albedo (degree of reflectivity) if it contains lots of aerosols (fine suspended particles) because these reflect solar energy back into space.

5. e Choices a to d are all true.

6. a Carbon, oxygen, sulfur, ammonia, phosphorous, and nitrogen also participate in biogeochemical cycles; it is a true cycle, happening over and over again, whereas a unidirectional change is one that precedes in one direction and never repeats; and the hydrologic cycle can maintain a steady state for long periods but can also change the proportions of distribution and thus produce global change.

7. b Air bubbles in ancient ice show the carbon-dioxide concentration of past atmospheres, and this is an indicator of atmospheric temperature.

8. b Dinosaurs became extinct at the end of the Cretaceous (K-T boundary), which was 65 million years ago and is marked by a thin clay layer rich in iridium.

9. d Although Earth's climate currently is experiencing warming, whether this will continue and whether it has all been caused by human activity are hotly debated issues.

10. c There have always been some naturally occurring radioactive materials in Earth's crust, locally concentrated around ore deposits, and therefore some radioactive pollution. Humankind's activities have increased the problem by changing the distribution of radioactive materials and by creating some new, nonnatural radioactive elements.

11. e Choices a to d are all true.

12. a Different computer models have produced conflicting scenarios of future climate. Choice a presents figures which are a worst-case scenario, not a proven fact.

13. c Sea-level changes are cyclical; what's happened in the past will no doubt happen in the future, and interior land will get inundated by seawater.

14. False They were the first life forms but they appeared 3.8 billion years ago.

15. True Earth's mobile asthenosphere and its surface temperature make it unique among the planets and influence many changes it undergoes.

16. False The term "Gaia" suggests that the Earth system (an interconnecting web of physical and biological phenomena on Earth) is analogous to a complex living entity; "anthropogenic change" means a human-induced change.

17. False "Steady-state condition" means that even though there may be a constant flow between reservoirs, the proportions of a chemical in different reservoirs remain fairly constant.

18. True The study of ancient glacial ice and its contents is one of the best tools scientists have to determine paleoclimates.

19. False They're scientists who study fossil pollen to decipher past climates.

20. False They are greenhouse gases that warm the atmosphere because they trap infrared radiation (heat) close to Earth's surface.

21. False The large amounts of aerosols volcanoes put into the air lower atmospheric temperature, as illustrated by 1816, which is known as the year without a summer.